Towards a Global 3G System

Advanced Mobile Communications in Europe

Volume 1

The Artech House Universal Personal Communications Series

Ramjee Prasad, Series Editor

CDMA for Wireless Personal Communications, Ramjee Prasad

OFDM for Wireless Multimedia Communications, Richard van Nee and Ramjee Prasad

Third Generation Mobile Communication Systems, Ramjee Prasad, Werner Mohr, and Walter Konhäuser, editors

Towards a Global 3G System: Advanced Mobile Communications in Europe, Volume 1, Ramjee Prasad, editor

Towards a Global 3G System: Advanced Mobile Communications in Europe, Volume 2, Ramjee Prasad, editor

Universal Wireless Personal Communications, Ramjee Prasad

WCDMA: Towards IP Mobility and Mobile Internet, Tero Ojanperä and Ramjee Prasad, editors

Wideband CDMA for Third Generation Mobile Communications, Tero Ojanperä and Ramjee Prasad, editors

For further information on these and other Artech House titles, including previously considered out-of-print books now available through our In-Print-Forever® (IPF®) program, contact:

Artech House
685 Canton Street
Norwood, MA 02062
Phone: 781-769-9750
Fax: 781-769-6334
e-mail: artech@artechhouse.com

Artech House
46 Gillingham Street
London SW1V 1AH UK
Phone: +44 (0)20 7596-8750
Fax: +44 (0)20 7630-0166
e-mail: artech-uk@artechhouse.com

Find us on the World Wide Web at:
www.artechhouse.com

Towards a Global 3G System

Advanced Mobile Communications in Europe

Volume 1

Ramjee Prasad

Editor

Artech House
Boston • London
www.artechhouse.com

Library of Congress Cataloging-in-Publication Data
Towards a global 3G system : advanced mobile communications in Europe / Ramjee Prasad, editor.
p. cm. — (Artech House universal personal communications series)
Includes bibliographical references and index.
ISBN 1-58053-138-5 (v. 1 : alk. paper)—ISBN 1-58053-139-3 (v. 2 : alk. paper)
1. Mobile communications systems—Europe. 2. Personal communication service systems—Europe. I. Prasad, Ramjee. II. Series.

TK6570.M6 T69 2001
384.5'3'094—dc21 2001022879

British Library Cataloguing in Publication Data
Towards a global 3G system : advanced mobile communications
in Europe
Vol. 1 Ramjee Prasad, editor. — (Artech House universal
personal communications series)
1. Global system for mobile communications
I. Prasad, Ramjee
621.3'8456

ISBN 1-58053-138-5

Cover design by Igor Valdman

International Standard Book Number: 1-58053-138-5
Library of Congress Catalog Card Number: 2001022879

10 9 8 7 6 5 4 3 2 1

"Nothing is done.
Everything in the world remains to be done or done over."

To my wife Jyoti, to our daughter Neeli,
and to our sons Anand and Rajeev.
—Ramjee Prasad

Contents

Foreword

Bringing together, in these volumes, the work performed in the scope of the European Union (EU)'s Advanced Communications Technologies and Services (ACTS) R&D Program (1995–98) in the area of mobile and personal communications, is both timely and purposeful.

Indeed, mobile and personal communications systems play a crucial role in building the wireless Information Society, where the user, no longer bound to a place or to a network or service provider, is effectively in control, and is always and everywhere connected. The information collected herein will certainly contribute to advancing the state of the art in this important area.

In this foreword, I will discuss the critical role played by European cooperative R&D in developing new concepts and systems and in the evolution of current systems, in the context of a European coordinated approach to Third Generation (3G).

This coordinated approach was devised against the backdrop of the tremendous global success of the Global System for Mobile communications (GSM), a success built around a three-pronged strategy:

- Europe chose to adopt a single, common, evolving[1], and open platform for Second Generation (2G), having learned the hard way from First Generation (1G) and the results of a fragmented market.[2]

---

[1] GSM evolution is built upon yearly releases, which were initially handled by the European Telecommunications Standardization Institute (ETSI)'s Technical Committee Special Mobile Group (SMG), but which have since been taken over by the 3G Partnership Project (3GPP). Currently, a 3GPP Technical Specification Group (TSG), having the responsibility for the release of a common version (covering both 3G and GSM) of a specification, will also be able to approve change requests to earlier releases of the corresponding GSM specification. This shows that GSM evolution will proceed even as 3G is specified, but will now proceed in a broader context.

[2] In fact, Europe created and launched the first cellular system, Nordic Mobile Telephone (NMT) in 1981, but soon the proliferation of incompatible analog systems at 450 MHz and later at 900 MHz

- Europe does not rely on the competition of standards, but on the competition in services and equipment within a mass market that developed on the basis of a common system platform, GSM, operating first at 900 MHz and later at 1,800 MHz.[3]
- Europe conducted a sweeping reform of telecommunications, fostering a liberalized and competitive market. The mobile sector was a pioneer in this respect.

As a result, with the capability to offer, for the first time, pan-European roaming, we saw an amazing growth: While in nine years the combined analog systems approached only 6 million subscribers, in half that time GSM surpassed 30 million subscribers, a factor of more than 10 in take-up. The growth has remained exponential, despite the fact that many countries have already reached penetrations above 70%.

The success of GSM has exceeded the most optimistic expectations: From a European standard considered to be ahead of its time, GSM achieved, and is set to maintain for quite a while, global market domination. This happened because GSM was able to grow quickly in Europe, due to a liberalized, highly competitive market, and was later able to build upon the European market to expand globally. Contributing to its growing acceptance are global roaming and the evolving nature of the system, which allows operators to deploy incrementally new services and capabilities (modular, phased deployment).

As a result of GSM's success, and with the same logic, Europe decided early on for a single 3G system, Universal Mobile Telecommunications System (UMTS). UMTS was conceived from the start to be a member of the International Telecommunication Union (ITU)'s International Mobile Telecommunications-2000 (IMT-2000) family, what was initially called Future Public Land Mobile Telecommunications System (FPLMTS).

A long-term perspective was considered essential in order to prepare the grounds properly for a successful deployment of UMTS:

- Certain activities needed to start early: This is the case notably of pre-competitive R&D, which brings together actors who cooperate at the stage of system definition and early technology development but will later compete in using the technology to develop products and services, yielding first system concepts and then standards.

(NMT-450 and NMT-900; Total Access Communication System, TACS-900, launched in the United Kingdom in 1985; Radiocom 2000, RC2000, a French system launched also in 1985; and C-Netz, a German system launched in 1986) led to lackluster growth.

[3] The introduction of the system, in both bands, was coordinated at the European level through the European Conference of Post and Telecommunications (CEPT) to facilitate deployment, namely across borders, but also to give manufacturers a pan-European market for which to fight.

- Other activities, notably those relating to regulatory issues, need to be addressed in a timely manner to provide legal certainty, promote the emerging markets, and enable new business cases.

To illustrate the long-term nature of our commitment to UMTS, the European Commission (EC) launched the concept in 1989, when GSM was still in the early stages of deployment in Europe. As a result, the Mobile Project in the Research into Advanced Communications in Europe (RACE) program effectively provided the ammunition to obtain the necessary spectrum at the ITU World Administrations Radiocommunication Conference (WARC) 1992: 230 MHz of spectrum on a worldwide basis in the 2-GHz band, including both terrestrial and satellite components. And as early as 1995, the EC granted to the European Telecommunications Standards Institute (ETSI) a general standardization mandate related to UMTS.

Regarding the coordinated approach to 3G, the need was identified to proactively link different areas of activity. For instance, spectrum planning is conditioned by systems concepts as they emerge from R&D and from standardization activities. Regulatory issues will be conditioned by systems and services, while policy, especially liberalization, will condition market developments.

R&D performed in the context of the RACE and ACTS programs contributed decisively to the standardization of UMTS and is now providing the basis for the harmonization of the use of the UMTS bands, as requested from CEPT. R&D also made a major contribution to the success of the ITU World Radiocommunication Conference (WRC) 2000 in all that relates to additional spectrum requirements.

In terms of policy, besides ensuring a liberalized and highly competitive market[4], the timely introduction of mandates to CEPT[5] and the publication of the UMTS Decision[6] send the right message to the industry.

Such a coordinated approach is only possible if all involved actors (industry, academia, users, spectrum managers, national administrations, and standardization organizations) get their acts together. A variety of mechanisms were and are used to facilitate this—the EU R&D framework programs (in this case, ACTS), the regular consultation with the national administrations and regulatory agencies, and the open consultations around ad hoc white/green papers.

In January 1995 the EC launched the UMTS Task Force with the objective of creating the conditions for a repeat of GSM's success by bringing together all of the above actors and effectively kick-starting a broader reflection on the new system, conceived to be capable of supporting, in particular, innovative multimedia services beyond the capability of 2G systems.

---

[4] Mobile Directive (96/2/EC) and Full Competition Directive (96/19/EC).

[5] Namely to harmonize frequency use.

[6] UMTS Decision (128/1999/EC).

As the concept matured, and the industry became convinced that 3G systems had a role complementary to GSM with the promise of bringing about mobile multimedia, the UMTS Task Force evolved into the UMTS Forum, established in 1996 as an industry association with the participation of manufacturers, operators, and regulators, as well as IT and media/content organizations, from all over the world.

In terms of R&D, two RACE II program projects, CODIT and ATDMA, began developing advanced high data-rate systems, one that was CDMA-based and the other that was TDMA-based. Then, in the scope of ACTS, a whole series of projects in the mobile and personal communications area started developing the concept of UMTS.

Without ignoring the significant contributions toward advanced services and applications and enabling technologies, two areas deserve particular mention in respect to UMTS:

- *Air interface:* at the origin of the two proposals retained in ETSI's historic agreement of January 29, 1998, a harmonized WCDMA/TD-CDMA scheme.[7]
- *Network:* provided major contributions to ITU-T, proposing the separation of radio dependent and radio independent parts of the network, and facilitating the design of the required interfaces and the interworking with 2G systems, namely, GSM.

Besides the contribution to UMTS, moving it from concept to reality, European cooperative R&D in the area of mobile and personal communications provided a determinate contribution in other aspects, all contributing to a more complete perspective of 3G:

- Broadband wireless, namely, developing the HiperLAN family;
- Digital broadcasting and its integration with mobile communications, which is called interactive broadcasting;
- Reconfigurable radio systems and networks, broadening the scope of software-defined radio.

It is now understood that both broadband wireless and interactive broadcasting will eventually be integrated with UMTS, while reconfigurable radio is regarded as an essential technology for the above integration, which goes well beyond the terminals to enable seamless service provision across heterogeneous networks.

---

[7] ETSI's breakthrough agreement on the UMTS Terrestrial Radio Access (UTRA) recommends the use of WCDMA in the paired portion of the radio spectrum and TD-CDMA in the unpaired portion. The coexistence of the two technologies requires optimization from the start; the maximum level of commonality must be sought with a subsequent effort to harmonize the two modes.

In order to cover the whole spectrum of system issues, the efforts of many projects, covering everything from radio access to backbone network, from terrestrial to satellite, from public to private systems, and from enabling technologies to services and applications, had to be brought to bear. The resulting mobile domain cluster worked particularly well in building up synergy among the participating projects, providing a discussion forum for the integrated perspective above as demonstrated by the mentioned results and the contents of these volumes.

Concertation was the mechanism for this clustering. To further enrich this cooperation, the projects were kept in close contact with the real world, exposed to new concepts, trends, technologies, and policy orientations through the organization of open thematic workshops and the annual Mobile Summit—certainly the largest European event in this area. To broaden the scope of the dissemination of the results, the projects are also heavily involved in major international conferences, both in terms of papers and participation in panels, as well as contributing to special issues of major international technical journals.

In conclusion, I thank my colleagues from all the ACTS Mobile Domain projects for their critical contribution toward 3G, their commitment to cooperative R&D and to making UMTS a reality, and, in particular, for their contributions to ETSI, CEPT, and ITU.

I also extend my thanks to the European Commission services that provided the framework for the cooperative R&D and animated the mobile domain, bringing in new ideas and launching new themes, and to those who were in charge of the regulatory framework, in particular, the UMTS Decision.

Robert Verrue
Director-General, DG INFSO
European Commission

Preface

राजविद्या राजगुह्यं पवित्रमिदमुत्तमम् ।
प्रत्यक्षावगमं धर्म्यं सुसुखं कर्तुमव्ययम् ॥२॥

rāja-vidyā rāja-guhyaṁ
pavitram idam uttamam
pratyakṣāvagamaṁ dharmyaṁ
su-sukhaṁ kartum avyayam

rāja-vidyā—the king of education; *rāja-guhyam*—the king of confidential knowledge; *pavitram*—the purest; *idam*—this; *uttamam*—transcendental; *pratyakṣa*—by direct experience; *avagamam*—understood; *dharmyam*—the principle of religion; *su-sukham*—very happy; *kartum*—to execute; *avyayam*—everlasting.

The knowledge is the king of education, the most secret of all secrets. It is the purest knowledge, and because it gives direct perception of the self by realization, it is the perfection of religion. It is everlasting, and it is joyfully performed.

The Bhagavad Gita (9.2)

The Advanced Communications Technologies and Services (ACTS) research program has launched many projects divided according to special thematic areas with the focus of this book on the mobile domain. The idea for the ACTS Success, Dissemination and Promotion (ASAP) project arose naturally from observation of the enormous efforts put into research and development programs funded by the European Commission, particularly in the mobile domain area. These involved many manufacturing companies, operators, and universities hoping, by joining their forces, to contribute to the progressive development of a global information society.

Collecting the results of the projects in the mobile domain area of the ACTS research program provides a permanent record and source of information about the achievements of the projects, which all contributed to the rapid and far-reaching successes of the European mobile communications industry which we have recently experienced.

At the start of the project, I was with the Delft University of Technology, where I was actively involved with the Future Radio Wideband Multiple Access Schemes (FRAMES) project. Naturally, being part of its activities and efforts could only convince me that it was a good idea to show how much dedication was needed before achieving a solution. Also, as a lecturer, I was aware that by summarizing the efforts of various projects, young students can learn exactly what research involves, that small details do matter, and that teamwork is of great importance.

Later on, I continued as the technical manager of the project at the Center for PersonKommunikation, Aalborg University, also an active participant in some ACTS projects. Here, the two volumes of this book were completed. The covered areas are depicted in Figure P.1.

The material of the book targets everyone interested in mobile communications, regardless of occupation. I hope that all readers will experience the benefits and power of this knowledge.

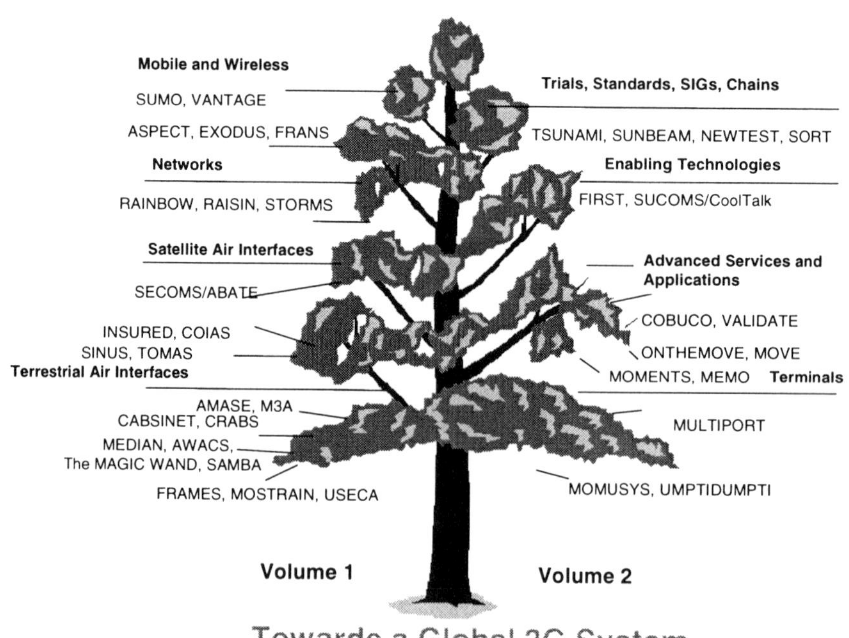

Figure P.1 Coverage of the book.

Acknowledgments

The material in this book originates from all the mobile domain projects conducted within the framework of the European-Union-funded ACTS program of pre-competitive, collaborative RTD. The idea was to provide a permanent record of the efforts and results of the projects, as well as a summary of their contributions towards the development of a new information society. The concept was developed further as a European ACTS project itself: The ACTS Success, Dissemination and Promotion in the Mobile Domain (ASAP). The editor, therefore, would like to thank the European Commission for providing this opportunity and, in particular, Jorge M. Pereira, who, as a project officer guided and fully supported the project until its completion.

The project started at Delft University of Technology, The Netherlands, and credit for the initial and significant efforts goes to Neill Whillans. He laid the groundwork for the project work, putting in many long days and evenings in the process.

Later the project's main technical work continued under the wing of the Center for PersonKommunikation (CPK) at Aalborg University, Denmark. I would like to thank the CPK for its support throughout the project. In particular, my thanks go to Rasmus Olsen and Martijn Kuipers, who offered a lot of support throughout the preparation of the CD-ROM and books.

Credit for her major efforts in helping me to put the material together and shape it into a final version goes to Albena Mihovska from CPK, who was involved in the technical management of the ASAP project as a research engineer.

In the scope of the ASAP project, two other individuals were involved in the manuscript preparation. Special thanks for her dedicated work go to Liljana Gavrilovska from the University of Skopje, Macedonia (currently with CPK) and Sudhanshu Jamuar from the Indian Institute of Technology in Delhi, India.

I would also like to thank Julie Lancashire from Artech House, who as part of the ASAP project was always available for fruitful discussions and whose immense support was indispensable.

Ramjee Prasad

Chapter 1

Mobile and Wireless

Mobile communications development is at a turning point. New services and technologies are being introduced, customer expectations are growing, and competition amongst operators and amongst service providers is intensifying. At the same time, the Internet is evolving quickly into a true information superhighway. Wireless services are also proliferating, while companies are starting to make way for the freedom that data networking and telecommunications offers their customers, partners, and employees. Simultaneously, telecommuting, videoconferencing, voice mail, paging, e-mail, cellular phones, personal digital assistants (PDAs), and laptops enable professionals to conduct business anywhere and at any time. Keeping abreast of new research and industry trends is key in developing a good strategy that is compatible with the fast pace of the information technology industry. When faced with making decisions, engineers need to be aware of current and emerging trends of the communications world. Many new ideas are not only creative and innovative but technically challenging as well. This calls for a very good system perspective and detailed knowledge in a number of key areas.

This introductory chapter tries to capture the most significant aspects of current and future telecommunications technology developments. It introduces the reader to the basic features of third-generation (3G) systems. Mobile communications in the global, wide, and local area will be described in the following sections.

1.1 EVOLUTION TOWARDS THIRD GENERATION: UNIVERSAL MOBILE TELECOMMUNICATION SYSTEM (UMTS) AND INTERNATIONAL MOBILE TELECOMMUNICATIONS-2000 (IMT-2000)

The universal mobile telecommunication system (UMTS) expected to be deployed in Europe in 2002 [1], will offer significant user benefits including high-quality wireless multimedia services through a convergent network of cellular and satellite components. It will deliver information directly to users and provide them with access to new and innovative services and applications beyond what current second-generation systems can provide. It will offer mobile personalized communications to the mass market regardless of location, network, or terminal [2].

Mobile and personal communications in the information society of the twenty-first century depend on a well-established, flexible, and evolving UMTS. The different aspects of UMTS and its deployment are considered in Sections 1.1.1–1.1.5.

1.1.1 UMTS

UMTS is part of the International Telecommunications Union's IMT-2000 vision of a global family of 3G mobile communication systems. The demand for high-quality wireless multimedia communications makes the development of UMTS key to the creation of a mass market expected to grow from one billion users worldwide in 2002 to two billion users by 2010. The goal of UMTS is to enable tomorrow's wireless information society by delivering broadband services (e.g., e-commerce, entertainment, and games) to mobile users. The introduction of UMTS will accelerate the convergence between telecommunications, media, and content industries, delivering new services and providing new revenue-generating opportunities. Users will be able to access information services for business, leisure, and education. UMTS will provide low-cost, high-capacity mobile communications, offering data rates up to 2 Mbps with global roaming and other advanced capabilities [3]-[6].

UMTS builds on today's significant investments in GSM mobile systems worldwide and has the support of hundreds of network operators, manufacturers, and equipment vendors [7]. It represents a unique opportunity to create a mass market for highly personalized and user-friendly mobile access.

The launch of UMTS services will see the evolution of a new, "open" communications universe, including providers of information and entertainment services, coming together harmoniously to deliver new communications services, characterized by mobility and advanced multimedia capabilities. Figure 1.1 illustrates the importance of UMTS in terms of a prognosis for its capacity and number of applications.

The development of UMTS is based on the success of the global system for mobile communications (GSM). Over 230 operators in more than 110 countries [8] have already adopted GSM. At the end of 1999, in Europe, 900- and 1,800-MHz GSM radio systems already accounted for 91% of the cellular market, with 140 million subscribers [9]. Figure 1.2 shows the segmentation of the worldwide cellular market.

Table 1.1 gives the penetration of cellular subscribers in different countries of Europe at the end of 1999. At the time of writing, the top 10 countries already have more than 50% penetration, with eight expected to be above that number by the end of 2000.

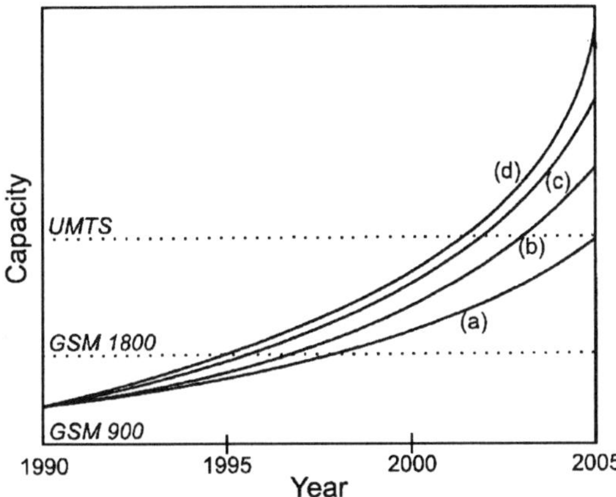

Figure 1.1 Importance of UMTS [9]: (a) current voice applications, (b) current data applications, (c) emerging applications, and (d) multimedia.

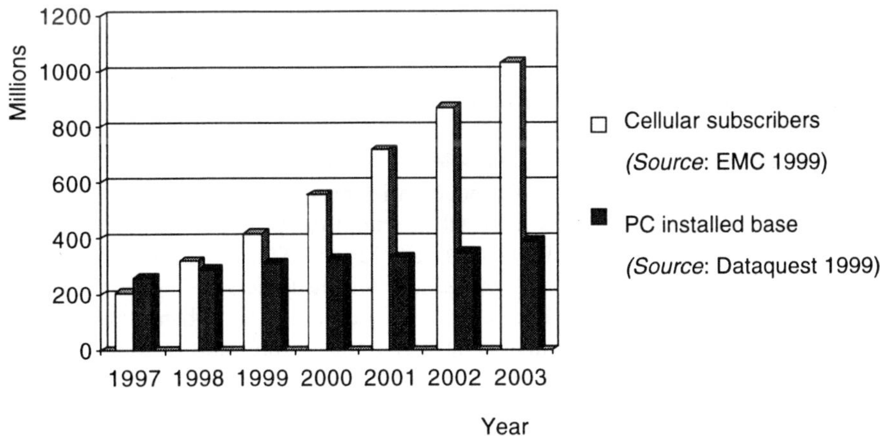

Figure 1.2 Segmentation of the worldwide cellular market.

Table 1.1

Penetration of Cellular in Europe (End of 1999)

EU-15	*Penetration of Population (percentage)*
Finland	65.22
Sweden	56.09
Denmark	49.75
Italy	49.61
Luxembourg	49.30
Austria	47.83
Portugal	44.57
The Netherlands	40.28
United Kingdom	36.34
Ireland	34.63
Greece	34.39
Spain	33.32
France	33.32
Belgium	29.36
Germany	25.65

1.1.2 Third-Generation Systems

The conviction that mobile and personal communications are a key driver for innovation and growth [10] in the next millennium has led to the launch of a number of European Union (EU)-funded research and development (R&D) projects. The Advanced Communications Technologies and Services (ACTS) program (1995–1998), with some projects extending beyond 2000, provided opportunities to further develop mobile and personal communications with the participation of academia and industry (operators, manufacturers, and service providers).

The definition of the third generation of mobile systems, called UMTS in Europe, was one of the most significant activities pursued by ACTS. The vision of UMTS had emerged from work undertaken within the program for research into advanced communications in Europe (RACE) [11] and ACTS. It called for support of all those services, facilities, and applications that customers had come to expect. It also provided the potential to accommodate future broadband multimedia services and applications with quality levels similar to the ones in today's fixed integrated broadband communications networks (IBCN). It was important that UMTS be designed in such a way to be perceived by customers as a broadband service evolution of second-generation technologies while ensuring a competitive service provision in a multioperator environment [10]. Figure 1.3 shows that UMTS will deliver narrowband and wideband services and support bandwidth-on-demand up to 2 Mbps [12].

UMTS represents a new generation of mobile communications systems in a world where personal services will be based on a combination of wireless/mobile and fixed services providing seamless end-to-end service to the user. This requires the following:

- An integrated offering of services, irrespective of the serving (wired/wireless) network;
- Mobile technology that supports a very broad mix of communication services and applications;
- Flexible, on-demand bandwidth allocation for a wide variety of applications;
- Standardization that allows full roaming and interworking capability but responsiveness to different types of markets.

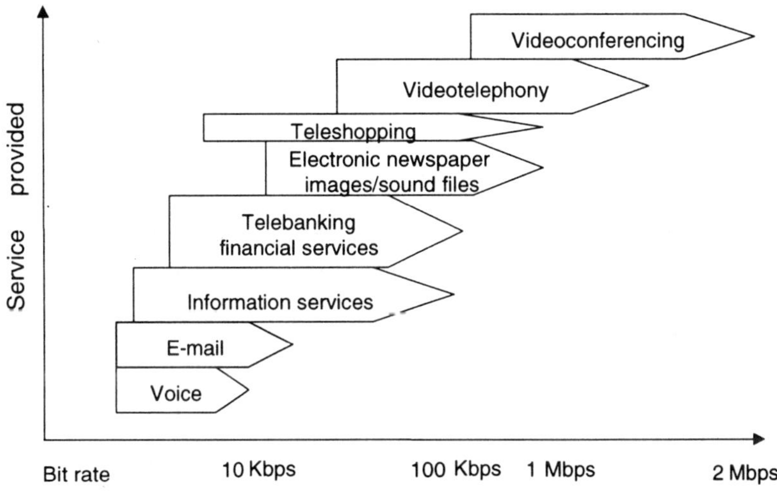

Figure 1.3 Support of bandwidth-on-demand for different types of services [12].

In that context, a good spectrum plan is of utter importance as spectrum is a limited and valuable resource that must be used in the most efficient way to satisfy the continuously increasing demand. Factors of influence to the overall spectrum demand, including UMTS/IMT-2000 needs, are population density and economic development [13]. To enable UMTS/IMT-2000 services to be offered globally spectrum bands need to be identified as relatively large blocks, if possible contiguous to the core band. Identification of additional bands for UMTS/IMT-2000 is related to the interest of the users worldwide. To facilitate roaming and the successful development of a mass market, worldwide harmonization of the frequencies is imperative. Figure 1.4 shows a possible solution for a flexible spectrum designation for UMTS/IMT-2000.

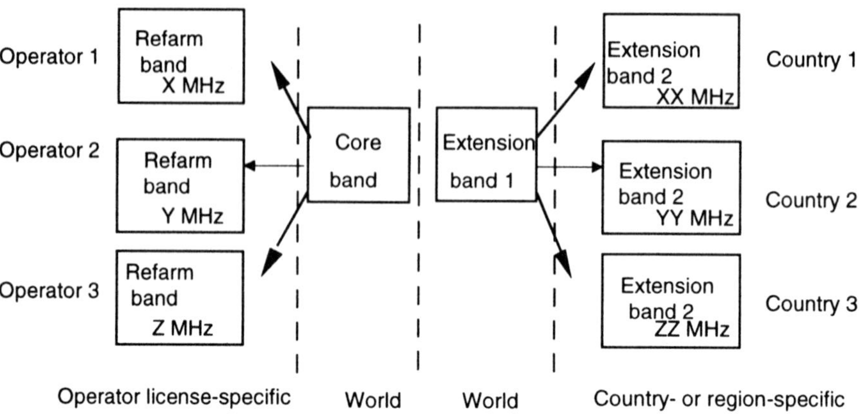

Figure 1.4 Spectrum vision [13].

The bands 1,885–2,025 MHz and 2,110–2,200 MHz, originally identified for IMT-2000, are now considered to be the core band where UMTS/IMT-2000 should initially be deployed. The identification of a worldwide core band gives a good opportunity for deployment of advanced mobile multimedia services on a worldwide basis.

For applications where there is a very large degree of asymmetry in the downstream and upstream traffic channels, UMTS, combined with digital audio broadcasting (DAB) and digital video broadcasting (DVB) techniques, can provide cost-effective solutions. The full exploitation of UMTS abilities as the integral mobile-access part of the broadband integrated switched digital network (B-ISDN) will lead to a major leap towards the provision of a technically integrated, comprehensive, and consistent personal communication system, supported by both fixed and mobile terminals. Figure 1.5 shows some of the services and applications to be offered by UMTS.

When determining the frequency spectrum requirements for UMTS/IMT-2000, however, the calculations cannot be limited to multimedia services and high-speed data services [13]. A number of the foreseen UMTS/IMT-2000 services, such as speech and low-speed data services, are currently delivered by second-generation systems, and this will remain unchanged at the initial introduction of UMTS/IMT-2000. In the satellite field, UMTS/IMT-2000–compliant systems and other mobile satellite systems (MSS) will share partially the same market. Therefore, to adopt multimedia applications sufficiently and to save spectrum resources, both circuit- and packet-switched radio access are assumed in the frequency spectrum requirements, while traffic calculations are carried out for both transmission principles. The flowchart of the calculation methodology is presented in Figure 1.6.

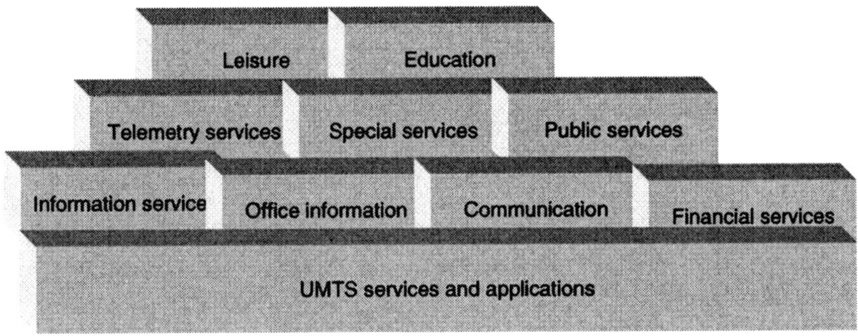

Figure 1.5 UMTS services and applications [6].

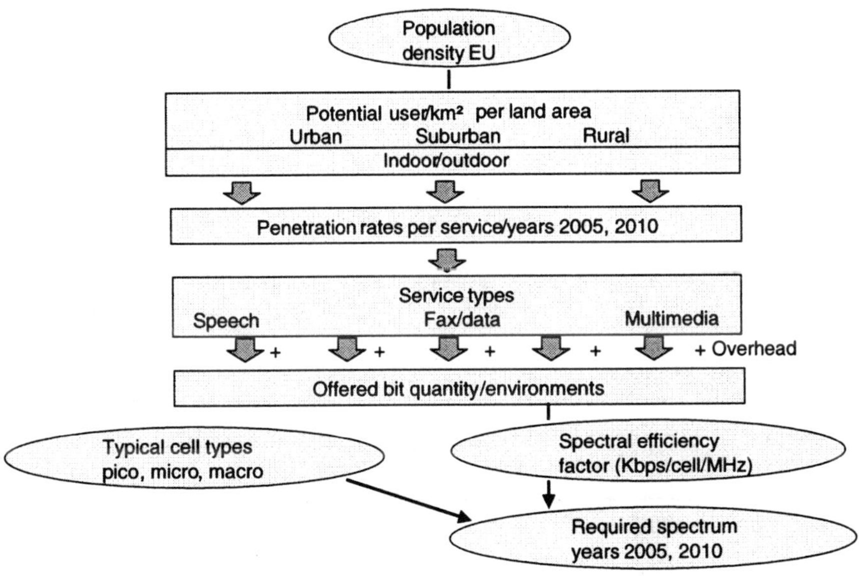

Figure 1.6 Calculation method for UMTS spectrum allocation [13].

The spectrum requirement for each service—speech, simple messaging, switched data, medium multimedia, high multimedia, and high interactive multimedia—are considered in each of six geographical operating environments: central business district (CBD), suburban, home, urban pedestrian, urban vehicular, and rural. Including the downlink and uplink, there are 2:6:6 = 72 cases to consider. The spectrum requirement calculation takes place independently for each case, with a final summation giving the total spectrum requirement.

The first stage in the calculation is to derive the total number of users per square kilometer for each of the six service classes. This can be readily derived by multiplying the population density by the service penetration. Each service in each environment generates a particular call rate (calls/hour), call duration (s), and bit rate (Kbps). Further multiplication generates the bit requirement (Kbps/km^2). The bit requirement must be increased to take account of a coding factor, overhead for signaling and packet retries, and blocking for circuit-switched services. This gives the figure for the *offered bit rate quantity*. It is reasonable to assume that the UMTS/IMT-2000 capability will be at least equivalent to that of a GSM-based system. To allow for improvements in system capability as UMTS/IMT-2000 evolves towards more sophisticated technology, some factor can be included in the calculations.

The terrestrial spectrum estimate for Europe (see Figure 1.7) as calculated by the UMTS Forum, for example, is aligned with the predicted growth of the number of mobile subscribers within the EU-15 countries. Table 1.2 shows the market forecasts for the number of physical users of mobile services and, out of them, the number of physical users of mobile multimedia (MM) services and the assumed penetration rates.

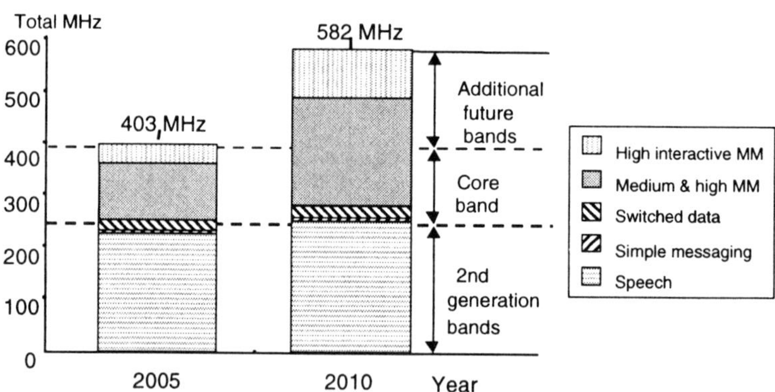

Figure 1.7 Terrestrial spectrum estimates for the years 2005 and 2010 in Europe (EU-15) [13].

The service revenues estimated for the total mobile market and based on the above calculations are € 104 billion per year, with € 24 billion for the mobile multimedia segment. The estimated traffic throughput is 6,320 million Mb/month and 3,800 million Mb/month, correspondingly. These figures include the use of enhanced second-generation services as well.

Table 1.2

Population of EU-15 and Penetration of Future Mobile Services [13]

Year	*Population in Millions*	*Physical Users in Millions*	*Penetration*	*Physical Users of MM Service in Millions*	*Penetration*
2005	385	200	0.52	32 (20 users of high MM service)	0.08 (0.05)
2010	387	260	0.67	90	0.23

1.1.3 Evolution of Fixed and Mobile Networks Towards UMTS

Today's fastest growing industry following the mobile telecommunications is the Internet. In a recently published report, the UMTS Forum predicts global revenues in excess of $164 billion for three forecasted services—customized infotainment, mobile intranet/extranet access, and multimedia messaging)—over the next 10 years. Customized infotainment is expected to produce the highest share of revenues thanks to its low delivery cost and high consumer appeal. The same report also predicts that more data than voice will flow over mobile networks as soon as 2005 [14]. The pressure placed upon spectrum because of MM applications growth is illustrated in Figure 1.8.

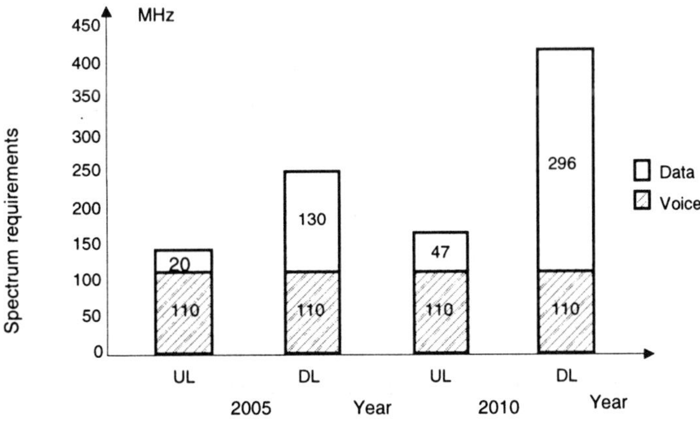

Figure 1.8 Mobile services types for 2005–2010 versus spectrum requirements [14].

UMTS will offer a range of mobile services (see Figure 1.5), increasing the number of subscribers in terms of high- and low-speed data users. Figure 1.9 shows how the new services will affect that number [15]. It should be noted that the forecast number of persons using mobile services is roughly equal to the speech service users but the use of the higher bandwidth multimedia services will increase over time, leading to a proportional decrease in the share of speech services [13]. The future mix of services should result in even more spectrum utilization than the current one, particularly through the use of delay in high-volume data applications.

UMTS will provide access to a converged market between the Internet and mobile telecommunications in terms of price, service, functionality, user services, and transmission technology, and at speeds up to 250 times higher than today's mobile telephones. UMTS operators will also be some of the first to implement end-to-end Internet protocol (IP)-based networks. In fact, while the general packet radio service (GPRS), starting at the time of the writing of this book and overlaid on GSM networks, is expected to use IPv4, UMTS is expected to be IPv6-based from the start [16] (see Figure 1.11).

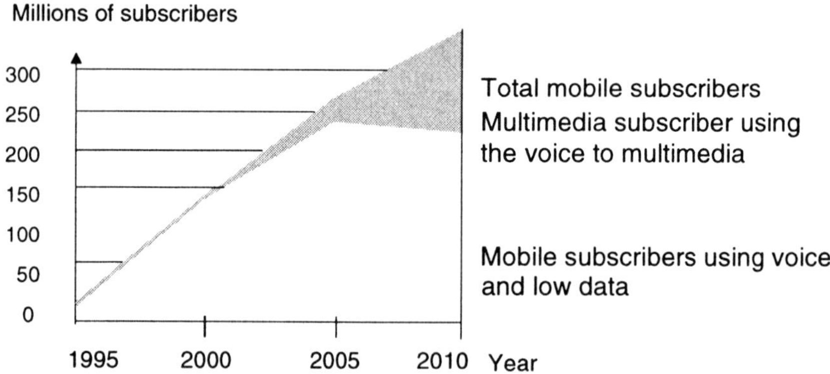

Figure 1.9 Prognosis for the total number of mobile subscribers through 2010 [15].

Figure 1.10 illustrates the growth of mobile Internet towards an all-IP–based network.

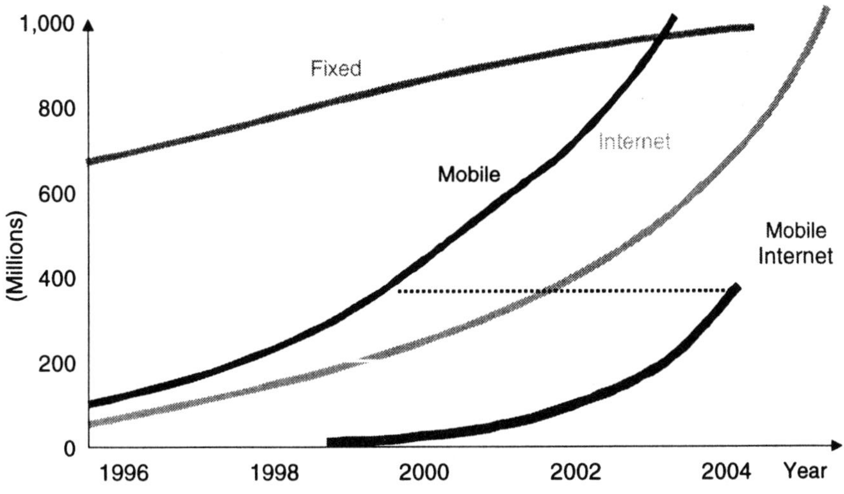

Figure 1.10 Growth of mobile Internet in terms of subscribers and prognosis by 2004 [16].

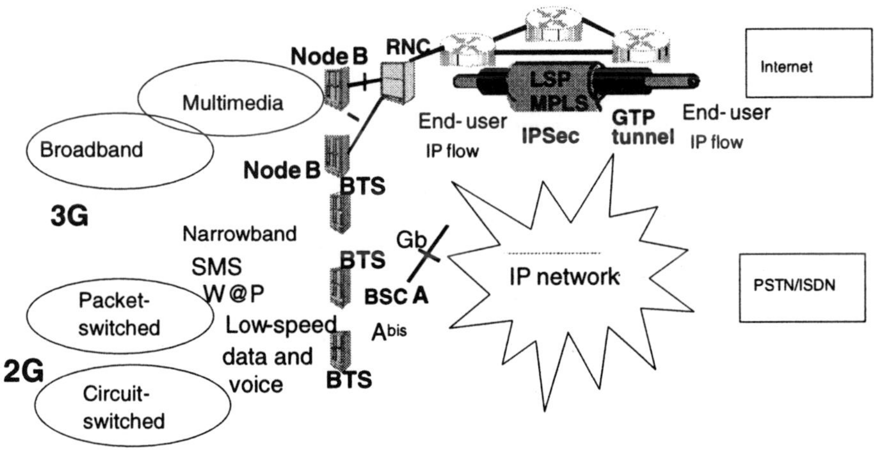

Figure 1.11 End-to-end IP–based mobile network interdomain architecture.

The UMTS Forum has predicted that UMTS/3G market acceptance will be driven by a multitude of applications [17]. The central element of the mobile information society will be the emergence of a powerful and user-friendly MM portal. On the other hand, the rapid, wide-scale introduction of IPv6 will play a vital role in overcoming problems related to end-to-end security as well as numbering, addressing, naming, and quality of service (QoS) for real-time applications and services [18]. The theme was echoed at a recent IPv6 summit in the United States, where wireless industry observers claimed that 3G mobile would represent the "killer application" for IPv6. The benefits of IPv6 to mobile network operators and mobile end users are briefly described as follows:

- Auto configuration;
- Embedded encryption support and authentication;
- Embedded mobility;
- Internet provider selection;
- Efficient packet processing in routers;
- Easy management and auto configuration;
- Efficient address allocation;
- Renumbering possible.

1.1.4 Mobile Satellite Systems

While second-generation (2G) MSSs are starting commercial operations, the scientific community, the telecommunications industry, operators, and regulatory bodies are working together towards the conception of new MSSs that fit UMTS/IMT-2000, thus allowing for considerable reuse of terminal hardware and software [19].

The concept of universal communications means information at any time, in any place, in any form. It is no longer a futuristic vision: three nongeostationary (NGSO) satellite systems were in the process of being introduced [20]–[22]. These systems were designed to provide global MSS, such as voice, low-rate data, and messaging to handheld terminals. They targeted a significant market for applications, such as cellular roaming, cellular nonserved areas, fleet management, rural telephony/data, and maritime/rail/aeronautical mobile applications [23]. MSS, which supply services from either moving terminals or portable/transportable terminals, will not try to replace or compete with the cellular services; they will mainly serve vast and remote geographical areas where the installation of terrestrial-based systems is not economically viable [24]. That will require some sort of coordination between cellular systems and MSSs. In the short term, integration at the service level will be acceptable [25]. MSS terminals will be able to work both as satellite terminals and as cellular terminals (e.g., ICO and GSM). In the longer term, a deeper integration of terrestrial and satellite networks will be necessary.

Current expectations are that less than 20% of the world's land area will be covered by terrestrial cellular networks within the envisaged timescales of UMTS/IMT-2000 [13]. Satellite systems are important to UMTS/IMT-2000 to provide complete coverage. On the other hand, in the EU-15 group, terrestrial UMTS/IMT-2000 can be expected to cover more than 80% of the population in 2010. MSSs are a solution to the problem of economically covering large areas and serving widely scattered or remote subscribers in rural areas in both developing and developed countries, as well as for ship and aircraft communications. Three levels of integration are identified to create transparent interworking between the terrestrial and satellite component:

- *Network integration.* Space and terrestrial networks operating as separate entities;
- *Equipment integration.* Requirements for common service standards, consistent transmission parameters, and common radio interface between satellite and terrestrial implementations;
- *System integration.* Satellite is an integral part of the terrestrial network and able to support hand-over between terrestrial and satellite megacells.

Assumptions used in the calculation of satellite spectrum requirements can be simplified by the conclusion that two types of terminals are expected to provide mobile satellite services: handheld terminals providing speech and low-rate data services and somewhat bigger terminals providing multimedia services [13]. The satellite spectrum estimates are presented in Figures 1.12–1.13, which also show the existing MSS allocations, totaling 229 MHz. In Europe, however, because the bands 2,520–2,535 MHz and 2,655–2,670 MHz are not expected to be made available for MSS, the total available MSS spectrum is 199 MHz. The comparison shows that there is a requirement for additional MSS spectrum of 50–60 MHz by 2010, with lower requirements in Europe [13].

UMTS, after being initially defined by ETSI SMG, is now under the responsibility of the 3G Partnerships Project (3GPP). The process is for the time being tentative. The UMTS/IMT-2000 system framework will provide for a wide range of services (e.g., toll-quality voice, low error-rate data services) to mobile, handheld and portable terminals in different environments (indoor, outdoor, public, private, cellular, cordless). In order to provide these services to a global market, both populated areas and remote regions worldwide must be covered by the new system; this will be achieved by integrating a terrestrial cellular radio network with a satellite-based system.

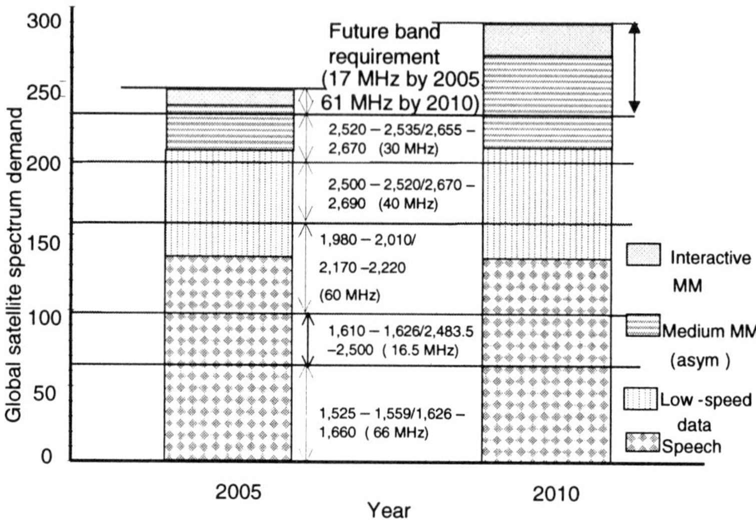

Figure 1.12 Satellite spectrum estimate for the years 2005–2010 (global hot spot) [13].

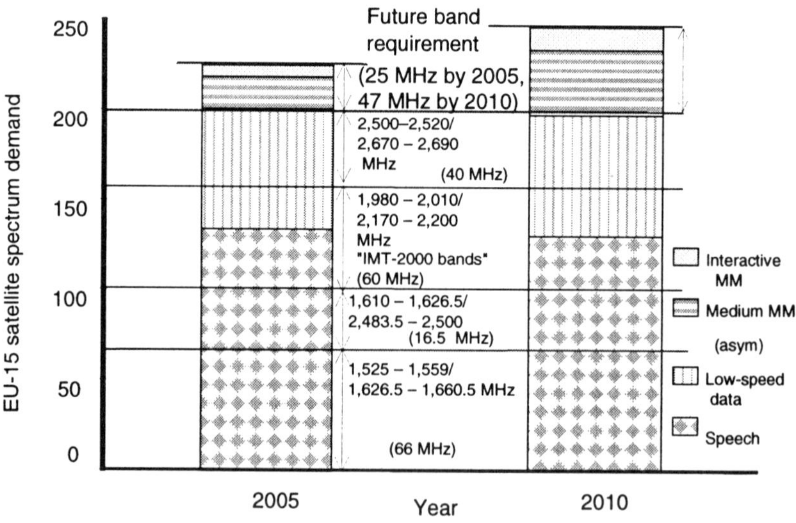

Figure 1.13 Satellite spectrum estimate for the years 2005–2010 in EU-15 [13].

New MSSs will then be conceived taking into account the definition of UMTS/IMT-2000. Code division multiple access (CDMA) seems to be a very good candidate because of its inherent properties of flexibility and robustness against interference.

Fixed and broadcasting satellite services are also undergoing a tremendous and rapid change. The situation is moving from a scenario of only a few operators to one in which commercial operators are leading the way in terms of innovation and establishment of new services (e.g., DVB, interactive television, and DAB). For this purpose, coded orthogonal frequency division multiplexing (COFDM) has been selected as the European standard for terrestrial DAB [38] and for DVB-T [39].

1.1.5 Technology Trends and Conclusions

To meet the UMTS-2002 deployment deadline [1], a number of steps are being undertaken by manufacturers, standardization bodies, operators, and regulators around the world. The goal is to create an adequate regulatory framework, to make licenses available, and to allocate adequate spectrum to operators. To stimulate uptake of services in a worldwide market, it is also very important to encourage the simultaneous uptake of UMTS in as many countries as possible, building upon the current GSM footprint. Table 1.3 presents the UMTS Forum prognosis for the number of users in the European mobile market [40].

Table 1.3

European Mobile Market in 2005 [40]

	Total Mobile	*MM*	*Percentage of MM*
Users (millions)	200	32	16
Service revenue (billion ECU)	104	24	23
Traffic (million MB/month)	6,320	3,800	60
Traffic (MB/users/month)	32	119	—

The transition to 3G represents a giant and expensive leap for GSM operators. An outstanding factor in this process is the ability of GSM to constantly reinvent itself to meet new requirements. Vendors claim that this leap can be made less costly and less frightening [41] by pursuing the evolutionary path through GPRS and enhanced data rates for global evolution (EDGE). The introduction of GPRS is one of the key steps in the evolution of today's GSM network to 3G because it offers an increase in data rates, from 9.6 Kbps to 115 Kbps. It is also attractive to users because of its capability to charge per data sent and received, and not per time. Using a packet data service, subscribers are always connected, so services are easy and quick to access. GPRS allows innovative services to be created enabling new and previously inaccessible market segments to be addressed. Machine-to-machine and person-to-machine communications become possible.

The implementation of EDGE is the next step towards 3G. EDGE will offer services and applications at speeds up to 384 Kbps, through high-level modulation. It will allow for the further exploration of the advantages of GPRS, with a fast-connection setup and higher bandwidth (E-GPRS). Often overlooked is the impact that the introduction of GPRS and EDGE will have on the core backbone network. At present, the GSM backbone operates at 2 Mbps, which is likely to be insufficient to handle the demand, but operators are understandingly reluctant to commit the considerable investment that will be needed to upgrade their backbone. The base-station system (BSS) is one of the most important elements in the GSM structure, and all the infrastructure manufacturers are continuously improving their base station portfolios to meet ever rising demand for higher capacity, greater coverage, and improved quality. With the latest challenge—the introduction of packet data and the need to handle higher data throughput—being overcome, the road ahead lies open for the greater challenge of UMTS.

Many communication industries are already presenting their vision on future developments, in particular on wireless Internet [42]. Combining the success of mobile communications and the Internet, wireless Internet is generally seen as the new frontier in the world of communications. Both hardware and software suppliers are developing products to prepare for the 3G of mobile communication. On the other hand, mobile network operators are investing in these new data technologies to pilot applications in the fields of mobile office, e-commerce, Web browsing, and so forth. An important aspect of this process is the "preparation" of the future users of the services of the 3G networks. This will require a proactive stance of the operators (at least) to stimulate a higher usage of current wireless data applications. It is crucial to evolve from the cumbersome, piecemeal provision of services to an integrated, user-friendly, personalized provision of enhanced services across networks [32]. Therefore, many operators should make the shift from focusing purely on new data technologies to applying marketing instruments with which to stimulate more user-friendly versions of current applications. Emphasizing lifestyle issues is considered very important for the future of mobile communications [42]. Mobile phone usage, especially among groups of adolescents and students, is very trendy. This potential target group has a high Internet usage and a positive attitude towards new technology. They also have an active and mobile lifestyle and spend much time within peer groups. Moreover, they carry newer handsets than the rest of the market and are the mobile phone generation of the future. This makes them very important for the acceptance of future wireless applications.

The European Telecommunication Standards Institute (ETSI) identified the following major trends in the process of evolving GSM and developing UMTS:

- From monopoly to competition;
- From hardware and technology to software and applications;

- From separate industries to convergence/integration;
- From telecommunications to Internet;
- From subscriber to client.

1.2 BEYOND UMTS: WIRELESS BROADBAND COMMUNICATIONS

Current wireless or mobile systems, despite their evolution, are still constrained in terms of the data rate they can offer and their flexibility to manipulate complex, yet user-friendly multimedia services to allow for the delivery of a rich variety of audio, visual, and text-based information in addition to simple voice services [29]. This need presents an opportunity to offer to users something new—a mobile system capable of managing and delivering a much wider range of information services to the mass market. In this context, UMTS becomes a very innovative platform to bring multimedia to the consumer. To ensure that UMTS completely develops its potential, IP, Internet, and other value-added service providers focus on extending their offering to UMTS.

UMTS, as a modular concept, takes full regard of the trend towards convergence of fixed and mobile networks and services, enabling the development of a huge number of applications. As an example, a laptop with an integrated UMTS communications module becomes a general-purpose communications and computing device for broadband Internet access, voice, video telephony, and conferencing for either mobile or residential use. As the number of IP networks and applications is growing fast, UMTS will become the most flexible broadband access technology because it allows for mobile, office, and residential use in a wide range of public and private networks. UMTS can support both IP and non-IP traffic in a variety of modes, including packet, circuit-switched, and virtual circuit, thus benefiting directly from the development and extension of IP standards for mobile communication. New developments such as IPv6 allow parameters like QoS, bit rate, and bit error rate (BER), vital for mobile operation, to be set by the operator or service provider. Developments on new domain name structures will increase the usability and flexibility of the system, providing unique addressing for each user, independent of terminal, application, or location. The challenge to the communications industry is to integrate the technologies needed for UMTS and thereby transform vision into reality.

1.2.1 Introduction to Wireless Broadband Communications

The development of telecommunications, with the growing convergence of technologies, can be characterized as a multimedia revolution. Third-generation licenses for advanced multimedia services are starting to be handed out (through auctions and beauty contests), and new technologies like GPRS and wireless access protocol (WAP) are being implemented in the more advanced networks. It is obvious that the world of mobile communications is undergoing the most radical change in its history. It is changing the way we work, learn, communicate, buy, and enjoy ourselves. At the same time arises the need for higher data rates.

As the popularity of the mobile phone has increased, so has the demand on such a scarce resource as the radio spectrum. In turn, this has driven the development of ever more efficient coding and modulation schemes. UMTS will offer global universal services up to 2 Mbps. For users demanding more capacity for high-quality applications, the MBS was capable of supporting advanced broadband services [43]. Interworking of broadband mobile technologies with a wireless local area network (WLAN), compatible with B-ISDN and asynchronous transfer mode (ATM) was seen as the key issue for future integrated interactive broadband multimedia communications [44]. Interworking of broadband mobile technologies with a wireless local area network (WLAN), compatible with B-ISDN and asynchronous transfer mode (ATM) was seen as the key issue for future integrated interactive broadband multimedia communications [45]. The objective of MBS is to give mobile users access to the range of broadband services available for fixed users of the IBCN. MBS is also meant to provide the enabling technology to support new applications that arise from the mobility offered by the system. It will offer data rates up to 155 Mbps to mobile users in various environments, which are necessary for future multimedia applications. MBS will be compatible within the limitations set by UMTS [43]. While UMTS might offer good coverage but limited capacity, MBS will provide fully broadband services in restricted areas.

UMTS and MBS must be locked into a convergence process toward access-independent, universal personal communications. Figure 1.14 illustrates the roles and interfaces that may be built upon the technology and standards solutions for UMTS. This is only one example of the arrangements that the highly flexible UMTS architecture and interfaces will enable.

The concept of MBS was born in the RACE program (1988–1992). Later, in the RACE II program (1992–1995), the MBS concept was developed in detail—architecture, services and applications, transmission techniques, protocols, and a demonstrator was built to prove the concept [45]. Additionally, the European Radio-Communications Office (ERO) provisionally allocated two paired frequency bands at 40 and 60 GHz [46]. The ACTS program (1995–1999) followed with different projects in the broadband area—SAMBA, MEDIAN, AWACS, and the Magic WAND.

The attributes "broadband" and "mobile" are key issues in wireless communications. Both depend on and influence the development of technology and its implementation. Broadband means in this context transmission capacity "beyond UMTS," namely in the range 10–155 Mbps with all possible communication types and services. Mobility includes features like coverage and operational environments, maximum speed, as well as size, weight, and power consumption of the mobile terminal.

Radio spectrum is essential to wireless broadband communications. It must be managed efficiently in order to achieve the best possible spectrum efficiency. For this purpose new bands beyond those assigned for UMTS [26]–[27] are needed—additional spectrum is necessary for new systems, but at the same time only millimeter-wave bands offer sufficient bandwidth for broadband transmission [28].

1.2.2 The Demand for High Data Rates

The demand for high data rates and high-capacity systems is acknowledged to be one of the "hot topics" in the area of telecommunications research [30]. In the communications industry, the ultimate goal is to provide customers with a variety of versatile broadband services, from multimedia to interactive and universal services.

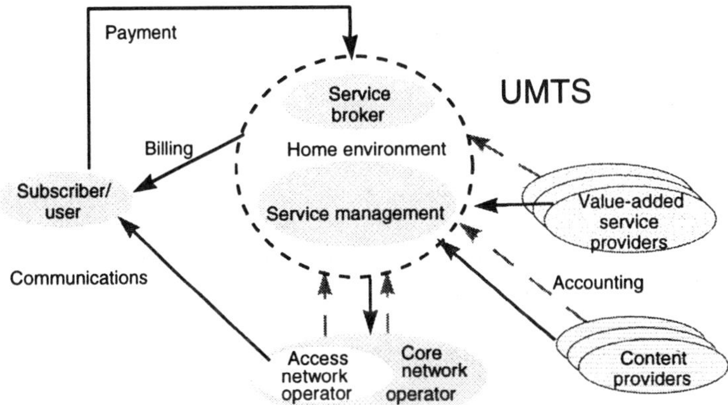

Figure 1.14 Evolution beyond 3G systems upon UMTS technology and solutions [29].

The evolution of technology has already made clear that the real broadband services will be first deployed and provided over IP-based networks. Future IPv6 networks will provide high-speed data transmission capability for full-fledged multimedia services while performing the following functions:

- Support sufficient throughput on an end-to-end basis;
- Ensure QoS for those applications;
- Provide a highly secure communication environment.

Because of advancements in information presentation and networking technologies, the Web has been promoting strong consumer demand for broadband services and is driving the emergence of global multimedia services. For example, there is a global demand for Web-based broadband services, obvious from the wide range of applications to many different businesses. There are many driving forces behind innovative broadband services:

- High-capacity/reliability links are available for the transfer of large amounts of information.
- The user sees technology as a means of augmenting senses in every field of telepresence.

In an evolutionary scenario, where one makes the most of existing and new systems, it is necessary to improve the efficiency of the three technological platforms—UMTS, MBS, WLANs—in order to offer new opportunities for advanced mobile services. The different system approaches are complementary. The wide-area cellular systems enable greater mobility through handoff and roaming, whereas WLANs and MBS offer higher data rates as a result of coverage area restriction, mainly indoors and outdoors, respectively, and reduced multipath delay spread. There is also a need to establish interoperability of these services; a major issue in this context is continuity of services across different radio environments: in-building picocells, outdoor microcells, and macrocells delivered by networks of widely differing capabilities. The technological feasibility of underlying mobile telecommunications technologies, for acceptable bearer capability, QoS, integration, and interworking requirements, should be identified through system-oriented demonstrators that take place using UMTS, MBS, and broadband WLANs system platforms. This is to perform a fine-tuning of the technologies that are to achieve the desired functionality.

1.2.3 WLANs

Mobility and proliferation of portable and laptop computers and PDAs, together with potential cost savings in avoiding the wiring and rewiring of buildings, are the driving forces behind the introduction of WLANs. WLANs have thus received a lot of attention from equipment manufacturers, because they represent the coming together of two of the fastest-growing segments of the computer industry, LANs and mobile computing. Ongoing research activities seek solutions that recognize application, environment, cost, performance, networking, and system architecture requirements, involving [28]:

- Frequency allocation/selection;
- Bandwidth-on-demand;
- Efficient coding schemes;
- Specification of medium access protocols and link control;
- Connectivity aspects related to the backbone wired or wireless communications networks.

Different working environments (e.g., office, production line, storehouse) have different requirements in terms of security, range, transmission rate, reuse of frequencies, cost, maintenance, and so forth. At present, commercially available WLANs are mostly specified to the IEEE 802.11 standard [31], but soon high-performance radio LAN (HIPERLAN) Type 1 equipment will be available. Similar to IEEE 802.11, the HIPERLAN Type 1 protocol is Ethernet-like and provides for a fully decentralized system allowing ad hoc networking [32]. Figure 1.15 illustrates both versions of the IEEE 802.11 standard.

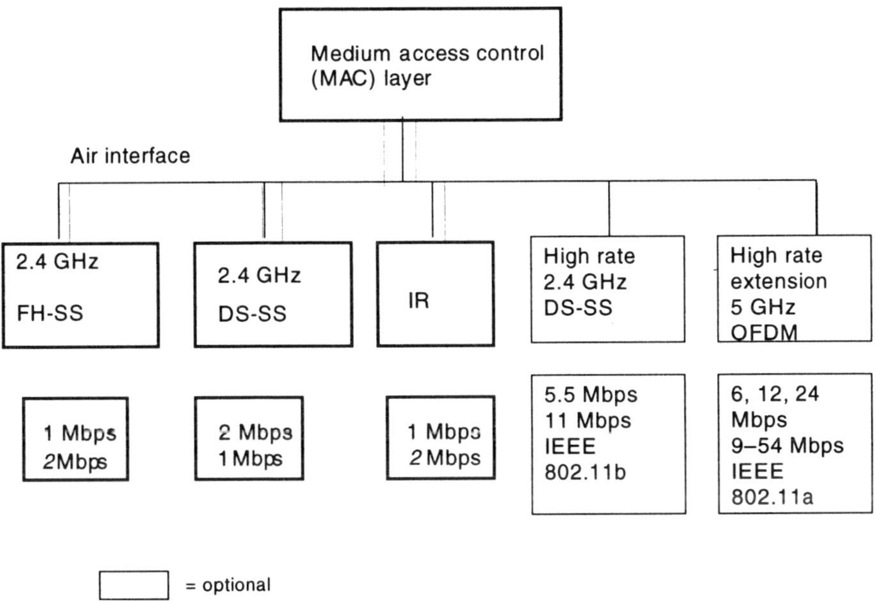

Figure 1.15 Scope of IEEE 802.11 standard.

The IEEE 802.11 has specified for WLANs high-speed extension, physical layers with bit rates in the range up to 11 Mbps at 2.4 GHz (IEEE 802.11B) and 24 Mbps or higher at 5 GHz (IEEE 802.11A) [33]. The techniques proposed for WLANs include orthogonal frequency division multiplexing (OFDM) for IEEE 802.11A and HIPERLAN Type 1 and complementary code keying (CCK) for IEEE 802.111B. The IEEE 802.11 architecture is illustrated in Figure 1.16.

The OFDM technique has recently been adopted not only by WLAN standards but also in DAB and DVB. OFDM has the ability to deliver high data rates without the requirement for explicit channel equalization and the flexibility to permit optimum control of individual subcarriers. Still in the process of development are timing and frequency synchronization issues, the knowledge of which is of particular importance for the full application of OFDM.

For the broadband fixed point-to-multipoint wireless access systems, examples are IEEE 802.16 and ETSI broadband radio access networks (BRANs) HIPERACCESS. The latter is a long-range variant following the HIPERLAN Type 1 standard and intended high-speed access (25 Mbps is the typical data rate) by residential and small business users to a wide variety of networks including the UMTS core networks, ATM networks, and IP-based networks. (HIPERLAN Type 2 might be used for distribution within premises.) Spectrum allocation in the 40.5–43.5-GHz band is also being discussed. The first batch of HIPERACCESS specifications is ready for publication in the first half of 2001.

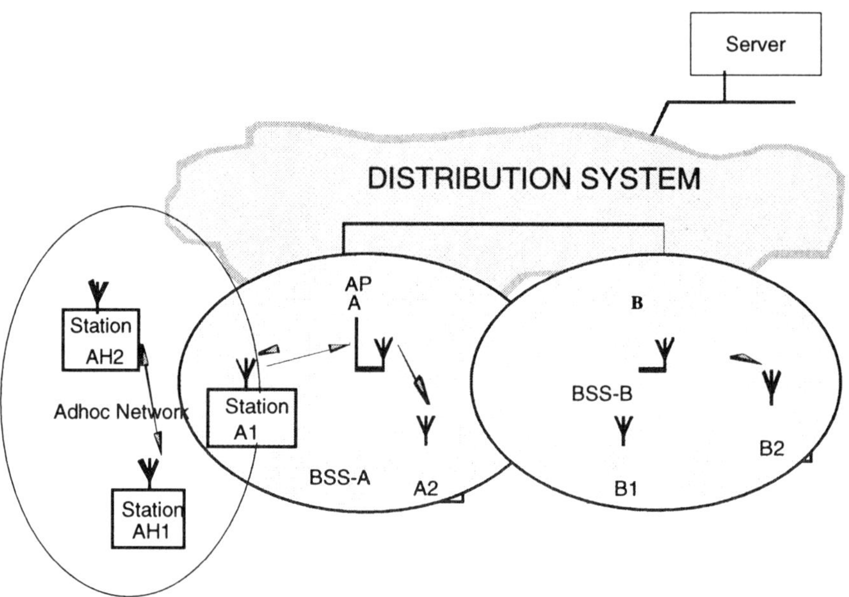

Figure 1.16 IEEE 802.11 architecture.

To support both asynchronous data and time critical services (e.g., packetized voice and video) that are bounded by specific time delays, and to achieve an acceptable QoS, ETSI BRAN developed the HIPERLAN Type 2 standard, which provides high-speed multimedia communications between different broadband core networks and mobile terminals [34]. HIPERLAN Type 2 provides a flexible platform for a variety of business and home multimedia applications that can support a set of bit rates up to 54 Mbps. In a typical business application scenario, a mobile terminal gets services over a fixed corporate/public network infrastructure. In addition to QoS, the network will provide mobile terminals with security and mobility management services when moving. In an exemplary home application scenario, low-cost and flexible networking is supported to interconnect wireless digital consumer devices.

The standard can support multibeam antennas to improve the link budget and to reduce interference in the radio network. It also defines a set of protocols (measurements and signaling) to provide support for a number of radio network functions: dynamic frequency selection (DFS), link adaptation, handover, and power control.

To cope with the varying radio link quality (interference and propagation conditions), a link adaptation scheme is used. It aims at keeping up the communications link at low signal-to-interference ratios (SNRs) in order to maintain the QoS and to trade off between the communications range and data rate. Based on link quality measurements, the physical layer data rate is adapted to the current link quality. Transmitter power control is supported in both mobile terminal (uplink) and access point (downlink). The uplink power control is mainly used to simplify the design of the access point receiver by avoiding automatic gain control at the access point. The main goal of downlink power control is to fulfill the regulatory requirements in Europe to decrease interference to other systems using the same 5-GHz band.

The 5-GHz band is open in Europe, the United States, and Japan. The current spectrum allocation at 5 GHz comprises 455 MHz in Europe, 300 MHz in the United States, and 100 MHz in Japan.

Parallel to the HIPERLAN Type 2 standardization work, the Multimedia Mobile Access Communications (MMAC) Association in Japan started to develop different high-speed radio access systems for business and home applications at 5 GHz. One of these systems for business applications in corporate and public networks is aligned with HIPERLAN Type 2 at both the physical layer and the data link control (DLC) layer. In addition, the PHY layer of IEEE 802.11 standard in the 5-GHz band is harmonized with that of HIPERLAN Type 2. With these alignments, the three communities succeeded to specify a unique radio platform at 5 GHz, which supports the development of cost-efficient multimode terminals for worldwide high-speed communications. The ETSI HIPERLAN Type 2 functional specifications encompass the physical (PHY) layer, the DLC layer, and the convergence layers (CLs) that perform service-specific functions between the DLC layer and the network layer. In other words, specific convergence sublayers are used on top of the HIPERLAN Type 2 PHY and DLC layers to provide access to networks such as IP, ATM, or UMTS. This makes HIPERLAN Type 2 a multinetwork air interface. Figure 1.17 shows the HIPERLAN Type 2 reference model [35].

Very recently the IEEE 802.16 [36] committee was formed to set standards for broadband fixed-point-to-multipoint systems, typically operating in bands between 10 GHz and 66 GHz and offering bit rates of several tens of megabits per second to customers.

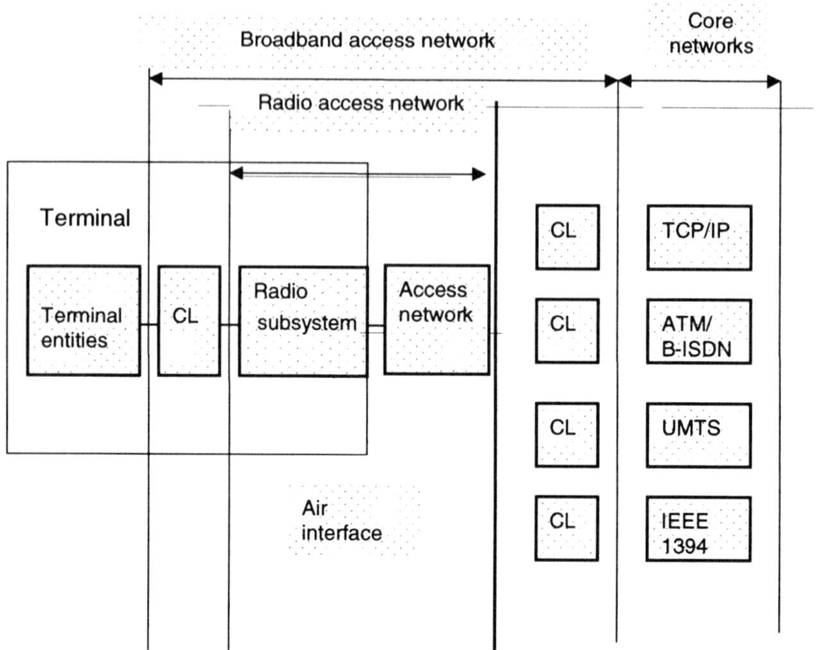

Figure 1.17 HIPERLAN Type 2 reference model [35].

Advances in coding, signal processing, and access technologies are extremely important for the further development of wireless systems since they help to increase the link capacity, while addressing the power, size, and complexity constraints of the portable units.

1.2.4 ATM-Based Wireless Mobile Broadband Multimedia Systems

ATM was identified as a prime candidate to provide high-speed multimedia communications [35]. Wireless ATM became a popular and challenging technical pursuit further intensified by the fact that future personal communication systems (PCSs) were once considered to be based on an ATM-backbone network infrastructure [37]. Currently, the perspective is mitigated by the rise of IP, but wireless ATM remains an important topic of research nevertheless. Table 1.4 presents the HIPERLAN Types family with associated spectrum allocation in the context of W-ATM [47].

Table 1.4
HIPERLAN Family with Associated Spectrum Allocation

Type 1: Wireless LAN		*Type 2: Wireless ATM*	*Type 3: (HIPERACCESS) Wireless Remote Access*	*Type 4: (HIPERLINK) Wireless Point-to-Point*
MAC (CSMA/CA)		W-ATM DLC	W-ATM DLC	W-ATM DLC
PHY: 5-17 GHz		PHY: 5 GHz	PHY: 5 GHz	PHY: 17 GHz
Up to 25 Mbps		Up to 25 Mbps	Up to 25 Mbps	Up to 155 Mbps
Approximately 50m		—	—	—
100 MHz common to all CEPT countries (+ 50 MHz on a national basis according to market demand)	200 MHz common to all CEPT countries	—	—	—
1W–100-mW transmitter		—	—	100-mW transmitter

ATM is based on packet transmission and switching, and differs from circuit switching in the utilization of network resources. Also, QoS can be guaranteed and variable rate data is supported. Because of its flexible bandwidth allocation, efficient multiplexing of bursty traffic, and provision of a wide range of wireless broadband services, ATM is a promising solution for the next-generation wireless communication systems.

Wireless broadband systems are mainly characterized by providing a much higher transmission rate than the primary rate interface of ISDN (2,048-Kbps) [46]; therefore, the spectral efficiency becomes a critical issue. One possible approach to achieve better spectral efficiency is to take advantage of the spatial dimension (i.e., use of array antennas) [48]; another is the integration of different services within a single network (multiplexing different streams).

These objectives have given rise to numerous proposals and standardization activities on how to integrate wireless links into the framework of ATM. One of these still ongoing activities, which originally started as a wireless ATM standardization initiative, is the HIPERLAN Type 2 system in the scope of ETSI's BRAN project [34]. Meanwhile, this system has become core-network-independent by introducing a convergence layer. It shall support mobility and be able to provide QoS and achieve the typically required data transfer rates of 25 Mbps that users expect from a wired or IEEE-1394 (FireWire) network.

The extension of broadband multimedia services to mobile wireless terminals means that wireless mobile ATM (wmATM) technologies will be able to provide innovative solutions to the construction of a generic broadband wireless core. wmATM will also act as a broadband open platform to support different air interfaces and hierarchical wireless systems. The packetized air link helps dynamic bandwidth allocation and therefore improves wireless spectrum utilization.

wmATM evolves from wireless ATM and is armed with a direct signaling and packet-based mobile control architecture that can be attractively applied in next-generation broadband wireless mobile and broadband wireless access networks [49]. This helps to construct a broadband wireless pipe and "IP over wmATM" open signaling suite to support various common air interfaces for the wireless industry.

Implementation issues involving the IP-over-wmATM solution for 3G, broadband wireless mobile-communication systems include the design of the wmATM medium-access controller, protocol stack, and wmATM module definition. The wide deployment of ATM and multimedia communications has created new value-added telecommunications services. A major issue is how to extend these attractive services to wireless and mobile applications.

IP technology is at the basis of the Internet explosion and effectively launched the information society. IP over ATM has been well accepted to provide global IP addressing and packet transmission as well as management in wired networks. Since 1992, wmATM technology has developed very quickly, and the first wmATM system was successfully demonstrated in late 1995. By using IP over wmATM, all current value-added services in wired networks can be extended to the mobile on a global basis, an objective common to 3G mobile communications.

1.3 PREVIEW OF THE BOOK

In summary, Chapter 1 introduced the ever-changing world of mobile communications including recent trends. It started with UMTS and was followed by wireless broadband communications and interactive broadcasting allowing readers to follow the evolution process towards 3G communications. Important aspects of that process are considered including the need for higher data rates and more spectrum efficiency and bandwidth, as well as some significant marketing issues.

Chapter 2 covers the mobile and personal communication networks area 4 of the ACTS program—the mobile domain. Section 2.1 presents an overview of the ACTS mobile domain. It starts by discussing the strategic importance of advanced communications. Among the addressed issues are the significance of the European Union–funded RTD and how the program results are being achieved and the participation in ACTS of industry and academia. It looks at ACTS as an integrated program, at ACTS concertation, at the emphasis on trials and testbeds, and the dissemination of results. Special attention is paid to the activities of the mobile domain, its scope and relevant projects, and achievements to date. Reflecting the integrated approach of the program, we will also refer, as appropriate, to projects from other domains, whose activity is relevant to the work performed in the scope of the mobile domain.

Chapter 3 details into terrestrial air interfaces. Section 3.1 discusses their essential features. Section 3.2 provides an overview of terrestrial UMTS and relevant projects, namely future radio wideband multiple access schemes (FRAMES), mobile services in high-speed trains (MOSTRAIN), and UMTS security architecture (USECA), in view of their main objectives, technical approach, key issues, and impact. Section 3.3 covers the projects related to terrestrial wireless broadband systems, focusing on the wireless ATM network demonstrator (The Magic WAND), the wireless broadband CPN oblique for professional and residential multimedia applications (MEDIAN), the wireless access communication system (AWACS), the system for advanced mobile broadband applications (SAMBA), and the ACTS Broadband Communications Joint Trial and Demonstration (ACCORD). Section 3.4 briefly discusses projects related to wireless access systems: CABSINET and CRABS (both from a related domain); AMASE, which covers agent-based mobile access to information services; and the M3A project related to mobile multimedia access using intelligent agents.

Chapter 4 presents the projects related to satellite air interfaces. Section 4.1 is an introduction. Section 4.2 discusses the main objectives, technical approach, and the key issues of projects related to satellite UMTS, including the novel satellite mobile applications (SINUS), integrated S-UMTS real environment demonstrator (INSURED), and intertribal testbed for mobile applications of satellite communications (TOMAS). Section 4.3 covers the satellite EHF communications for mobile multimedia services (SECOMS/ABATE) project, which is related to satellite broadband communications and a project related to the convergence of Internet-ATM-satellite (COIAS).

Chapter 5 discusses the projects related to the evolution of existing and future networks. The radio access independent broadband on wireless (RAINBOW) project looks into a generic UMTS access infrastructure that is able to cope with different innovative radio access techniques. The project RAISIN, a continuation of RAINBOW, was concerned with trial integration and is also presented in this section. In addition, another network planning project that considers software tools for optimization of resources in mobile systems (STORMS) is described. Related projects from other domains about advanced security for personal communications (ASPECTS); the implementation and validation of enhancements on ATM-based B-ISDN (EXODUS); and the demonstration of a hybrid radio-optical fiber loop for interactive broadband services (FRANS) are also discussed.

Section 5.3 covers interoperability between complementary satellite systems and terrestrial networks (SUMO). The service platform for other ACTS trials and the applications project (VANTAGE), originally from the Service engineering, security, and communications management domain are also discussed.

REFERENCES

[1] UMTS Decision: Decision No 128/1999/EC, "On the Coordinated Introduction of a Third-Generation Mobile and Wireless Communications System (UMTS) in the Community," European Commission, January 1999, p. 1.

[2] UMTS Forum 1997, " The UMTS Vision," p. 2.

[3] Prasad, R., *CDMA for Wireless Personal Communications,* Norwood, MA: Artech House, 1996.

[4] Prasad, R., *Universal Personal Communications,* Norwood, MA: Artech House, 1998.

[5] Ojanperä, T., and R. Prasad, *Wideband CDMA for Third Generation Mobile Communications,* Norwood, MA: Artech House, 1998.

[6] Prasad, R., W. Mohr, and W. Konhäuser, *Third Generation Mobile Communication Systems,* Norwood, MA: Artech House, 2000.

[7] Pereira, J. M., " The European Approach to Third Generation," *Proc. EC Workshop,* Bejing, China, September 1999.

[8] Heath, M., " Preparing for the Licence Bid," *Mobile Communications International,* September 1999, p. 63.

[9] Verue, R., and J. Pereira, "Insuring the Success of Third Generation," *European Commission,* 1999, p. 4.

[10] Schwarz da Silva, J., et al., " Evolution Towards UMTS," *European Commission,* p. 14.

[11] RACE Vision of UMTS, *Workshop on Third Generation Mobile Systems,* DG XIII-B, European Commission, Brussels, Belgium, January 1997.

[12] Mohr, W., et al., "UTRA Concept for Third-Generation Mobile Radio Systems," Tutorial, *Proceedings VTC-Fall,* Amsterdam, September 19–22, 1999.

[13] UMTS Forum Report No 6, "UMTS/IMT-2000 Spectrum," December 1998.

[14] UMTS Forum Report, No 9, "The UMTS Third-Generation Market-Structuring the Service and Revenues Opportunities," October 2000.

[15] UMTS Forum, *Report,* No. 5, September 1998.

[16] IPv6 Forum Report, 1999.

[17] UMTS Forum, "Shaping the Mobile Multimedia Future—An Extended Vision from the UMTS Forum," *Report,* No. 10, October 2000.

[18] UMTS Forum, "Enabling UMTS/Third-Generation Services and Applications," *Report,* No 11, November 2000.

[19] Colzi, E., "Digital Signal Processing Techniques for Personal and Broadcasting Satellite Communication Systems," Ph.D. Dissertation, ESA, October 1999, p. I.

[20] Sterling, E., Harleid, J. E., "The Iridium System—A Revolutionary Satellite Communication System Developed with Innovative Applications of Technology," *Proceedings of the IEEE Military Satellite Communication Conference,* McLean, VA, 1991, pp. 436–440.

[21] Widemann, R. A., A. B. Salmasi, and D. Rouffet, "Globastar: Mobile Communications Wherever You Are," *AIAA 14-th International Communication Satellite Systems Conference,* Washington DC, 1992.

[22] Singh, J., "Project-21/Inmarsat-P: Putting Reality into the Hand-Held Sat-Phone Vision," *Proceedings of the Mobile Satellite Communications Conference,* Paris, France, October 1993.

[23] Balduino, P. R. H., "Latin America Goes Wireless Via Satellite," *IEEE Communications Magazine,* September 1995.

[24] Abrishakmar, F., and Z. Siveski, "PCS Global Mobile Satellite," *IEEE Communications Magazine,* September 1996.

[25] Del Re, E., "A Coordinated European Effort for the Definition of a Satellite Integrated Environment for Future Mobile Communications," *IEEE Communications Magazine,* February 1996.

[26] UMTS Decision, "Decision of 30 June 1997 on the Frequency Bands for the Introduction of UMTS," *CEPT/ERC/DEC(97)07,* The Hague, The Netherlands, June, 1997.

[27] UMTS Decision, "ERC Decision on the Harmonized Utilization of Spectrum for Terrestrial UMTS Operating Within the Bands 1900–1980 MHz, 2010–2025 MHz, and 2110–2170 MHz," CEPT/ERC/DEC(99)25, Oslo, Norway, December 1999.

[28] Final Report, "Mobile Broadband System," *ERO,* July 1997.

[29] UMTS Forum, "The Path Towards UMTS—Technologies for the Information Society," *Report No. 2,* 1998.

[30] "Communications for Society—Visionary Research," *European Commission,* DG XIII/B, February 1997.

[31] Ariyavisitakul, S. L, et al., "Broadband Wireless Techniques," *Guest Editorial IEEE Journal on Selected Areas in Communications,* October 1999, Vol. 17, No. 10, pp. 1,709–1,710.

[32] Pereira, J. M., "Mobile Multimedia Perspectives: From GSM to UMTS and Beyond," *IEEE TUDelft Student Chapter Symposium,* Delft, 1998.

[33] IEEE 802.11 Web site, http:// grouper.ieee.org/groups/802/11.

[34] http://www.etsi.org/.

[35] ETSI, "High Performance Radio Local Area Network (HIPERLAN) Type2; Systems Overview," *TR 101 683*, Oct.1998.

[36] IEEE 802.16 Web site, http://grouper.ieee.org/groups/802/16.

[37] Lu, W. and Q. Bi, "Wireless Mobile ATM Technologies for Third-Generation Wireless Communications," *Guest Editorial, IEEE Communications Magazine,* November 1999, Vol. 37, No. 11, p. 36.

[38] ETSI-European Telecommunications Standards Institute, "Radio Broadcast Systems; Digital Audio Broadcasting (DAB) to Mobile, Portable and Fixed Receivers," ETS 3000 401, February 1995.

[39] ETSI-European Telecommunications Standards Institute, "Digital Terrestrial Broadcasting Systems," ETS 300 744, February 1997.

[40] UMTS Forum, "A Regulatory Framework for UMTS," *Report No 1,* June 1997.

[41] Channing, I., "Evolution of the Base Station," *Mobile Communications,* September 1999, p. 24.

[42] Helme, S., "What's Your Proposition?" *Mobile Communications*, September 1999, p. 45.

[43] Prögler, M., C. Evci, and M. Umehira, "Air Interface Access Schemes for Broadband Mobile Systems," *IEEE Communications Magazine*, September 1999, p. 106.

[44] Evci, C., "HIPERLAN: A Viable Solution for the Future Market of High Resolution Multimedia Services," *MBS Conference*, London, UK, March, 1997.

[45] Fernandes, L., "R2067-MBS, A System Concept and Technologies for Mobile Broadband Communications," *RACE Mobile Telecommunication Summit,* Cascais, Portugal, November 22–25, 1995, pp. 62–72.

[46] ERO, "Results of the Detailed Spectrum Investigation, 1st Phase-3400 MHz to 105 GHz," *European Commission,* DG XIII/B, Brussels, Belgium, February 1997.

[47] Pereira, J. M., "Indoor Wireless Broadband Communications: R&D Perspectives in Europe," *TUDelft Workshop*, 1997.

[48] Vornefeld, U., et al., "SDMA Techniques for Wireless ATM," *IEEE Communications Magazine,* November 1999, Vol. 37, No. 11, pp. 52–57.

[49] Hu, J., "Applying IP Over wmATM Technology to Third-Generation Wireless Communications," *IEEE Communications Magazine,* November 1999, Vol. 37, No. 11, pp. 64–67.

Chapter 2

The ACTS Mobile Domain

The extent to which system and technology, as well as service and application aspects of mobile communications has progressed in the past few years is breathtaking. This evolution is due to a wide set of social, technological, and market drivers, with the ACTS projects playing a fundamental role. The ACTS demonstration/trial-oriented program for technology development, with its clear system objective, brought together in precompetitive, cooperative research the major players from industry and academia, as well as the users. The RTD activities, launched in the program framework, have been able to anticipate, develop, and promote many of those steps of innovation. The principal objectives addressed by the current ACTS projects in the mobile domain include the development of third-generation platforms for the cost-effective transport of broadband services across various radio environments and under different operational conditions [1]. Since the scope of mobile communications definitely encompasses multimedia, the objective is to progressively extend multimedia capabilities and the provision of high-performance services and enable their integration and interworking with the future broadband-core network [2]. This chapter details how ACTS and the projects in the mobile domain contributed to the further development of mobile communications. It starts by discussing the mobile domain, its scope and organization, and the interrelationships between mobile domain projects and relevant projects from other areas.

2.1 THE ACTS PROGRAM

ACTS (1994–1998) was a European Union–funded program of precompetitive, collaborative RTD, in the scope of the fourth framework of European RTD. The program was the successor to the earlier RACE program.

ACTS was organized in the following domains by associating projects in the same area [3]:

- Interactive multimedia services, including interactive digital television;

- Photonic networks;
- High-speed networking, including ATM;
- Mobile and personal communication networks;
- Intelligence in services and networks, including network management.

In fact, the effectiveness of the action of the ACTS projects resides in three main ingredients [4]:

- The balanced mix of diverse and complementary players in the consortia;
- The systematic contribution to the standardization process;
- The high level of teamwork achieved within the ACTS program and the extensive dissemination of results.

Cooperation in the ACTS program has further reinforced worldwide development in key areas of telecommunications, essential for the establishment of a true information society [5]. It has led to wide industrial consensus on common open platforms for mobile multimedia communications (as in the case of UMTS), digital television, and interactive video services (namely DVB), fostering rapid and coherent deployment of new working practices and electronic commerce [6].

Each area of the ACTS program contributed with major achievements but the highest added value has come from the coherence across the whole program, ensured through regular concertedness involving all projects, and the world-class level of all work, ensured through annual self-assessment and independent technical auditing of every project.

The ACTS program provided a first opportunity to test advanced mobile and personal communications services and technologies [5]. It involved service providers, communications operators, equipment manufacturers, and users. From the user's perspective, the ACTS program strived to ensure that current mobile services are extended to include multimedia and broadband services. Another objective was to make access to services possible regardless of the underlying networks and to provide for an automatic adaptation of convenient, lightweight, and power-efficient terminals to whatever air-interface parameters are appropriate to the user's location, mobility, and desired services.

RTD in mobile and personal communications is believed to be essential because of its multiplicative, enabling effect in all areas of the information society. Mobile and personal communications are recognized as a major driving force of social-economic progress and are crucial for fostering European industrial competitiveness and sustained economic growth. RTD's impact extends well beyond the industries directly involved and enables a totally new way of life and business [6].

2.2 THE MOBILE DOMAIN

2.2.1 Organizations and Activities

The objective of the mobile domain was to accommodate the foreseen demand for mobile personal mobile communications well beyond 2000 and to permit the European industry to retain its development in this area. The issues addressed included the following.

- The integration of fixed and mobile broadband networks, including the development of components and subsystems to carry broadband services over radio, and the standardization of interfaces to allow for universal mobile telecommunications to integrate a range of services;
- The concept of "personal communication space" through the development of technologies for individual authentication and privacy, the management of location/registration databases, and man-machine interfaces;
- Advanced technologies and systems for satellite communications;
- Advanced services and applications;
- Enabling technologies and tools for mobile communications including software radio and high-temperature superconductivity technologies.

The scenario is depicted in Figure 2.1, which shows the work spanned in picocell in-building coverage to global coverage by satellite. The applications range from indoor wireless with user bit rates approaching 155 Mbps to full terminal mobility and much lower bit-rate requirements (9.6 Kbps).

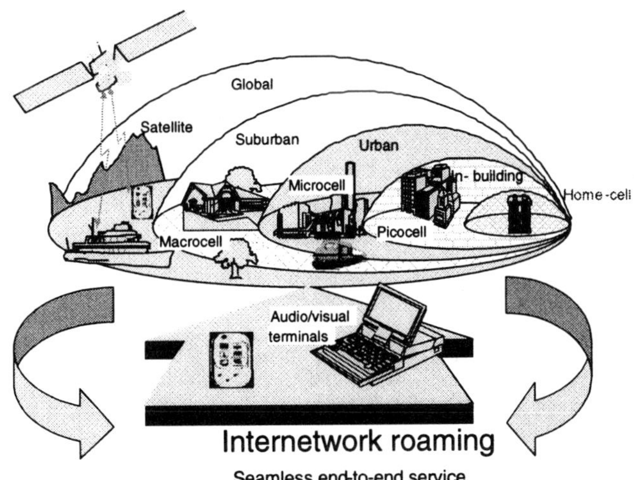

Figure 2.1 Hierarchical cellular structure.

Three network platforms are the focus of work: UMTS, WLAN, and MBS. The satellite dimension was also taken into consideration. All projects included a demonstration/trial phase. In 1998, 20- and 23-Mbps WLAN systems operating at 5.2 GHz and 19 GHz, respectively, were demonstrated. A satellite platform in the Ka-band was also trialed.

Work in the ACTS mobile domain has been instrumental in achieving the historic agreement on 3G radio technologies for multimedia services (UTRA) [7], which constituted the cornerstone of UMTS developments.

3G mobile communication systems aim at integrating all different services of 2G systems, while providing for true mobile multimedia capabilities. The target is competitive service provision to a majority of users and a wide range of broadband services (voice, video, data, and multimedia) that are consistent and compatible with technology developments within fixed telecommunication networks [5].

UMTS, the European 3G mobile system, is conceived as a multifunction, multiservice, and multiapplication digital mobile system that will provide personal communications at rates ranging from 144 Kbps up to 2 Mbps. The minimum-performance capabilities for circuit- and packet-switched data that have been agreed upon by ITU TG 8/1 are listed as follows [8]:

- Vehicular environment (144 Kbps);
- Pedestrian and outdoor to indoor environment (384 Kbps);
- Indoor office environment (2048 Kbps).

It will support universal roaming and provide for broadband multimedia services. UMTS is designed to have terrestrial and satellite components with a suitable degree of commonality between them, including the radio interfaces [9]. The RTD effort concentrated on the UMTS support of ATM and IP and the compatibility of UMTS and existing fixed networks. The allocation of intelligent functionality, the level of integration of the satellite component of UMTS, and the multiservice convergence philosophy of the UMTS radio interface were also included in the RTD activities. Emerging concepts and technologies include software radio, technology for smart antennas, high-temperature superconductive components and subsystems, effective software tools able to assist, and support operators to design and plan UMTS networks.

Application-development projects have provided insight into what the users should expect from future generations of mobile communications networks. Some demonstrated applications include the following:

- Distribution of advanced multimedia products by wireless means;
- Wireless services in a hospital environment using high-performance WLANs;

- Advanced mobile communications for the construction sector;
- Usability of services and equipment by all, including people with special needs;
- Wireless TV studio applications and support to emergency services using MBS;
- Virtual mobile office supported by WLANs;
- Telemedicine and entertainment supported by a satellite in the Ka-band.

2.2.2 Scope of the Domain

A key feature of the ACTS program was the use of practical experimentation and trials, many of them involving real users in real working conditions, to test and validate the technology developed in the projects. Figure 2.2 describes the organization of the ACTS mobile domain and the relation with other domains. Projects in the mobile domain address three aspects of RTD in 3G systems:

- Advanced mobile/wireless services and applications;
- Mobile/wireless platforms;
- Enabling technologies.

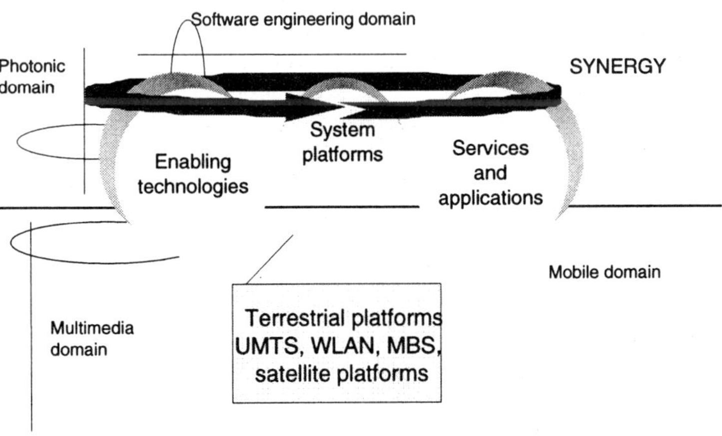

Figure 2.2 Scope of the mobile domain and relevance to other domains.

A very good coverage of platforms and applications was established. End-user applications were included, as well as the core technology issues. The regulatory, licensing, standards technology, and industrial factors for success are critical for a successful UMTS investment and market launch [6]. If the target for mass-market introduction is to be achieved, then most of initiatives need to be firmly on their way to resolving the remaining problems and issues. A major achievement of the ACTS program was its timely definition of the UTRA air-interface for the 3G mobile communication, which launched many projects into action [4].

Table 2.1 describes the project portfolio of the mobile domain. The chapters where the projects are covered are also included. For more general information on the ACTS and mobile projects, the interested reader should refer to [10]. The reader is also directed to the Proceedings of the Annual Mobile-Domain Summits for results on the RTD effort [11]–[15].

The common factor of all projects is that they contain a strong and clearly addressed thread of user involvement and innovative technology and that they share a common effort toward technology, service, and application trials.

The demonstration and assessment of novel services and applications takes into account the full implications of user environment, system characteristics, service provision, and control. The projects aim at proving the validity of novel components or subsystem technologies, including multimode transceivers and tools for network planning.

2.2.2.1 Enabling Technologies

Several access technologies are available and will become available in the near future [16]. These evolving and emerging technologies represent a very flexible and powerful platform to support future requirements of services and applications. At the core of the seamless support network to these different access technologies are a series of enabling technologies, whose role is essential for permitting such networks to meet the capacity and QoS requirements at a cost/performance level attractive to operators, service providers, and users. These enabling technologies range from those usually related to hardware issues (antennas, adaptive wideband radio front-ends, and so forth), to those related to software/algorithmic issues (authentication, security, network planning and control, novel equalization, compression, and coding/decoding algorithms). Therefore, software-defined radio concepts will be a key technology [16].

Software radio concepts for adaptive and flexible transmission schemes were the main subject of the project FIRST, aiming to demonstrate that it is feasible and cost-effective to develop and deploy intelligent multimode terminals capable of operation with UMTS as well as with multiple 2G standards, and with the ability to deliver multimedia services to mobile users. Software radio was also in the core of the project SORT.

Table 2.1

Project Portfolio of the Mobile Domain

Project	*Target*	*Reference*
ACCORD	Wireless broadband systems	Chapter 3, Volume 1
AMASE	Wireless access systems	Chapter 3, Volume 1
ASPECT	Terrestrial network	Chapter 5, Volume 1
AWACS	Wireless broadband terrestrial air interfaces	Chapter 3, Volume 1
CABSINET	Wireless access air interfaces	Chapter 3, Volume 1
COBUCO	Case study/services and applications	Chapter 3, Volume 2
CRABS	Wireless access air interfaces	Chapter 3, Volume 2
EXODUS	Terrestrial network	Chapter 5, Volume 1
FIRST	Practical promotion of UMTS/software radio	Chapter 2, Volume 2
FRAMES	Terrestrial UMTS/air interfaces	Chapter 3, Volume 1
FRANS	Terrestrial network	Chapter 5, Volume 1
INSURED	Demonstration of S-UMTS services	Chapter 4, Volume 1
The Magic WAND	Wireless ATM	Chapter 3, Volume 1
MEDIAN	Wireless broadband terrestrial air interfaces	Chapter 3, Volume 1
MEMO	Multimedia services/services and applications	Chapter 3, Volume 2
MICC	Case-study/services and applications	Chapter 3, Volume 2
MOMENTS	Multimedia services/services and applications	Chapter 3, Volume 2
MOMUSYS	Mobile multimedia system/terminals	Chapter 4, Volume 2
MOSTRAIN	Terrestrial UMTS air interfaces	Chapter 3, Volume 1
MOVE	Middleware/services and applications	Chapter 3, Volume 2
M3A	Wireless access systems	Chapter 3, Volume 1
MULTIPORT	UMTS in healthcare/services and applications	Chapter 3, Volume 2
NEWTEST	Link design aspects in UMTS satellite components/enabling technologies	Chapter 2, Volume 2
ON THE MOVE	Middleware/services and applications	Chapter 3, Volume 2
RAINBOW/ RAISIN	Terrestrial networks	Chapter 5, Volume 1
SAMBA	Advanced mobile broadband applications/ wireless broadband terrestrial air interfaces	Chapter 3, Volume 1
SECOMS /ABATE	Broadband satellite air interfaces	Chapter 4, Volume 1
SUCOMS/Cool Talk	Enabling technologies	Chapter 2, Volume 2
SINUS	Satellite UMTS air interfaces	Chapter 4, Volume 1
SORT	Software radio/enabling technologies	Chapter 2, Volume 2
STORMS	Terrestrial networks	Chapter 5, Volume 1
SUMO	Satellite networks	Chapter 4, Volume 1
SUNBEAM	Software radio/enabling technologies	Chapter 2, Volume 2
TOMAS	Satellite UMTS air interfaces	Chapter 4, Volume 1
TSUNAMI II	Adaptive antennas for satellite UMTS/enabling technologies	Chapter 2, Volume 2
VALIDATE	Broadcasting/services and applications	Chapter 3,Volume 2
VANTAGE	Satellite networks	Chapter 5, Volume 1
UMPTIDUMPTI	UMTS implementation for disabled people/terminals	Chapter 4, Volume 2
USECA	Security in UMTS	Chapter 3, Volume 1

The project TSUNAMI II sought to develop further the technology for smart antennas and to demonstrate a way to deploy adaptive antennas within the infrastructure of 3G mobile systems, such as UMTS. The adaptive antennas developed by the project are a new technology with the potential to provide large increases in capacity.

The introduction of high-temperature superconductivity (HTS) components in the base stations of mobile communications systems is a recent "hot topic" for research worldwide. HTS, applied to the subsystems of mobile communications networks, will offer enhanced performance as a result of the lower losses in filters, delay lines, and resonators, as well as lower noise figures. The project SUCOMS was involved in the design, manufacture, and assessment of communications transceiver components and subsystems built to a predefined specification. The NEWTEST project was about advanced algorithms based on neural-network techniques to combat the signal impairments that satellite nonlinear channels traditionally encounter.

It should be noted that, although an interesting and necessary challenge, this only represents a fraction of the overall support and technology required to realize the potential of the concept of a seamless network [17]. Other necessary developments include network/terminal cooperation for seamless interstandard handoff, QoS management, a secure software download mechanism, a terminal software architecture supporting reconfiguration, configuration management, and capability negotiation. All these issues have been undertaken by the projects in the frames of the new IST program as a continuation of the efforts of the ACTS program.

2.2.2.2 Mobile/Wireless Platforms

The essential platforms of relevance to mobile communications are those that meet the requirements of a variety of environments (public and private, urban and rural, indoor and outdoor) and different levels of service and availability. These can be broadly classified as cellular public networks (UMTS), private and public LANs with full or reduced mobility (MBSs and WLANs), and global low mobility networks (satellites). The system design of some of these technologies is mainly based on the traditional vertical approach to support a certain set of services with a particular technology. Depending on future user requirements and economic trends, new radio access schemes such as wideband radio access for high mobility might be developed [16]. Their integration in a common flexible and expandable platform will provide a multiplicity of possibilities for current and future services and applications to users in a single terminal. The user expectations to more sophisticated services with a QoS comparable to wireline access based on the current development of the Internet are increasing. Figure 2.3 gives an overview of these evolving and emerging radio access technologies [18].

The convergence point for the supply of information and entertainment and the demand from end users is expected to form the mobile multimedia portal [19]. As the end user's preferred point of entry into all-IP–based services and content, the portal is where the customer interacts with the entity that provides the services.

Using an intelligent IP-access platform with dynamic service-selection capabilities, the portal owner can provide personalized location-dependent services that are tailored to the mobile users' individual requirements and choices. The harmonization of the UMTS standards in the IMT-2000 framework with the standards on the Internet side is necessary to make the mobile multimedia portal a workable solution in an international networking environment, especially for the roaming user.

Most of these concepts were at the core of the work of various ACTS projects. The Magic Wand project was involved in the definition of HIPERLAN Type 2 and HiperLink, while the project AWACS contributed to HiperLink and HiperAccess [20]. Wireless customer premises were also studied; the project MEDIAN concentrated its work on providing 20–155 Mbps at 60 GHz with a wATM approach. The project SAMBA targeted the 155-Mbps data rates with high mobility, considering the outdoor, mobile scenario with MBS.

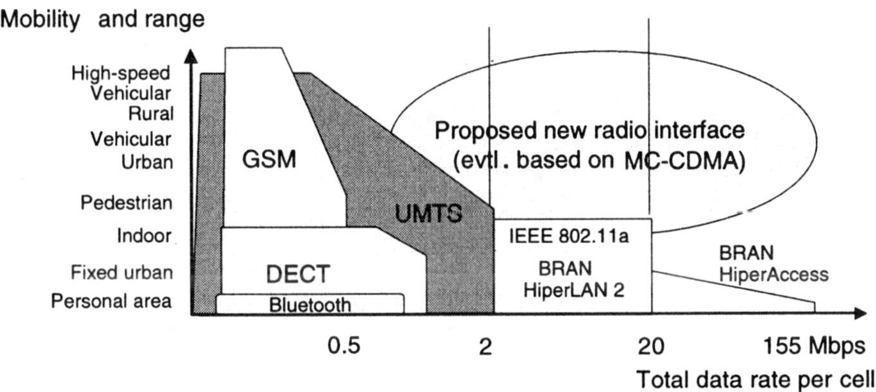

Figure 2.3 Existing and proposed [18] wireless systems to support data rate, range, and mobility.

The project RAINBOW, and its extension, the project RAISIN, focused on the flexibility approach to a common platform. They developed the generic radio access network (GRAN) approach (see Chapter 5 for more details).

The project EXODUS aimed at identifying a smooth transition from fixed and 2G mobile systems towards UMTS from the fixed-network perspective, while the project ACCORD anticipated the importance of solutions for the IP mobility problem. Mobility management will be a part of a new media access system as an interface between the core network and the particular access technology to connect a user via a single number for different access systems to the network [16] (see Chapter 1). Basic parts of this concept are currently being investigated in the

Broadband Radio Access for IP-Based Network (BRAIN) as part of the information society technology (IST) program [21].

2.2.2.3 Advanced Services and Applications

The key points in the development of future-generation systems are the interworking and integration of different access systems on a common IP-based platform and the necessary multimode or adaptive and multiband terminals for different access systems to support a wide range of multimedia services [16]. The services and applications area is another technical challenge that requires extensive research activities together with the software architecture and supporting technologies areas in order to make the mobile multimedia happen.

In general, basic service concepts are common between 2G and 3G [19]. The service delivery mechanisms and the user device attributes and interfaces are expected to improve with 3G. The applications that have been developed for the Web and the intranet will be a key source for 3G mobile applications. Many people will prefer to access the information via mobile devices rather than personal computers. Most mobile users will utilize the Internet differently than today, more for needful purposes than leisurely browsing. The emerging m-commerce offers a multitude of services, such as m-broking, m-shopping, and m-auctioning.

The degree to which networked multimedia services enter into common usage will have a significant influence on users' attitudes to mobile multimedia services [22]. A high level of adoption will make mobile services highly attractive as users come to feel comfortable with the concept of relying on electronic services. Furthermore, the high levels of IT literacy that a strong market for fixed networked multimedia will create will simplify the service creation process for mobile multimedia service providers, make it easier for new services to be accepted, and increase the level of service and terminal functionality that the marketplace will find attractive.

All this was included in the basic objectives of a number of projects contributing to the ACTS program and undertaken by the mobile domain area. The projects found technical solutions concerning multimedia services and applications. The concept of a basic service platform was the focus of the project MOMENTS, while the project MEMO considered functions as voice over IP (VoIP) as part of the UMTS services. Services targeted at residential customers and, therefore containing a great deal of audio and video, online newspapers, and so forth, triggered research in the area of broadcasting multimedia services (see projects VALIDATE and MEMO).

In 1999 expectations were that less than 20% of the world's land area will be covered by terrestrial cellular networks within the envisaged timescales of UMTS/IMT-2000 [22]. Satellite systems, which provide worldwide service, are therefore important to provide for complete coverage. In turn, this is expected to result in further demand for terrestrially based UMTS/IMT-2000 services. Price differences between terrestrial and satellite services will play an important role in the usage of mobile satellite services. Within the ACTS program, the focus became the realization of low-speed services provided by handheld terminals and their integration into UMTS (see projects SINUS and INSURED), while the project TOMAS developed a testbed for high-speed services with a focus on the realization of highly portable or even mobile terminal equipment and their application in field trials. The integration of terrestrial (see project FRAMES) and satellite UMTS involved joint research activities between the latter project and the project SINUS, both major projects within the ACTS program.

3G communications will lead to a growing demand of 3G terminals. In the next one to two years, 3G terminals are most likely to use existing designs and be based on voice-centric applications and supply access to the Internet via WAP or an IP-based microbrowser [19]. For Internet access, the early adopters will link their existing PDA or notebook computer to their phones for access to new applications. As mobile audio streaming technology advances, specifically designed devices may become available to support MP3 applications for the audio market. In Europe, broad implementation of mobile Internet browsing is not expected before 2002 with the launch of 3G networks combined with the availability of browser enhanced 3G terminals.

The project UMPTIDUMPTI's main objective was to ensure that emerging broadband and mobile services and equipment are usable by all. The project MultiPort focused on database access using PDAs in the medical care sector. Research of the project MoMuSys was based on the MPEG-4 standard that supports audio-video coding with spatial and temporal scalability.

All mobile domain projects sought to describe UMTS in the context of future developments in telecommunications, information technology, and media, bearing in mind that today's concept of separate mobile and fixed telecommunications networks was not sufficient to explain the realities of tomorrow's business environment. A clear focus was kept on the potential of UMTS to embrace new technologies, concepts, and services.

2.2.2.4 Projects from Other Domains and Programs

There are several programs related to information and communications technologies (ICTs) within the framework of EU research. Relevant to the area of mobile/wireless communications, there were ACTS projects from other domains, as well as some projects in the ESPRIT "IP for Mobility and Microelectronics" program.

2.2.3 Concertation and Dissemination

The overall theme of the domain concertation is to build a "road map" describing the future evolution of mobile communication networks and technologies.

Concertation is a key process that ensures that the value of the ACTS program as a whole exceeds that of the individual projects. The concertation process has two main components, the domains and the chains [6]. In the second volume of this book, special attention is paid to these "inside" special interest groups (SIGs) and chains activities because they are key to successful research.

Domains were formed around the main technical areas of the whole ACTS program. Here, projects report results, undergo an informal review, exchange ideas, prepare workshops, and carry out other activities of common interest. This increases the synergy between projects and helps to avoid the overlap and duplication of effort. Specific issues of common interest to mobile projects are tackled by the SIGs with the aim of reaching results that are applicable to the widest possible set of projects, while contributing to the standardization bodies work. Table 2.2 illustrates the basic objectives and contributions of each of the SIGs.

Table 2.2

Objectives and Outputs from the Mobile Domain SIGs

Air interface (SIG 1)	To investigate the aspects affecting the choice and design of air interfaces for 3G mobile communications systems (UMTS/FPLMTS, MBS, WLANs), including satellite aspects	Common deliverable on air interface options
Joint trials cluster (SIG 2)	To provide a forum for the exchange of views with the aim of using the maximum possible of H/W system resources provided by the projects developing H/W platforms in order to support the largest possible number of application demonstrators	Joint trials
Mobile terminals (SIG 3)	To determine the implication of the user requirements and technologies implications on the specification and design of terminals for the 3G mobile communication systems	Common deliverables in preparation will be submitted to ETSI
Network issues (SIG 4)	To perform work on all aspects affecting the network architecture of future mobile systems (UMTS< MBS< WLANs) and their satellite components; this work will primarily cover network architecture and planning, service definition and provisioning, impact of radio-interface features on network aspects, protocol architecture, and networking aspects	Analysis of the UMTS transport chain extended to satellite-based services, wireless ATM systems, and the transport chain for signaling information New ATM adaptation layer, type 2, proposed by ITU-T for the transport of low bit rate services. A comparison with the use of AAL5 has been performed First version of the deliverable "Preliminary Framework for the UMTS Network," released

Chains consist of a number of projects drawn from different parts of the program, which mutually support a defined objective. They contribute to a specific result, such as the achievement and dissemination of a guideline or realization of a complex demonstration, acknowledged to be useful to a wider community outside the ACTS program. Chains are formed as the need arises and are dissolved when their goals are achieved. Individual projects are frequently active in more than one chain at any given time. The different chain groups are listed in Table 2.3.

Table 2.3

ACTS Chains Relevant to the Mobile Domain

Chain Groups	*Related Projects*
Group BA	Broadband access networks: economics and evolution
Group GA	Generic access to applications (user perspective)
Group NI	Network level interoperability and management
Group SI	Global service integration
Group XB	External: broadening and awareness

ACTS concertation may be broadly defined as the unification of people and their organizations or projects. In unifying, they can profit from each others' knowledge and experience, coordinate or encourage the convergence of ongoing work on the most important issues, and build a widely based consensus on the ways to realize advanced communications.

2.3 ACTS AND THE IST PROGRAMS IN THE SCOPE OF FUTURE MOBILE COMMUNICATIONS

The impact of ACTS is ultimately on the acceleration of the emergence of an information society by speeding consensus on the nature of advanced communications networks and services that could benefit all economic sectors of activity, not only in Europe but also worldwide. A key factor to success will be the ability to articulate long-term perspectives in technological developments with the necessary flexibility in program management to reorient work, as the environment changes. This requires an ongoing effort at keeping updated the vision of future societal and technical trends. The goal can be achieved by monitoring social-economic trends relevant to IST developments, by bringing views together through consultation mechanisms, and by periodically assessing the validity of parts of the program. Thus, ACTS can build a platform of common understanding of technical and market trends and develop ways to meet customer and citizen needs. It is also essential to encourage project orientation. Figure 2.4 illustrates the scope of ACTS guidelines toward their progressive endorsement.

Beyond the present ACTS program, communications research is expected to form the core of a single, broader research, aimed at creating a user-friendly information society. This includes the unification of all future research into a single integrated program.

Recently, the IST program of research, technology, and demonstration under the fifth framework of European research has been deployed. It focuses on the global networked society (1999–2000), taking into consideration the technology, policy, and regulatory framework, and new applications and skills. Expressed in commercial terms, both the ACTS and IST programs contribute to the techno-economic feasibility of alternative technological solutions under specific circumstances and thus help to achieve a focus on the most viable options for product and service deployment.

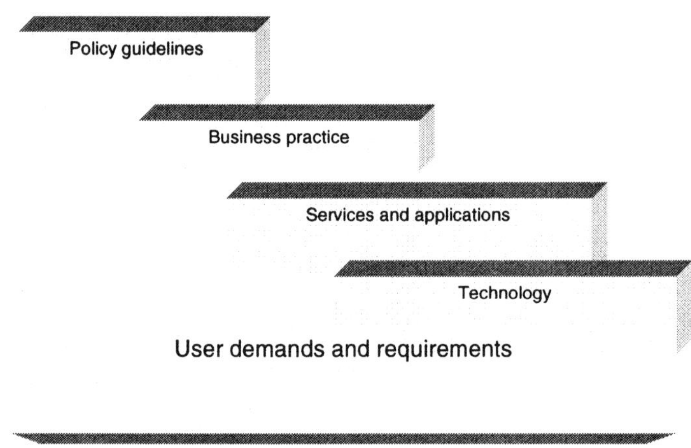

Figure 2.4 Scope of ACTS guidelines.

This reduces risks and uncertainties, thereby creating a favorable climate for organizations to invest in advanced communications infrastructures and for new services to rapidly gain the adoption necessary for long-term viability.

REFERENCES

[1] Schwarz da Silva, J., et al., "Mobile and Personal Communications: ACTS and Beyond," *Proc. PIMRC Conference,* London, 1997.

[2] Schwarz da Silva, J., et al., "Evolution Toward UMTS," *HF Communications,* 1997, p.12.

[3] ACTS 98 CD-ROM, *European Commission, DG XIII,* Brussels, 1998.

[4] Verrue, R., *IST Mobile Summit,* Galway, Ireland, October 2000.

[5] Pereira, J., et al., "From Wireless Data to Mobile Multimedia: R&D Perspectives in Europe," *European Commission,* Brussels, 1999, pp. 2–26.

[6] ACTS Independent Report of the Monitoring Committee on the ACTS Program, *European Commission,* Brussels, 1998, p. 14.

[7] ETSI Resolution, January 1998.

[8] http://www.itu.int/.

[9] Da Silva, J. S., et al., "Mobile and Personal Communications: ACTS and Beyond," *Proceedings PIMRC,* Helsinki, 1997.

[10] ASAP Project Web Site, www.cpk.auc.dk/asap/.

[11] *Proc. RACE Mobile Telecommunications Summit,* European Commission, Cascais, November 1995.

[12] *Proc. ACTS Mobile Telecommunications Summit,* European Commission, Granada, November 1996.

[13] *Proc. ACTS Mobile Telecommunications Summit,* European Commission, Aalborg, October 1997.

[14] *Proc. ACTS Mobile Telecommunications Summit,* European Commission, Rhodes, October 1998.

[15] *Proc. ACTS Mobile Telecommunications Summit,* European Commission, Sorrento, Italy, June 1999

[16] Mohr, W. and R. Becher, "Mobile Communications Beyond Third Generation," *Proceedings VTC'2000,* Boston, MA, September 2000, CD ROM.

[17] Drew, N. J., et al., "Compelling Needs and Technologies to Support Reconfigurable Terminals," *IST Mobile Summit,* Galway, Ireland, October 2000, CD ROM.

[18] NTT DoCoMo, "The Fourth Generation of Mobile Communications Systems," *ITU-R, Document 8F/INFO1-E,* February 2000.

[19] UMTS Forum, "Shaping the Mobile Multimedia Future," Report No.10, September 2000.

[20] Pereira, J., "A Unified Open Architecture To Deliver Mobile Multimedia," in *Wireless Multimedia Network Technologies,* Norwell, MA: Kluwer Academic Publishers, 2000.

[21] IST Web site, www.cordis.lu/ist/projects.

[22] UMTS Forum, "The Future Mobile Market," Report No. 8, March 1999.

Chapter 3

Terrestrial Air Interfaces

3.1 INTRODUCTION

With the new millennium, we are on the threshold of a new era—the mobile information society. As we become ever more mobile, everything we are used to doing at our desktop, we want to do when we are on the move. The demand for access to information at any time and any place are the challenges that scientists and industries are facing on the way to establishing the mobile information society.

The vision of the future includes total mobility (i.e., having access to all kinds of data, regardless of time or location). For instance, by means of a mobile phone-like device one will be able to access full flight, bus, and train schedules from anywhere—or hold a videoconference by using a terminal no bigger than a mobile phone of today.

These are the perspectives of the next step in the evolution towards the mobile information society. Unlike first- and second-generation systems (e.g., NMT and digital GSM, respectively), 3G systems, like IMT-2000 and UMTS, are based on much higher transfer speeds and a convergence of IS95, 136, and the Internet. More advanced data and seamless multimedia services will become available in addition to mobile telephony, and eventually the user will not need to be aware of the underlying technology.

3G represents a shift from "voice to data to multimedia" in all its forms to provide us with high-speed Internet access, mobile LAN, videoconferencing, and so forth. All these services will be accessible from mobile terminals.

The need for higher and higher data rates while on the move has forced research activities beyond the 2 Mbps offered by UMTS [1]. Data rates will go up to 155 Mbps and provide truly broadband wireless access. Also, because of the dominating role of IP-based data traffic in the future, the networks and systems will have to be designed for efficient packet data transfer.

This chapter provides an overview of the actions that took place in the field of radio air interface design within the frames of the ACTS program, to support multimedia services in different environments. First, the project FRAMES is discussed. This was an ACTS project dealing with the terrestrial component of the

UMTS radio interface and its goal was to develop a radio-interface proposal, fulfilling the requirements for terrestrial 3G mobile radio systems. Similarly, the project MOSTRAIN was developed with the objective of developing mobile multimedia communications technology for the very HST environment and its inclusion in the UMTS specification.

UMTS security was the main objective of the project USECA. It was concerned in particular with support of the specification process by means of developing a viable and complete UMTS security architecture, also to be used by ETSI as the basis for standardization [2]. The main issues under investigation were the security features and requirements, security mechanisms, security architecture, public key infrastructure (PKI), and terminal security.

Consequently, future mobile systems must somehow incorporate WLAN capability in order to maintain "universality" [3]. As a result, WLAN type systems, as HIPERLAN type 2 and broadcast systems as DAB and DVB-T have been standardized and are now available. The WLAN type systems are designed in particular for high data rate access, low range, and in general low mobility [4]. WLANs are related to broadband communications and aim at services more promising than the 2-GHz band currently allocated for UMTS. Interworking of broadband mobile technologies with a wireless LAN compatible with B-ISDN and ATM was considered the key issue for future integrated, interactive broadband multimedia communications [5]. The core networks, however, are moving towards more transparent transport techniques without any distinction between circuit-switched and packet-oriented networks to support real-time and non-real-time services in the same network [6]. IP is then the most promising solution from this perspective.

Different frequency bands have already been allocated to WLAN type systems, also in the millimeter frequency range, to combat the scarcity of available frequency. Within the ACTS program, several research projects were dedicated to investigation in this important area. These included the project MEDIAN (60-GHz range), the project SAMBA (30–40 GHz), and the projects AWACS and the Magic WAND (17–19 GHz). The very important integration between the terrestrial and satellite component of UMTS was investigated by the project ACCORD. Other projects were related to the problems of high mobility and high uplink capacity (e.g., for interactive broadcasting), and large coverage provided by infrastructure. Such issues are referred to in the ACTS projects CABSINET and CRABS. Later, the project AMASE focused on high user mobility among various networks, and devices regardless of physical location. The change of the mobile market from a voice-dominated environment to one in which mobile Internet and enhanced data services will be equally important, requires new service concepts and an in-depth understanding of the new user [7]. Addressing the mobility theme and integration issues for the near-term technology was the main objective of the project M3A. Figure 3.1 gives an overview of the terrestrial air interface–oriented projects and their scope.

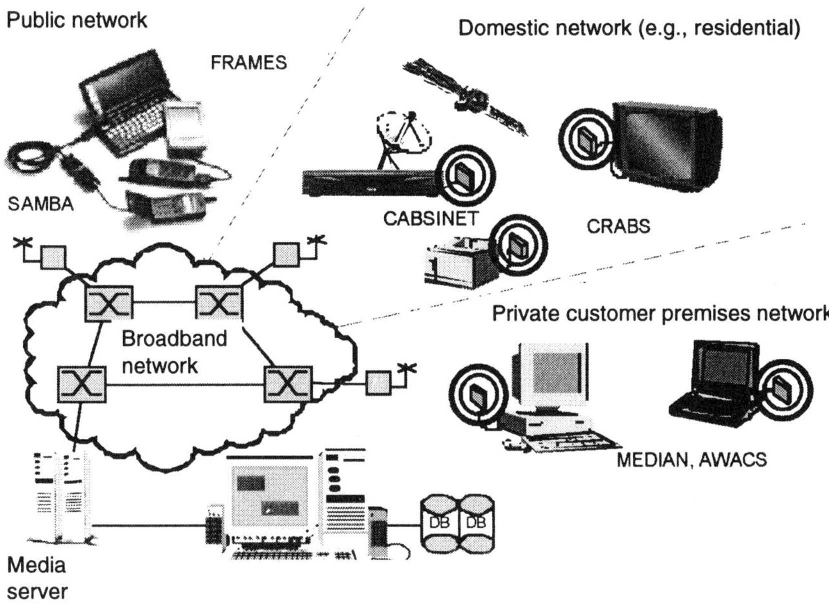

Figure 3.1 Overview of the projects related to terrestrial air interfaces.

3.2 TERRESTRIAL UMTS

Although Marconi's innovative perception of the electromagnetic waves and the air interface in 1897 was the first milestone on the significant road to a shared use of the radio spectrum, wireless communication started to take off only a century later [8]. The first-generation cellular and cordless telephone networks, which were based on analog technology with FM modulation, have been successfully deployed throughout the world since the early and mid 1980s. A typical example of a first-generation cellular telephone system is the advanced mobile phone service (AMPS). 2G wireless systems employ digital modulation and advanced call-processing capabilities. Two advantages have been offered in view of the processing complexity. One is the use of spectrally efficient radio-transmission schemes, such as time-division multiple access (TDMA) and CDMA, in comparison to the analog frequency-division multiple access (FDMA) schemes, previously employed. The other one is the provision for implementation of a wide variety of integrated speech and data services, such as facsimile, paging, and low-data-rate network access. Examples of 2G wireless systems include GSM in Europe; TDMA IS-54/IS-136, and CDMA IS-95 digital cellular standards in the United States; and, in Japan, personal digital cellular (PDC), which is the second largest digital cellular system after GSM, DECT, personal access communications system (PACS), and personal handyphone system (PHS).

3G wireless systems are evolving from mature 2G networks, with the aim of providing universal access and global roaming. These systems are also expected to support multidimensional (multi-information media, multitransmission media, and multilayered networks), high-speed wireless communications—an important milestone towards achieving the grand vision of ubiquitous personal communications. The introduction of wideband packet-data services for wireless Internet up to 2 Mbps is probably the main attribute of 3G systems, which are intended to unify the diverse systems we see today into a seamless radio infrastructure capable of offering a wide range of services in many different radio environments. The quality we have come to expect from wireline telecommunications networks will be maintained.

Since the mid 1980s, studies on 3G systems have been carried out within the International Telecommunication Union (ITU), and particularly within TG8/1, where it was called future public land mobile telecommunication systems (FPLMTS), lately renamed International Mobile Telecommunications-2000 (IMT-2000) [9]–[10]. In Europe, research and development on 3G technology, commonly referred to as UMTS and MBS, have been conducted under the RACE and ACTS programs. Key elements in the definition of 3G systems are the radio-access system and radio transmission technology (RTT). By the end of June 1998, 10 proposals from Europe, the United States, Japan, China, and Korea, were submitted to ITU radio communication sector (ITU-R) for candidate RTTs on the IMT-2000 terrestrial component [11]. In almost all the proposals, direct-sequence CDMA (DS-CDMA) was chosen as a leading multiple-access technique. Some important air interface specifications of these proposals are summarized in Table 3.1. Common to almost all the CDMA-based proposals, the following system specifications/implementations characterize 3G CDMA systems:

- Wideband CDMA;
- Pilot-aided coherent reverse link;
- Fast closed-loop power control on the forward link;
- Antenna diversity in the BS;
- Seamless interfrequency handoff to support hierarchical cells.

These CDMA-based RTT proposals can be categorized into three classes based on their system specifications:

- UTRA (Europe), W-CDMA (Japan), W-CDMA/NA (United States), CDMA II (Korea), WIMS-WCDMA (United States);

Table 3.1

Air Interface Specifications for 3G CDMA-Based Proposals [8]

Idea	*UTRA*	*cdma 2000*	*W-CDMA/ NA*	*WIMS W-CDMA*	*W-CDMA*	*TD-SCDMA*	*CDMA II*	*CDMA I*
MA	FDD: DS-CDMA	FDD: DS-CDMA	FDD: DS-CDMA	FDD: DS-CDMA	FDD: DS-CDMA	TDMA/ CDMA	DS-CDMA	DS-CDMA
	TDD: T/	TDD: T/	TDD: T/	TDD: DS-W-CDMA	TDD: T/			
	CDMA	CDMA	CDMA	DS-S-TDMA	CDMA			
Duplex scheme	FDD/ TDD	FDD/ TDD	FDD/ TDD	FDD/ TDD	FDD/ TDD	TDD	FDD	FDD
Frame length	10 ms	20/5 ms	10 ms	10 ms	10 ms	5 ms	10 ms	10 ms
Data mod.	QPSK	QPSK	QPSK	QPSK	QPSK	QPSK	QPSK	QPSK
	dual-channel	BPSK	dual-channel		dual-channel	16QAM	BPSK	BPSK

- cdma2000 (United States), CDMA I (Korea);
- TD-SCDMA (China).

The key differences between the groups are chip rate, synchronous/-asynchronous BS operation, and the pilot structures. Various international-standards development organizations have agreed to cooperate in the preparation of globally applicable technical specifications for 3G mobile systems under 3GPP and 3GPP2 partnership projects [2]–[12].

In most of the CDMA-based RTT proposals, wideband CDMA (W-CDMA) was proposed for the system with a bandwidth of 5–20 MHz, because of its inherent advantages in terms of multipath resolution capability, lower transmission power, and support of high bit-rate service [13]. Among the W-CDMA-based proposals, a typical chip rate of 4.096 Mcps is used for the 5-MHz band allocation. The choice of the chip rate is based on consideration of backward compatibility with GSM and PDC [8].

Frequency-division duplex (FDD) mode is considered the major duplexing technology in almost all proposals. Nevertheless, time-division duplex (TDD) mode is also supported by proposals such as W-CDMA, UTRA, and cdma2000. The introduction of a TDD mode is mainly because of the asymmetric frequency bands designated by the ITU. Also, the asymmetric nature of the data traffic on the forward and reverse links anticipated in the next-generation wireless systems (e.g., Internet applications) suggests that TDD mode might be preferred over FDD on certain occasions when efficient utilization of the available resources is necessary. Thus, an air interface with provision of both FDD and TDD can provide sufficient flexibility of spectrum allotment. Therefore, in most proposals the TDD mode was designed as similar to the FDD mode as possible. It is expected that this will facilitate easy implementation of dual-mode FDD/TDD phones as well as reuse of integrated circuits (ICs) in single-mode TDD phones.

In TDD mode, the code, frequency, and timeslot define a physical channel. In FDD mode, a physical channel is defined by its code and frequency and possibly by the relative phase. In general, physical channels can be broadly categorized into two basic classes: dedicated physical channel (DPCH) and common physical channel (CPCH). It should be noted that the design of forward- and reverse-link physical layer focuses on different aspects. In the forward link, efficient spectrum utilization and high traffic throughput are the prime concerns. An important factor in the reverse-link design is the power consumption for longer battery-life terminals. These considerations affect the configuration of physical channels for the forward and reverse links.

In the European UTRA proposal, dual-channel quadrature phase shift keying (QPSK) modulation is adopted on the reverse link, where the reverse-link DPDCH and DPCCH are mapped to the I and Q channels, respectively. The I and Q channels are then spread to the chip rate with two different channelization codes and subsequently complex-scrambled by an MS-specific complex code. For multicode transmission, each additional reverse link DPDCH may be transmitted on the I- or Q-channel. Either short or long scrambling codes should be used on the reverse link. The modulation and spreading scheme of UTRA is depicted in Figure 3.2.

Besides 3G activities, GSM, IS-95, and IS-136 continue evolving towards IMT-2000/UMTS requirements [14]–[15]. This evolution makes it already possible to introduce a subset of IMT-2000 services into current 2G systems; the innovative 3G wideband access then provides the full set of IMT-2000 services with increased spectrum efficiency and flexibility [8].

In particular, the evolution of GSM in its second phase is currently under way. Key enhancements include the high-speed circuit switched data (HSCSD) service facilitating data rates up to 57.6 Kbps using four timeslots, advanced speech call items (ASCI), intelligent network (IN) technologies introduced via customized application for mobile enhanced logic (CAMEL), enhanced short-message services (SMS), and high-speed GPRS. As mentioned in Chapter 1, GPRS is an important data service for GSM that allows full mobility and wide-area coverage with data rates up to 115.2 Kbps and supports both IP and X.25.Beyond Phase 2+ of the GSM. Another future evolution of GSM that is developed by ETSI is EDGE. EDGE uses the same spectrum allocation as today and has commonalities with

GSM in terms of carrier bandwidth (200 KHz), symbol rate (270.833 ksymbolsps), and frame format (8 TDMA slots/4.6-ms frame) that are beneficial to the design of multimode terminals. Finally, the high-level modulation implemented in EDGE will further extend data rates for high-link quality conditions, building upon GPRS and HSCSD [16].

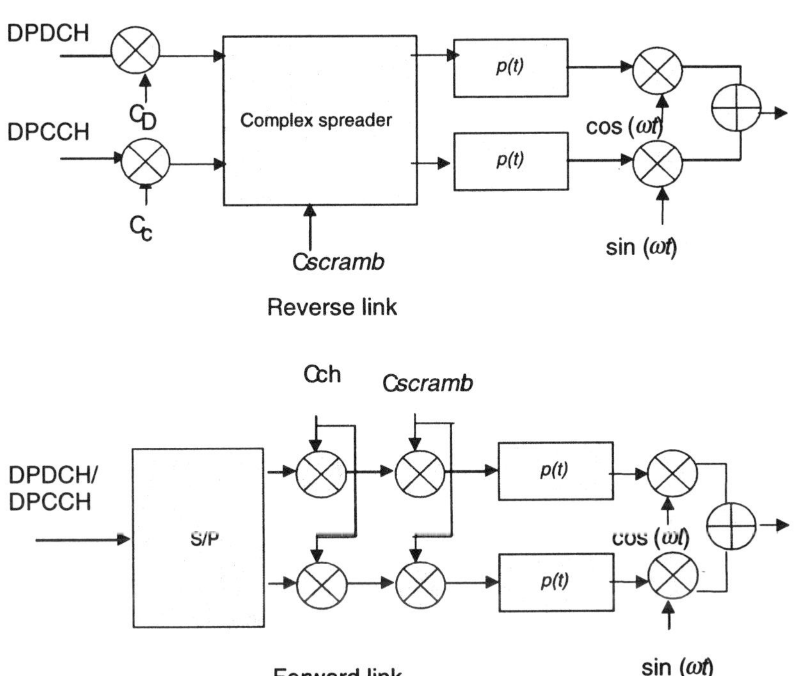

Figure 3.2 UTRA modulation/spreading scheme.

3.2.1 Future Radio Wideband Multiple Access Schemes—The FRAMES Project

The FRAMES project was the major ACTS project that addressed the problem of the terrestrial component of the UMTS radio interface. FRAMES continued preceding research programs within the frame of RACE I and II programs that investigated advanced TDMA- and CDMA-based radio interfaces [17]. The main goal of FRAMES was to develop a UMTS radio interface that fulfilled the requirements of terrestrial, 3G mobile radio systems.

Defined by the ITU, the FRAMES multiple access (FMA) concept lent itself to the introduction of novel technologies, such as joint detection, multiuser detection for interference rejection, advanced coding schemes (e.g., turbo coding), and adaptive antennas that would contribute to more efficient utilization of the UMTS frequency bands. The results of the FRAMES project were submitted to ETSI for use on future standardization work on the next-generation, radio-air interface systems.

3.2.1.1 FRAMES Contribution to Standardization Decisions

During its four-year duration (1995–1999), FRAMES definitely contributed to the standardization of 3G in Europe and other regions. A lot of the work was devoted to contributions to standardization bodies, mainly ETSI SMG2, on the radio-transmission technology. The main achievement of the FRAMES project in terms of the ETSI standardization process was the input to the consensus decision on UTRA. In addition, the project put substantial effort on the architecture and protocol design of the radio interface and on radio resource management techniques. In parallel, a hardware demonstrator for validation and demonstration purposes was developed.

ETSI SMG was responsible for the standardization process of IMT-2000 in Europe. In January 1998, after an extensive debate, the decision was reached by a consensus within ETSI on the radio interface. The solution was called UTRA and was based on research within the FRAMES project and ETSI requirements. The UTRA proposal agreed on the use of TD/CDMA in the unpaired portion of the UMTS spectrum in TDD mode, and on the use of WCDMA in the paired portion with FDD mode.

The input from FRAMES was substantial in the following decisions:

- FMA2 (FRAMES multiple access scheme mode 2) formed the WCDMA concept (UTRA FDD);
- FMA1 (FRAMES multiple access scheme mode1) with spreading formed the TD/CDMA concept (UTRA TDD);
- FMA1 without spreading was included for high bit rates in the U.S. proposal for IMT-2000 based on IS-136.

The UTRA FDD (WCDMA) was based on FMA2 only for the uplink, because FMA2 proposal employed code multiplexing of data and control information and it fitted better in the uplink, as power amplifier limitations are very important there. For the downlink, it was based on the Association for Radio Industry and Business (ARIB) proposal. The UTRA TDD (TD-CDMA) was based on FMA1 with spreading, but the parameters were modified for TDD application and harmonized with UTRA FDD.

The combination of FDD and TDD supported the different service needs in a spectrum-efficient way. The motivation for having different schemes for the FDD and TDD components of the UTRA concept was to provide the schemes that were best suited for FDD and TDD, respectively. The main characteristics of the UTRA proposals are shown in Table 3.2.

The harmonization between the modes was performed for the chip rate, pulse shape, and frame length, to facilitate the production of dual-mode terminals. In addition, the proposed solution aimed at supporting spectrum allocation as small as 2.5 MHz.

Table 3.2

Key Characteristics of the UTRA Concept

	UTRA FDD	*UTRA TDD*
Multiple access	WCDMA	TD-CDMA
Duplex method	FDD	TDD
Carrier spacing	5/10/20 MHz	5 MHz
Chip rate	4.096 Mcps	4.096 Mcps
Slot structure	16 slots per frame	16 slots per frame
Spreading	Spreading factor 4-256, short codes for DL and UL; long optional for UL	Orthogonal, spreading factor 1, 2, 4, 8, and 16 chips/symbol
Frame length	10 ms	10 ms
Multiple rates	Multicode, variable spreading factor	Multislot, multicode
Modulation	DL: QPSK; BPSK	QPSK
Pulse shaping	Root raised cosine roll-off = 0.22	Root raised cosine, roll-off = 0.22
Handover	Mobile-controlled soft handover	Mobile-assisted hard handover
IF handover	Mobile-assisted hard handover	Mobile-assisted hard handover

Figure 3.3 illustrates the process of defining a proposal for UMTS radio access and the subsequent process of evaluating and further development of the FRAMES scheme toward an UTRA concept.

FRAMES produced significant results on the physical layer for coding, modulation, and receiver algorithms and some of them were submitted within the UTRA concept to ITU-R for IMT-2000 [18]–[19]. The project contributed not only to the definition of the protocol structure and architecture of the radio interface but also to the development of GRAN. The main feature of the GRAN concept was the separation of the core network from the radio-access network and the description of radio bearers in a generic, parametric manner. Thus, the preservation of the radio-dependent functionalities within the GRAN was made possible. The GRAN concept was incorporated in the work of standardization bodies on the UMTS terrestrial radio-access network (UTRAN), while FRAMES contributed to the standardization processes in ARIB (Japan), TIA and T1P1 in the United States, and TTA in Korea. In the scope of the FRAMES project, a system demonstrator was built to validate and demonstrate the basic functionality required for the justification of the FRAMES specifications. The scope of the project required integration with other ACTS projects (e.g., RAINBOW).

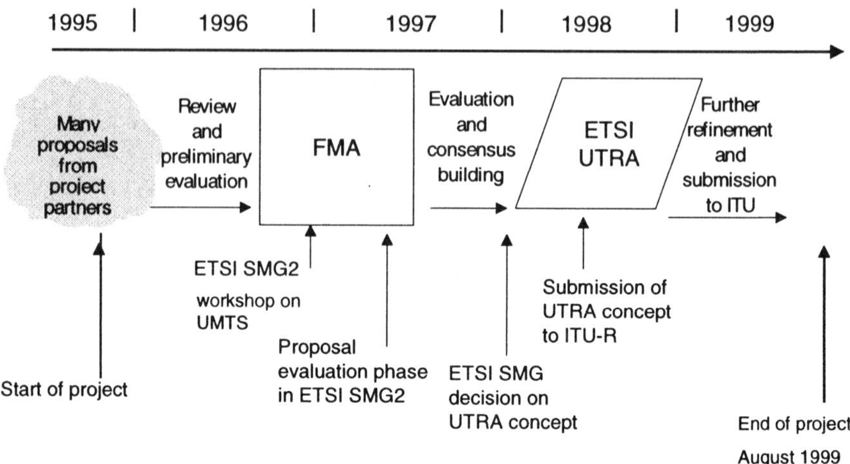

Figure 3.3 Adaptation of concepts to ETSI decision.

3.2.1.2 Original FMA Concept

The development of FMA

FRAMES had as a technical objective the definition, development, and evaluation of an efficient wideband multiple-access radio scheme that fulfilled the UMTS requirements. During the initial phase of the project various radio-access concepts were proposed and the effort was spent on combining the input into complete radio-access schemes [20]. Table 3.3 provides a list of the initial concepts and a brief explanation of their characteristics. Some of the concepts were complete radio-interface proposals; others regarded only certain key aspects. The evaluation of the proposals knew different phases. It started with a quantitative and qualitative evaluation and the initial concepts were the starting point. The project supported the definition of UMTS requirements within SMG; these requirements were later used as an evaluation basis. One strong requirement was to allow easy dual-mode (GSM-UMTS) terminal implementations and to maximize the reusability of GSM infrastructure; another one was the provision of efficient handling of packet services.

The ultimate goal of the evaluation process was the definition of an FMA concept. The FMA needed to be efficient in accordance with a wide range of applications, communication services, and different deployment scenarios. It had to offer an open platform to cater to current and future developments of UMTS radio interface standards.

Table 3.3

Initial Concepts Proposals

Abbreviation	*Concept Description*
CDMA-CC	CDMA using complementary code sets; combines TDMA and CDMA by using complementary code sets
CTDMA	CDMA; combines TDMA and CDMA, uses inverse filtering receivers
DS/SFH-CDMA	Direct sequence/slow frequency hopping CDMA; combines direct sequence and frequency hopping CDMA
JD-CDMA	Joint detection CDMA; combines TDMA and CDMA using joint detection receivers for intracell interference rejection
MC-CDMA	Multicarrier CDMA; DS-CDMA concept with frequency-domain spreading using OFDM
MC-TDMA	Multicarrier TDMA; TDMA scheme with multiple carriers for high bit rates
MCo-CDMA	Multicode CDMA; CDMA scheme with multiple codes for high bit rates
MUD-CDMA	Multiuser detection CDMA; CDMA scheme with multiuser detection for intracell interference rejection
TDMA/SFH-DSR	TDMA/slow frequency hopping with dual signal receiver; TDMA scheme using special receiver for simultaneous reception of two signals
TDMA -OFDM	TDMA with OFDM; TDMA concept with OFDM for multiplexing different users

The development of the FMA knew two stages [21]. At the first stage several initial candidate schemes were compared and schemes with similar characteristics combined. This stage resulted in two multiple access schemes, each with three options:

- Multicarrier TDMA (multiples of 200 kHz), single wideband TDMA (WTDMA, bandwidth 1–2 MHz), and hybrid CDMA/TDMA (bandwidth 1.6 MHz);
- Asynchronous CDMA (WCDMA, bandwidth 6 MHz), OFDM/CDMA and synchronous CDMA.

At the second stage, a more extended-criteria evaluation was performed. Provision of various data rates in different environments and bearer-service flexibility, spectrum efficiency, coverage, support for adaptive antennas, hierarchical cell structures (HCS), duplex method, support for public and private environments, and handover were considered. The criteria also addressed the terminal impacts (power consumption and complexity, GSM/UMTS dual mode terminals), and the BSS impacts (evolution from existing systems, BSS complexity and cost). The evaluated quantitative performance measures included coverage and capacity spectrum/efficiency; that was done via computer simulations and the results were generated back to the standardization bodies (mainly SMG2), suggesting some improvements.

The evaluation showed the weak and strong points of the suggested schemes. WTDMA proved to be flexible for TDD with asymmetric services and an optimization for high-bit-rate services was possible. The main problem was the complexity of the equalizer in a large cell. CDMA/TDMA provided good performances both for voice and data, but it was quite complex because of the joint detection for low-bit-rate services. WCDMA was suitable for circuit-switched variable rate services because of its good multirate capabilities. It proved, however, less suitable for asymmetric services in TDD mode and showed some difficulties related to power control in the case of packet-switched services.

Multicarrier options were dropped from further consideration because of complexity and performance reasons, including a too complex RF stage of the multicarrier TDMA, low performance of the OFDA/CDMA when considered only for the downlink, and low performance of the synchronous CDMA because of lack of fast power control. Unlike complex multicarrier TDMA, a single carrier scheme (200-kHz carrier) was implemented combining EDGE with 16-QAM modulation rather than conventional Gaussian minimum-shift keying (GMSK).

All these proposals were once again carefully examined and compared. The previous conclusions suggested the combination of WTDMA and hybrid CDMA/TDMA into mode 1 (WTDMA with and without spreading) and the harmonization of some WCDMA parameters into mode 2. So, during the first two years the FRAMES project developed the harmonized FMA platform consisting of two modes [22]:

- Mode 1 (FMA1): WTDMA with and without spreading;
- Mode 2 (FMA2): WCDMA.

The basic FMA multiple access schemes are illustrated in Figure 3.4. Both modes fulfilled the UMTS requirements and were harmonized to each other and with GSM. The harmonization between FMA modes involved the harmonization of radio parameters in terms of carrier spacing and clock rates, and protocols whenever possible. The harmonization with GSM was mainly concentrated on an evolved GSM core network, on a common frequency grid (200 kHz), and clock-based on the GSM clock with the objective of facilitating dual mode GSM-UMTS phones.

The main achievement of the FRAMES project was the decision on a harmonized dual-mode access platform. Through combinations of access schemes the two proposed modes offer an optimized solution for the implementation of UMTS requirements.

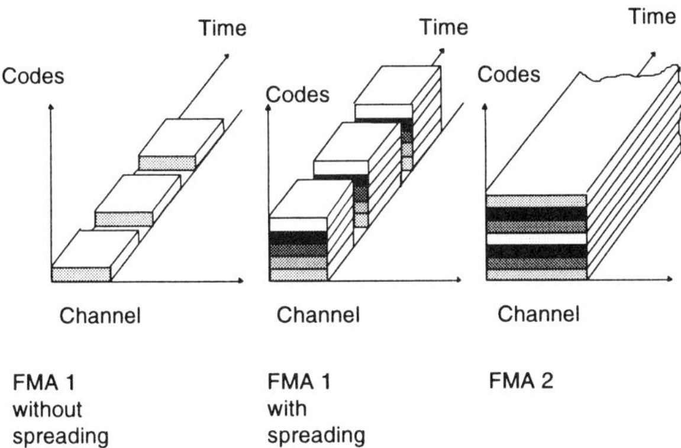

Figure 3.4 Basic FMA multiple access schemes.

Both modes support bit rates up to 2 Mbps. In mode FMA1 variable bit rates with low granularity are supported by resource reassignment together with adaptive coding. The slotted structure of FMA1 is well suited for the bursty, packet type of traffic. Multislot/multicode possibilities lead to higher flexibility in service provision of different bearer services to one user by packing them into different slots/codes. In FMA2 variable bit rates are supported by adaptive coding and power assignments Because of this power sharing and long initial synchronization, this mode is more suitable for moderately varying circuit-switched services. For both modes, different combinations of coding and link adaptation obtain different operating points for the services. Mixed bearer services for one user can be multiplexed and have different operating points.

Spectrum efficiency is an important parameter in wireless systems. It is strongly affected by the receiver algorithms. In the FRAMES project, multiuser receivers both for uplink and downlink were studied. By applying the interference-suppression/cancellation technique, significant performance and capacity improvements were obtained. Joint detection was used to eliminate either intracell or intercell cochannel interference or both, which led to increased spectrum efficiency. Adaptive antennas also introduced further improvements.

Hierarchical cells were supported by interfrequency handover, which in mode FMA1 was inherent part of the system and in mode FMA2 was supported through discontinuous uplink transmission and dual receivers in the downlink. The research showed that mode FMA1 was better suited for operations in the private environments. It can support asymmetric data services in TDD mode, as the number of slots in the uplink and downlink can be varied. In mode FMA2 flexible allocations of resources between up and downlink proved difficult, so it was more applicable for FDD mode. Coverage evaluation was carried out for both FRAMES modes.

The two major goals of wireless systems are high capacity and high QoS. Both of these goals can be dominated by the choice of modulation and channel coding. In FRAMES, intensive research was carried out toward improving modulation and coding because spectral efficiency, performance, and BER probability can be strongly affected by these techniques. Turbo coding was evaluated carefully for UTRA/TDD and the results showed that performance close to theoretical limits could be reached.

For both FMA modes, simple divisions and multiplication could generate the clock from a 26-MHz clock (same frequency as for GSM). This is one of the most important criteria for the simple design of dual-mode terminals (UMTS and GSM).

The FRAMES concept was modeled to a layered approach [23]–[25]. A protocol stack was designed to support multilayer connections and various QoS requirements, while harmonization between the two FMA modes was essential in the design. The FMA protocol stack consisted of mode-specific elements and common mode-overlapping functionalities (see Figure 3.5).

There were basic differences on the radio link layer (RLC) and media access control (MAC) layer for both modes, but they were harmonized as far as possible. The logical link control (LLC) layer was mode independent, while the radio network layer (RNL) provided some fundamental differences for soft handover in FMA2 and hard handover in FMA1. Otherwise, most of the protocols were unified, and for the RNL bearer the handling procedures were generic. Both modes could use the same "core" protocols.

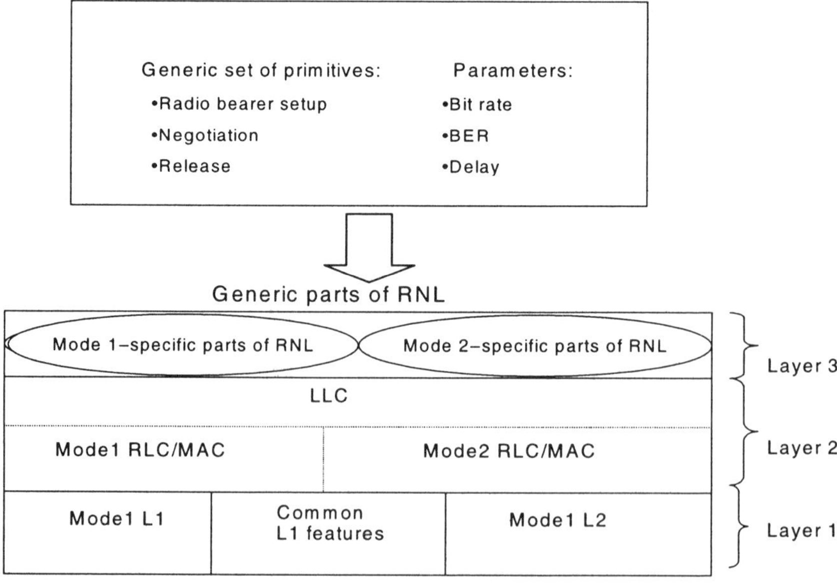

Figure 3.5 Harmonized protocol stack of the FRAMES concept.

The main contribution to the protocol concept was the definition of transport channels within FMA2. Transport channels were defined as intermediate steps between physical channels and logical channels, and they modeled the data transfer services that the physical layer offered to the MAC layer. They represented the pipes for information transfer between peer MAC entities. The concept of a transport channel was included in UTRA transport control functionalities, first for UTRA FDD and later for UTRA TDD [22]. The main properties of the physical layer are presented in Table 3.4.

Table 3.4

Basic Parameters of FMA 1 and FMA2

	FMA1 *(TDMA with and without spreading)*		*FMA2* *(CDMA)*
	Without spreading	With spreading	—
Multiple access method	TDMA	TDMA CDMA	WB-CDMA
Carrier Chip/bit rate	2.6 Mbps	2.167 Mcps	4.096 Mcps
Bandwidth	1.6 MHz		4.4–5 MHz
Duplex method	FDD		FDD TDD for further study
Interference reduction	Joint detection supported	—	Multi user detection supported
Spreading codes	N/A	Orthogonal of length 16 chips	Spreading factor 4 to 256, short codes
Multirate concept	Multislot	Multislot and multicode	Variable spreading and multicode
Detection	Coherent, based on midamble		Coherent (reference symbol or pilot)
Handover	Mobile assisted hard handover		Mobile controlled soft handover
Interfrequency handover	Supported		Supported
Frequency hopping	Frame by frame / slot by slot		N/A

FMA1 mode was based on a carrier spacing of 1.6 MHz (eight times GSM), a frame of 4.615 ms (same as GSM), and within the frame the split of three different slot lengths: 1/8, 1/16, and 1/64 of the frame length with the 1/8 slot used with spreading and the 1/16 and 1/64 slots used without spreading. Depending on the radio environment and service, the frame and burst structure can be dynamically adapted. The FMA1 frame structure is shown in Figure 3.6.

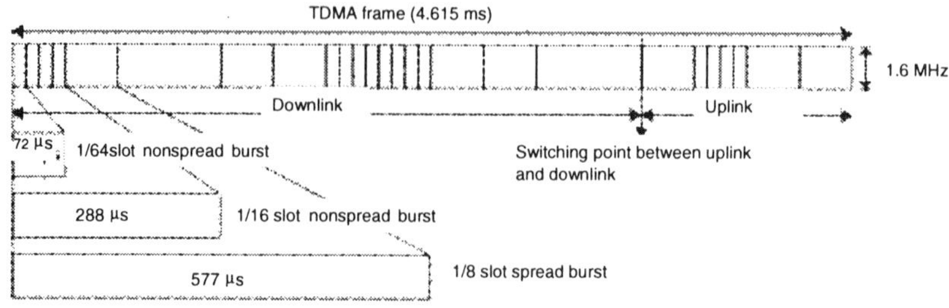

Figure 3.6 Unit TDMA frame structure of FMA1.

A basic physical channel is one timeslot in the case of FMA1 without spreading, and one timeslot and one spreading code in the case of FMA1 with spreading. In FMA1, user bit rates from a few kilobits per second up to 2 Mbps can be achieved by allocating a different number of basic channels to a user. The physical contents of the timeslots are bursts of corresponding length. Within each 1/64 and 1/16 slot, one burst of corresponding length can be transmitted. Within the 1/8 slot, up to nine or even more bursts with corresponding length, separated by different spreading codes can be transmitted. Within the same timeslot, these bursts can be allocated to one or more users. There are five types of traffic bursts, three nonspread and two spread (Figure 3.7 and Figure 3.8).

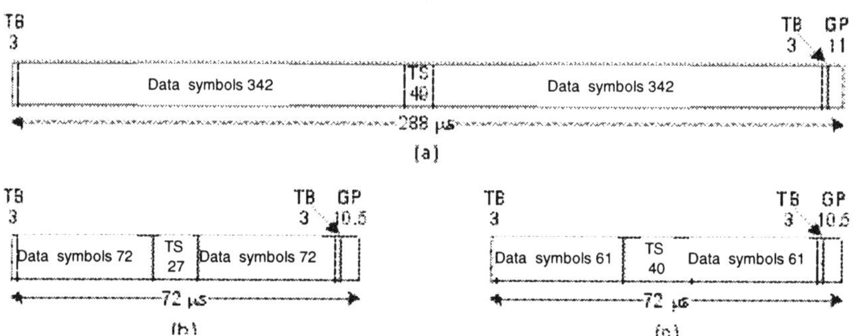

Figure 3.7 Nonspread data burst for different number of data symbols: (a) 342, (b) 72, and (c) 61.

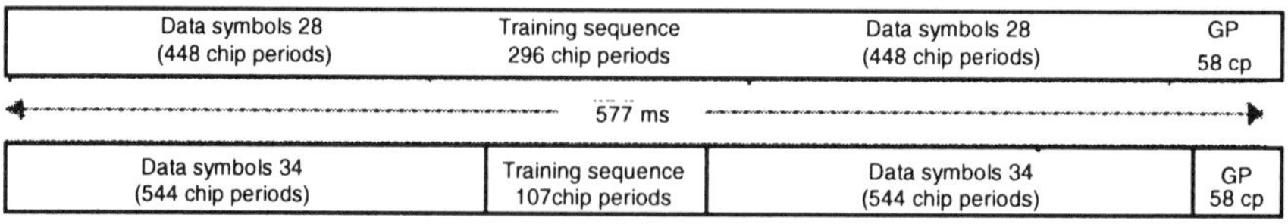

Data symbols 28 (448 chip periods)	Training sequence 296 chip periods	Data symbols 28 (448 chip periods)	GP 58 cp

577 ms

Data symbols 34 (544 chip periods)	Training sequence 107chip periods	Data symbols 34 (544 chip periods)	GP 58 cp

Figure 3.8 Spread speech data burst.

Each burst consists of a training sequence, two data blocks, and a guard period. The nonspread bursts differ in the length of the burst and in the length of the training sequence (27 and 49 symbols), leading to different numbers of payload symbols and different tolerable multipath delays. The spreading bursts differ in the length of the training sequence (107 and 296 chips), leading to different applications (uplink or downlink, different numbers of users per timeslot, and different tolerable multipath spreads).

FMA1 without spreading was designed to provide large flexibility in terms of radio-resource management and very good service flexibility, with bit rates from a few kilobits per second to 2 Mbps gradually. The training sequence needed to be optimized to fully support adaptive antenna techniques and interference cancellation techniques. It was optimized for high bit rates and small- to medium-sized cells. The highest bit rate can be achieved in rather small cells. For larger cells FMA1 with spreading can be used because it can better tolerate large delay spreads. The multiframe structure of FMA1 was designed to provide a means for efficient realization of a packet-access protocol and to provide compatibility with the GSM multiframe structure to simplify handovers between FMA1 and GSM. The spread option has two different training sequences, one that can be used for a short delay spread, the other for large. Both options (with and without spreading) and their combinations cover the whole range of 2G and UMTS service requirements and provide efficient and flexible use of the radio resource. The spreading in FMA1 is fixed, with a spreading factor of 16. The flexibility in bit rates is achieved with multicode and/or multislot principles.

FMA2 used a 10 ms frame length. Although different from GSM, it allowed handovers between the systems, since 12 FMA2 frames were equal to a single GSM multiframe of 120 ms. FMA2 defined two types of dedicated channels on both the uplink and downlink. Control information was transmitted on the dedicated physical control channel (DPCCH), and the user data is transmitted on the dedicated physical data channel (DPDCH) [17]. Figure 3.9 shows the principal frame structure of the uplink DPDCH. Each 10-ms DPDCH frame on a single code carries $160*2^k$ bits, where $k = 0.1 \ldots, 6$ corresponding to the spreading factor of $256/2^k$ with the 4.096 Mchips basic chip rate. Multiple variable rate services could be provided on a single physical code resource by using time multiplexing within each PDCH frame. The overall DPDCH bit rate varied on a frame-by-frame basis with the data rate indicated on the DPCCH.

In most cases only one DPDCH is allocated per connection, and services are jointly interleaved sharing the same DPDCH. However, multiple DPDCH can also be allocated (to avoid a low spreading factor at high data rates). The dedicated DPCCH was used to transmit the control information generated at layer 1 (pilot symbols for coherent detection, power control signaling bits, and frame control feeder (FCF) with rate informations. The control informations ware multiplexed within a slot structure within each DPCCH frame. The length of each slot may vary in multiples of smallest slot length of 0.625 ms.

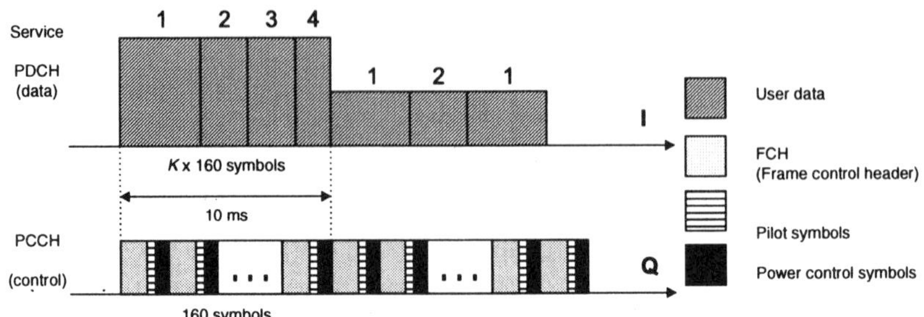

Figure 3.9 WCDMA uplink multirate transmission.

FMA1 supports both real-time (RT) and non-real-time (NRT) packet services [23]. For RT services error protection forward error correction (FEC) codes are used. As advanced solution turbo codes and rate-compatible punctured convolutional codes (RCPC) are implemented to meet the QoS requirements. Concatenated codes are used to achieve very low BER. NRT services use type-2 hybrid ARQ scheme. Interleaving in FMA1 is done on frame-by-frame basis for low-rate services, and slot by slot for higher rate services.

The data modulation technique used in FMA1 without spreading is binary offset QAM, which has good performance while keeping the requirements on power amplifier linearity on a reasonable level. Quaternary offset QAM is used for 2 Mbps. In FMA1 with spreading, the data modulation is linear (QPSK/16QAM), but linearized GMSK is used for spreading modulation. FMA1 has slow power control (with 50-dB dynamic range) with an option to use fast power control on a burst basis.

In FMA mode, because of coherent detection, hard decision parallel interference cancellation (HD-PIC) receiver is used only in the uplink. IC techniques can be applied to adaptive antennas, which additionally enhance the system performance and capacity. FMA1 uses midambles for channel estimation. FMA1 with spreading uses joint detection in the uplink and in the downlink to cancel intracell interference. In FMA1 without spreading, it is used to cancel intercell interference.

In FMA2 interleaving can span over one or several 10-ms frames. FMA2 is based on 3.84-Mchips CDMA, with 4.8-MHz channel spacing, and using optimized 10-ms frames.

Both modes use coherent detection. For channel estimation FMA2 in the upper link uses a varying number of reference symbols and a continuous pilot channel in the downlink.

The FMA2 mode has a variable spreading factor, from 4 to 256. Different user bit rates are supported using multiple codes. The choice of multiple codes depends on direction of transmission and bit rate requirements. The multiple codes with smaller spreading are used to keep the amplitude variations low. In the downlink multiple codes are more convenient. The optional long codes can be used in the uplink to reduce the correlations between the cells, when no multiuser detection is employed. When the long codes are turned off, efficient and low-complexity multiuser detectors can be used to upgrade the performances.

Error control for both modes in the initial stage was based on traditional convolutional coding. More advanced coding techniques like concatenated coding (Reed-Solomon coding), sequential decoding, and several ARQ schemes were suggested in later stages. FMA2 mode uses TCH codes. For FMA2 offset QPSK is used in the uplink, and ordinary QPSK is used in the downlink. FMA2 mode has fast power control (80-dB dynamic range) based on open- and closed-loop control.

In view of spectrum allocation issues within UMTS/IMT-2000, the proposed UMTS frequency allocation consisted of several parts. The main part of terrestrial applications is 1,920–1,980 MHz, and 2,110–2,170 MHz. In FRAMES, it was assumed that these bands should be used for FDD. As an additional band for terrestrial applications the 1,900–1,920-MHz band which cannot be used for FDD was allocated. FMA1 mode was designed to also allow the use of TDD. FMA2 mode was designed only for FDD.

3.2.1.3 System Demonstrator

Within the FRAMES project a hardware demonstrator was developed to help the validation and evaluation of the suggested solutions in FMA1 mode, both for spread and nonspread options. The main objectives of the demonstrator were to allow the evaluation of the FMA1 mode in a real-world environment and to provide joint trials with other ACTS projects.

The FRAMES FMA1 demonstrator is shown in Figure 3.10. The communication between the base stations and the control terminals was realized through Ethernet networks. A detailed overview of the demonstrator can be found in [26]–[27].

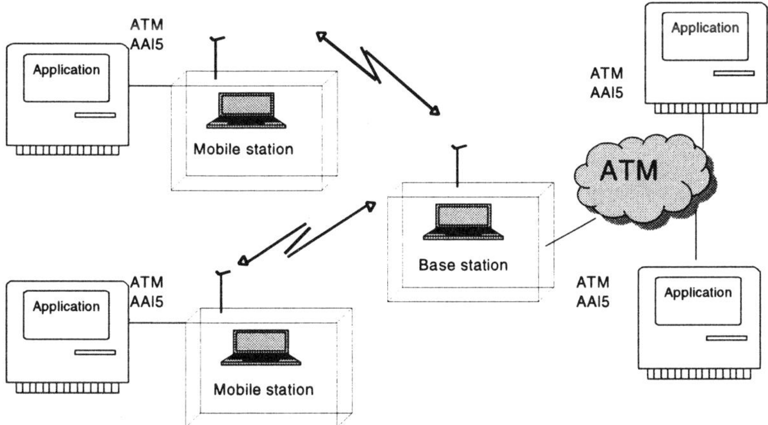

Figure 3.10 FRAMES demonstrator.

The MS and BS platforms each consist of three units (control, baseband, and RF) with very strong software support for the control unit structure. The control unit is connected to a central control terminal (CCT). The CCT provides the man-machine interface to the demonstrator. It is composed of bearer (BCT) and analysis (ACT), which are the control and analysis interfaces to the demonstrator. The BCT provides the demonstrator with the different operation mode configurations for both mobile and base stations, together with monitoring the system operations. The functionality of the mobile and base stations was designed on the layered principle and was presented with the FRAMES demonstrator application software that provides functionality within each of the protocol layers (see Figure 3.11).

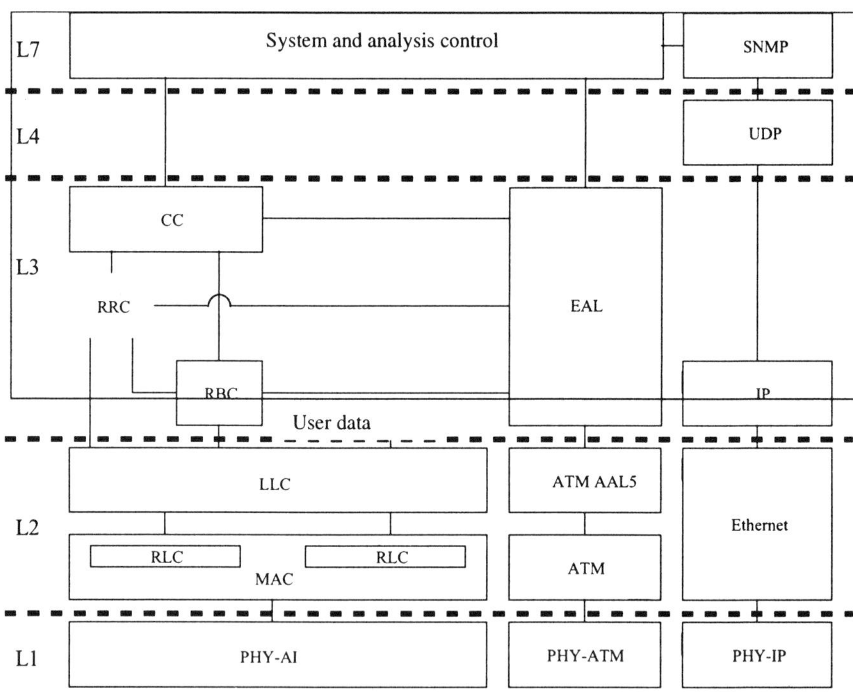

Figure 3.11 FRAMES software protocol stack.

The frame structure is depicted in Figure 3.12. Both BS and MS have two timeslots active in the downlink (TS 0 and 1) and in the uplink (TS 5 and 6). Two active timeslots are needed to support high bit rates. The demonstrator could process up to eight codes per slot. The switching point in the uplink and downlink was fixed midway through the frame. The parameters correspond to the FMA concept (1.6-MHz carrier, spreading factor of 16, QPSK and 16QAM modulations, slow power control—60-dB dynamics on uplink UL, BRE=10^{-3}, and a maximum bit rate of 512 Kbps). Circuit- and packet-switched services were investigated.

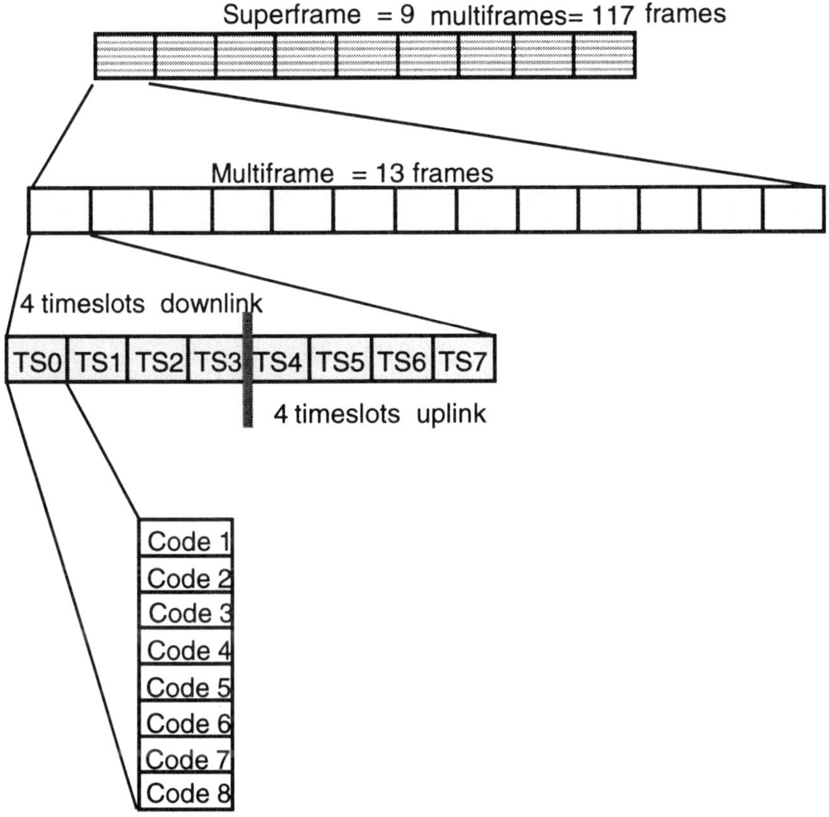

Figure 3.12 Frame structure of FRAMES demonstrator.

During the establishment of the testbed demonstrator, intensive laboratory activities are undertaken. In the laboratory environment, the channel simulator was used as the transmission medium (both at the RF and base-band level) and many quantitative performance measurements in terms of BER, transmission delay, and so forth, were completed. The performance evaluation of the spreading option. The quality of communication in various environments (picocells, microcells), within different bearer classes (data rate, BER, delay) and various operating conditions (connection establishment, traffic) was analyzed. The FRAMES testbed was able to provide high bit-rate services supported by both circuit and packet-switched bearer services. It also demonstrated bearer flexibility.

The demonstrator was used for indoor and outdoor trials. The indoor trials were performed in a laboratory environment or in large conference rooms, with limited user mobility. The outdoor trials were performed in a microcell-type environment with either pedestrian or vehicles as users for the FRAMES.

In order to meet the UMTS and ETSI requirements, the FRAMES demonstrator was adapted to the TDD TD-CDMA, as proposed in the Delta Concept Group evaluation report from ETSI SMG2 [28]. The idea was to demonstrate the feasibility of the proposed radio interface and to gain credibility for the proposal. This demonstration involved application terminals that provided multimedia applications through ATM/AAL5 link. Videoconference service with high bit rates (around 400 Kbps) and high QoS was demonstrated as well. Other typical Internet applications (videoconference, Web browsing, white board, data file transfer, e-mail, and so forth) were supported both locally and through remote connections to the Internet network. The demonstrator provided a flexible and powerful signal processor platform for bench and field trials [29]. The tests were concentrated on joint detection with respect to power control, real outdoor and indoor simulator supported trials, and service demonstration.

Joint demonstrations were performed with the projects RAINBOW and FIRST (see Chapter 5). The interface between the projects was an AAL5 interface. The network demonstrations with RAINBOW were performed with simple speech service, provided by the RAINBOW demonstrator.

The network demonstrations with RAINBOW were performed to demonstrate advanced call and bearer functionality as controlled by the RAINBOW access network. The interface was set between the FRAMES radio interface and the RAINBOW network. The information passed over the interface was formatted into ATM cells and routed by allocation of VCI/VPI addresses. Some messages were addressed to the FRAMES control processor, and the others were passed over the interface unaffected by the FRAMES. Figure 3.13 shows the overall structure of interconnection between FRAMES and external applications/networks. RAINBOW network demonstrators implemented a FRAMES adaptation layer (FAL) to interface with FRAMES through the ATM/AAL5 standard. FRAMES implemented a common external adaptation layer (EAL) to interface with RAINBOW demonstrators. A WCDMA evaluation system was built for experimentation with the WCDMA radio access network and ATM and to apply these technologies to different wideband services [22]–[30].

3.2.1.4 Further Activities Beyond FRAMES

In order to harmonize and standardize the similar solutions that come from different standardization bodies (ETSI, ARIB, TTA, T1P1, and TIA), new international standardization bodies were established: 3GPP and 3GPP2 [31]. 3GPP is involved with the standardization and harmonization of similar decisions from ETSI, ARIB, TTA, and T1P1, WCDMA, and the related TDD proposals [32]–[33]. 3GPP2 concentrates on the harmonization of cdma2000-based proposals from TIA. These international bodies and ITU-R continued to work on the finalization and further development of standards for UMTS/IMT-2000. To enable harmonization amongst the many different proposals, major international operators initiated activities within the Operators Harmonization Group (OHG) [34]. They started the harmonization process originally for the CDMA-based proposals, and created a global 3G, consisting of three modes: DS-CDMA, MC-CDMA, and TDD-CDMA.

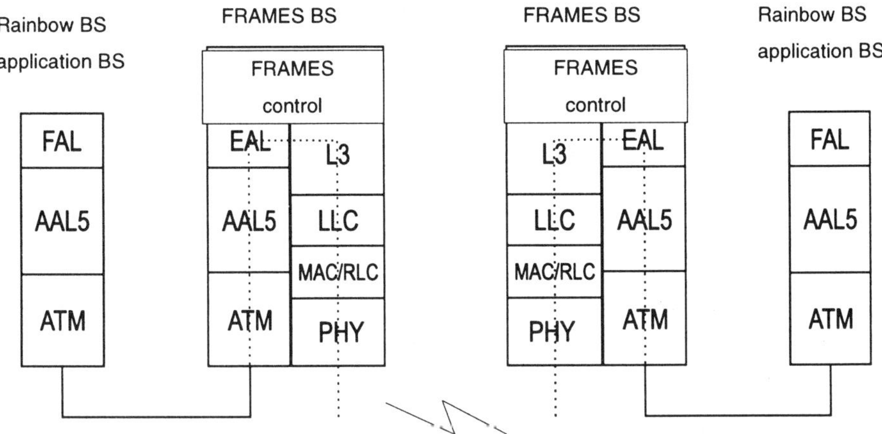

Figure 3.13 Overall structure of the interface for the joint trials.

3.2.2 Mobile Services for HSTs—The Project MOSTRAIN

MOSTRAIN addressed the development of 3G mobile communication technologies in a demanding environment concerned with the delivery of services to and from HSTs [35]–[36]. HSTs were seen as transport mode for medium-distances travel with increasing social, economic, and environmental benefits. Even if the number of the users is limited, providing access to passengers on board trains was an important aspect of the introduction of UMTS.

The project did not fulfill its goals, but it addressed the major problems concerning railway users. Recognizing the speed of the moving trains as the most crucial parameter, the main interest in the MOSTRAIN project focused on the special features of the application to high-speed trains (HSTs). The MOSTRAIN consortium included a number of partners from railway and network operators and leading manufacturers in railway communication systems.

3.2.2.1 Communications in the Railway Environment

Provision of mobile communications facilities everywhere could accomplish one of the main objectives of UMTS. The current velocity of HSTs is about 300 km/h. It is expected that in the next 20 years it will grow up to 500 km/h. These speeds were not addressed by 2G systems (GSM, DCS-1800) or in GSM-R, and they require a different approach to users' requirements and to operators' needs. Thus, a decision for signaling system based on mobile radio.

The diverse railway environment can be grouped into the following groups:

- HSTs (average traffic levels are relatively low);
- Traditional rural railways (moderate and slow speed—traffic varies from low to only occasional);
- Urban or metropolitan railway lines (lower speed, significantly higher traffic level);
- Station areas (low-speed, large traffic load).

HST operates in all four environments with the requirement that they were seamless to the user. The developed communication system considered the problem of a large number of simultaneous handovers; priorities ranging from noncritical data to incub-signaling, specific fading, and shadowing conditions; predictive handovers due to known position of the trains; penetration losses due to galvanized windows; and so forth.

The main objectives of the project MOSTRAIN for the provisioning of UMTS services in a railway environment were to ensure that 3G technologies could provide services, appropriate architecture structures, and radio channels to the railway user. To achieve these objectives, the project identified a set of requirements for passengers and railway users. The radio-channel and system architecture for the support of services in HSTs in various railway environments was also determined. MOSTRAIN examined appropriate access, transmission, and processing techniques for optimum UMTS transmission.

MOSTRAIN planned to develop a simulator and UMTS "generic kernel" to examine optimum architecture, parameters, and resource allocation. It investigated the development and improvements of existing antennas for "on-train" and "in-train" use. The idea was to combine results into a demonstrator, which would show UMTS-type applications in high-speed environments.

MOSTRAIN established links with other ACTS and railway projects, and the ETSI standardization committees. One of the key issues of the project MOSTRAIN was to deal with the split between the shared and dedicated infrastructure. Shared infrastructure assumes communications provided by "public" operators/providers utilizing equipment that is also used for nonrailway communications. Base stations in this scenario are not assumed to be located near the track, and propagation models do not assume line-of-sight (LOS). Dedicated infrastructure uses dedicated equipment for railway communications, combined with specialized techniques. It was supposed that the train users would be able to operate their UMTS mobiles. This scenario created the possibility of utilizing the two-link architecture. The architecture had separate links from the track side to the train and within the train to the mobiles, with a baseband element in between. The link within the train to the mobiles needed to be based on standard UMTS. The trackside-to-train link can transmit UMTS calls using alternative transmission techniques optimized for the HST environment.

The technical work within MOSTRAIN was carried out within the areas of services, architecture, propagation channel, physical layer, and demonstration.

Services

The project recognized groups of users as passengers, railway operators, and focus groups. The main interest of the passenger group was considered the application areas of the mobile-office and informed-traveler concepts. The mobile office was primarily of interest to business travelers. Accordingly, priority applications were split in a few groups recognized as follows.

- Essential services (mobile phones, fax, credit card transactions, e-mail, railway information, and so forth);
- Useful services (data file transmission/reception, publicity services, information on other transport systems, video-on-demand);
- Entertainment services (hotel reservation, weather report, paging, and so forth);
- Irrelevant services (videotelephony, teleconference).

Railway operators were demanding more safety, group calls, and higher QoS requirements. Public operators had mostly critical requirements for the safety of critical data (in-cab signaling) and the need of dedicated infrastructures.

MOSTRAIN developed a traffic model that recognized passenger and railway-specific categories. The traffic model was based on three types of matrix relations between user services and bearer classes, bearer classes and train types, and train types and the environment.

Propagation Channel

One of the main initial tasks of the project MOSTRAIN was the characterization of the railway propagation channel to ensure that it fitted in the UMTS 3G concept [37]. The results influenced the architecture concept and the physical layer design.

The problem of radio propagation was formulated in the context of the dedicated and shared infrastructure. It considered different environments (tunnels and cuttings, stations, along the track, penetration into the train, in-train propagation). The dedicated infrastructure located the base station trackside (in order to provide LOS) and the shared infrastructure assumed that the base station was located away from the track (without LOS).

The characterization of the propagation model in MOSTRAIN was based on theoretical calculations with the ray-tracing method and narrowband and broadband measurements in the 2-GHz range. In the case of base stations along the track, LOS could be obtained with consequential low to moderate delay spread and feeding. Very rapid transients required additional predictive handover, while the Doppler shift was high. Base stations located away from the track could rarely achieve LOS. The radio channel experienced deep and rapid fading, with moderate transients and Doppler shift. Penetration into trains could vary from few decibels (France TGV with large nonmetallic windows) to 30 dB for trains with galvanized windows. This raised the feasibility of positioning the onboard repeaters.

Architecture

The architecture concept needed to consider various implementation options for the delivery of UMTS services on the railway, both for dedicated and shared infrastructures. Two major architecture structures were suggested: single-link and two-link. In the single-link architecture there was direct UMTS communication from a ground base station (BS) to a UMTS mobile on the train. This could be either direct or via a repeater. The repeater should handle loss in trains with a high propagation loss into the carriage (see Figure 3.14 a and b). In the two-link architecture, separate links from the trackside to the train and within the train to the mobiles were established with a baseband element in between them (see Figure 3.8 c and d).

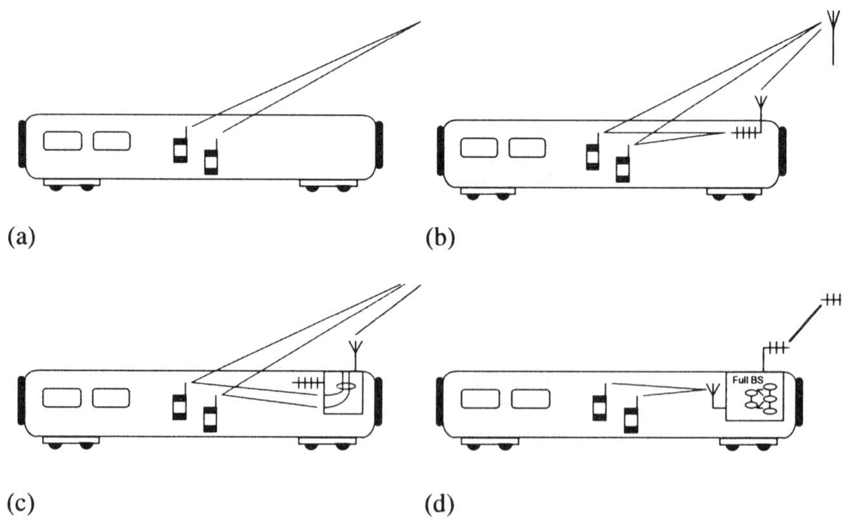

Figure 3.14 (a, b) Single-link communication and (c, d) two-link communication.

The two-link architecture was found to be the most appropriate for the dedicated infrastructure, while the single-link architecture was only applicable to the shared infrastructure.

The benefit of the two-link architecture was that it allowed a truncated radio link that could be optimized for use on the railway. It did not need to be a standard UMTS, but the cost and the availability of equipment had to be considered. The handover was required only for truncated radio link and not on individual calls. It allowed for the scheduling of calls and was more appropriate for in-cab signaling. The single-link architecture had to deal with the possibility of mass handovers at high speeds.

Physical Layer

The work performed in MOSTRAIN on the physical layer aspects has input from the works performed on the architecture and propagation channel. The physical layer concept was analyzed separately for the dedicated and shared infrastructures. The project performed an analysis of generic techniques such as diversity, modulation, equalization, and error correcting codes under different channel conditions (fast and slow fading, ISI, attenuation, Doppler perturbations). Within the dedicated infrastructure, the physical layer considered optimization of the trackside to the train link. In the shared infrastructure, the standard UMTS system was used for communication with the train. Research was concentrated on potential enhancements to standard UMTS that could improve QoS for onboard users.

In parallel, MOSTRAIN developed antenna schemes for the railway environments. Ground base station (GBS) and train base station (TBS) antennas for the 2-GHz range were investigated, and the design for the prototype demonstrator was decided. For HST and for both infrastructures, leaky feeders (LF) were suggested for radio coverage on cross sections and tunnels.

Macro- and microdiversity schemes were analyzed in order to define the diversity methods that could be implemented in the HST environment. Amongst the many diversity schemes (space diversity, fields component diversity, polarization diversity, angle diversity, frequency diversity, and time diversity), space diversity offered some gains in the HST environment, except for the tunnel environment.

A few candidate techniques (ATDMA, RNET, DS-CDMA, and OFDM) were carefully researched under different environmental conditions. After comparison and selection, ATDMA and RNET were chosen for the two-link architecture as appropriate candidates. Advanced TDMA is a classical solution, already addressed in the RACE II project. It is based on a fixed-frequency carrier during the slot duration. RNET is a novel access method, based on a new transmission technique, a frequency ramp that sweeps over the frequency bandwidth (during the slot duration). The slot in this case is called a cell. ATDMA-based techniques should be combined with coherent detection with equalization. RNET, due to the sweeping ramp, is more applicable for noncoherent detection.

Equalization work provided some valuable results that could be used in other ACTS projects considering the feasibility of UMTS equalizer design. The research was more focused on shared infrastructure, where the propagation channel had a fast, time-varying nature. Two types of equalizers were analyzed: K-VA (consisting of MLSE and the Kalman-type channel estimator) and SBS-MAP (consisting of symbol-by-symbol maximum–a-posteriori decoder and by a square-root Kalman-type channel estimator). SBS-MAP offers better performance with slightly increased complexity.

Analysis of error-correcting schemes in relation to their applicability to the railway environment included block and convolutional codes, trellis-coded modulation (TCM) schemes, and short-frame turbo codes. The designed physical layer hardware prototype was implemented in the demonstrator.

Demonstration

The project MOSTRAIN was terminated before the originally planned deadline. Thus, finalization of the demonstrator was not completed, and final results are not available. The significance of this work, however, lies within the extensive planning and demonstrator architecture design. Two demonstrator entities were devised, the simulator demonstrator and the in-train demonstrator. The simulation demonstrator itself was built of another two entities, the HSTB simulator and the off-line demonstrator.

The simulation demonstrator allowed the project to run many reference scenarios in a simulated environment in a long simulation period. It supported both dedicated and shared infrastructures. The HSTB simulator simulated the behavior at high-level modes (burst and call level simulation). The off-line demonstrator simulates behavior at a third-level mode (the bit-level simulation mode) on a short timescale. It uses HSTB results as inputs and provides input for the in-train demonstrator. The in-train demonstrator supported the MOSTRAIN field trials. It was used to validate physical transmission between an HST and a ground infrastructure.

Hardware realization of the demonstrator was proposed only for the in-train demonstrator for parts that were recognized as common with the other one. So, the RF transmission, equalization, and channel coding were either in hardware or simulated. Such a demonstrator platform was recognized as "emulated." The emulated architecture of the demonstrator is presented in Figure 3.15. The emulated demonstrator can be used to compare an actual air interface with a simulated air interface and provide for relevant system performances.

The project MOSTRAIN terminated before achieving all of its intended objectives. Nevertheless, the research work within the framework of this project had significant benefits both to UMTS and the railway community.

The project performed a detailed investigation of services for railway users, with special attention on UMTS multimedia requirements. The project MOSTRAIN undertook a comprehensive study of the railway propagation channel. Both theoretical modeling and measurements were accomplished. An important achievement was the investigation on the effect of directive antennas.

In the architecture domain innovative works included some key design decisions, including infrastructure options, two-link architecture, choice of ATDMA, and RNET for the dedicated infrastructure option.

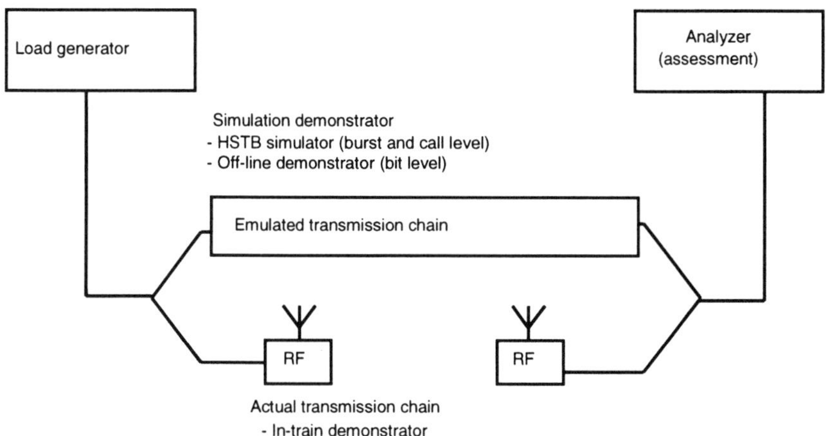

Figure 3.15 Demonstrator-emulated architecture.

The work on the physical layer and the demonstrator, although not completed within the project, provided many significant results that were of benefit to other projects.

3.2.3 UMTS Security Architecture—The USECA Project

The key benefits that UMTS/IMT-2000 promises include improvements in quality and security, the incorporation of broadband and networked multimedia services, the flexibility of future service creation and introduction, and ubiquitous service portability [38]. The security methods already standardized for GSM make it the most secure cellular telecommunication standard currently available. Although the confidentiality of a call and the anonymity of the GSM subscriber are only provided on the radio channel, this is a major step towards achieving end-to-end security. Subscriber anonymity is ensured through the use of temporary identification numbers. The confidentiality of the communication itself on the radio link is performed by the application of encryption algorithms and frequency hopping that can only be realized using digital systems.

Within the ACTS program the need for security in 3G networks was recognized in a timely fashion. The project USECA presented a comprehensive review of GSM security mechanisms with the objective of defining a path for achieving security in UMTS networks [39]. GSM terminal security relies on the integrity of the international mobile equipment identity (IMEI), a 15-digit number aimed at identifying the mobile terminal [40]. It is composed of the type approval code (TAC—six digits), the final assembly code (FAC—2 digits), the serial number (SNR—six digits), and a spare digit that has the value of zero when the IMEI is transmitted. A central body issues the TAC, while the terminal manufacturer issues the FAC. Unique IMEI is securely stored in each terminal during the manufacturing process as random data in distributed format. From the network side, these numbers are stored in a central database, the equipment identity register (EIR), which contains three lists, white, gray, and black. The white list contains all the terminals to be admitted for use on the network. The gray list contains all the terminals that need to be monitored when they access the network. The black list contains barred terminals that have no rights to access the network—for example, stolen terminals. The following main problems, however, may occur in such implementations:

- The network operators and service providers are not willing to support this service because of high cost. On a global scale, it would require the cooperation of all network operators and service providers.
- The secure storage of the IMEI in the terminal has not helped. Manufacturers have attempted to distribute the digits randomly, so as to make it very difficult to modify. This, however, has not entirely prevented cloning or modifying the IMEI. Any solution that increases the production cost is doubtful since the pricing of the equipment is increasingly competitive.
- The modification of the IMEI is not forbidden by law, so devices can be obtained on the market to modify the terminal identity.

As soon as one of the mechanisms is compromised, the security of the system is threatened.

Within the USECA project, a viable and complete UMTS security architecture was built that is currently used by ETSI as basis for standardization. The first task of USECA was to determine a current set of requirements for UMTS security. To compile them, existing ETSI technical standards and reports, FPLMTS/IMT-2000 documents, and other material were used as sources to determine and develop the appropriate security mechanisms required for the UMTS security architecture. The relevant domains considered were USIM, terminal, access network, serving network, home network, and application network.

Public key cryptography allows parties to communicate securely without prior exchange of secret key material. It is expected to be an essential part of UMTS security. The supporting infrastructure for public key cryptography consists of a network of trusted third parties (TTPs) that provide key management and certification services.

As part of the development of UMTS, the functional capabilities of the USIM that are expected to control access to networks and services need to be defined. In defining these, the USIM is expected to have a broader use than just authentication and storage of some personal data; it may also be a host to electronic payment systems and other applications. The project proposed required USIM specifications accordingly.

The main objective of the project was to ensure that a viable and complete UMTS security architecture was developed as a basis for standardization by ETSI. It was clear that UMTS could only operate in a commercially successful way and meet users' acceptance if reliable and effective security measures were implemented from the very beginning. A lot of the work had been done in critical parts of this area by other projects, namely ASPeCT (ACTS), MONET (RACE), and "3GS3—3G Mobile Telecommunications System Security Studies" (UK LINK program). However, it had a long way to go to establish security architecture covering all of the relevant aspects. This was partly because other aspects of the UMTS specification were not sufficiently advanced at the time for all the necessary details of security to be specified. The timely introduction of the security solutions to the overall UMTS architecture was of critical importance with work progressing faster in resolving other aspects to lead to the introduction of 3G systems. Work in USECA concentrated on the following activities:

- Provision of a focal point for UMTS security work;
- Technical basis for the definition of UMTS security standards by ETSI;
- Building on the work of relevant ACTS projects to provide the required security expertise;
- Review of security requirements arising from the set of services defined for UMTS and definition of a comprehensive set of security features for UMTS;
- Definition of a comprehensive set of security mechanisms, protocols, and procedures (with the exception of encryption algorithms) for UMTS;
- Definition of a complete functional and physical security architecture for UMTS;
- Definition of a public key infrastructure for UMTS;
- Definition of security features and procedures involving the USIM;
- Validation of critical concepts in demonstrators.

The project work actually started with the results obtained from previous and ongoing collaborative research projects and standards groups. The results from security- and nonsecurity-related work were considered in the study. Results from another ACTS project, ASPeCT, were the main source of information for the security-related work. The security-related results were compared with the results in the areas of service definition and network architecture to find possible inconsistencies. The work on UMTS security in the project proceeded after the following trends:

- *Security features and requirements.* The first task was to determine a current set of requirements for UMTS security. Furthermore, new UMTS services, whose impact on security had not yet been thoroughly studied, were identified. The multiple registrations of a user on several terminals or of several users on one terminal are examples of such service. This resulted in the definition of new or modified security features to contribute to a consolidated description of the UMTS security features.
- *Security mechanisms.* Security mechanisms provide the building blocks for the realization of security features. Using the list of security features and requirements, this work area determined and developed the appropriate security features required for UMTS security architecture. The list of security mechanisms available from previous work was incomplete, even for the existing list of security features. The existing ones were reviewed and evaluated. Mechanisms providing security for the new UMTS services and concepts were defined.
- *Security architecture.* A stable UMTS architecture was not available for earlier projects. Therefore, the security mechanisms that were proposed could not be integrated into the UMTS security architecture. In addition, new concepts were introduced in UMTS that were not available in GSM and whose impact on security had not been studied (i.e., macrodiversity and the air interface). The air interface was studied from a security point of view because it can considerably impact the confidentiality mechanism.
- *Public key infrastructure (PKI).* Public key cryptography allows parties to communicate securely without prior exchange of secret key material. It is expected to be an essential part of UMTS security. The supporting infrastructure for public key cryptography consists of a network of TTPs that provide key management and certification services. An effective PKI also requires availability of appropriate standards. Thus, results were used to propose a full set of standards against which the UMTS PKI was developed. In addition to the technical issues surrounding PKI, the legal aspects of PKI were also studied, with particular reference to the use of digital signatures.
- *USIM.* USIM is a key component of UMTS security. As part of the development of UMTS, the functional capabilities of the USIM that were expected to control access to networks and services had to be defined. In defining these, the USIM was expected to have a broader use than just authentication and the storing of some personal data, as it might also be host to electronic payment systems and other applications. Clearly, the specifications of smartcards at the time were too restrictive for such a

proposal, but major manufacturers have announced memory ranges of up to 70 KB within the next two years. More flexibility regarding the use of security mechanisms will require a new approach to the interface between the USIM and the terminal.

- *Terminal security.* The work on terminal security concerned logical and physical security. Logical security relates to division of authentication and other functionality between the USIM and the terminal, while physical security concerns theft, terminal cloning, terminal abuse, defective terminals, and nontype approval. The work addressed the issues related to the terminal side of the interface with the USIM, as well as the issue of how to bar stolen or cloned equipment.
- *Demonstrations.* Critical concepts developed in the project USECA were validated by means of proper demonstration that focuses on mobile/USIM interaction.
- *Standardization.* Close collaboration with standardization bodies was needed with the participation of project partners in the relevant ETSI groups [2]. The project provided a sound and validated technical basis for the definition of UMTS security standards by ETSI.
- *Collaboration with other ACTS projects.* A number of relevant ACTS projects working on issues concerning various parts of the UMTS architecture did not address security issues. The closest collaboration was with the FRAMES project.

Work focused on the development of the comprehensive UMTS security architecture and specific issues or parts of the system were addressed. Each member of the first group provided input into the development of those specific parts of UMTS under consideration. In addition to the main areas of technical work, a demonstrator was also developed to provide an initial validation of the results of the USIM and terminal security groups. The following issues required further work in the UMTS architecture:

- An incomplete list of security requirements and features for UMTS;
- Undefined security mechanisms for UMTS;
- Undefined security architecture for UMTS;
- Undefined PKI for support of UMTS;
- No clear definition of the role of the USIM within UMTS.

The terminal was registered at a specific location area after the network had authenticated the SIM. The terminal could either remain on standby or the user initiated a call setup. The administrative use of the IMEI enabled the operator to check the terminal identity at call setup. The purpose of this feature was to make sure that no stolen or unauthorized terminal was used in the system. The equipment identification procedure consisted of the following steps:

- The MSC/VLR requested the IMEI from the terminal to send it to the EIR.
- Upon reception of the IMEI at the AUC, the EIR made use of a database consisting of three lists:
    - A white list containing all the number series of all authorized equipment identities that the different participating GSM countries were allocated;
    - A black list containing all equipment identities considered to be barred;
    - A gray list containing (at the operator's discretion) faulty or nonapproved terminals under observation but not barred from service.

Although the GSM specification recommended using the IMEI check for every call, the frequency of identification depends on the operator's policy. The equipment identification procedure is illustrated in Figure 3.16. The procedure started with the MSC/VLR requesting the terminal for its IMEI. The IMEI request was initiated as an MSC/VLR combination as a result of the terminal requesting for the call setup. In response, the terminal sent its IMEI, which was then sent to the EIR for verification and checking against the stored values. A positive result permitted the terminal to proceed further with the call, otherwise permission was denied and the terminal was barred.

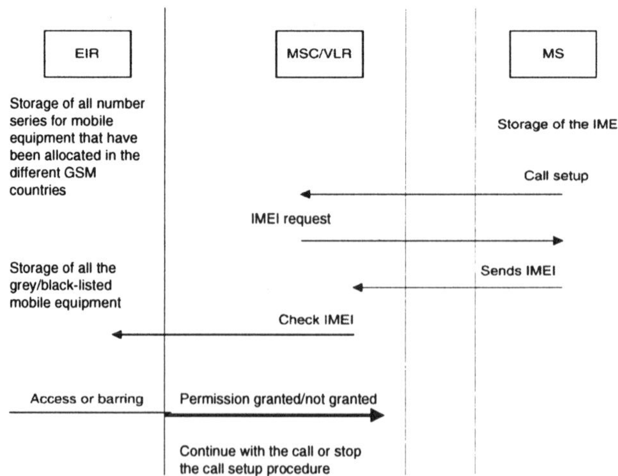

Figure 3.16 Terminal identification.

3.2.3.1 GPRS Terminal Security

The GPRS is a data service providing high-speed packet radio access for GSM terminals and packet-switched routing functionality based on the GSM infrastructure [41]. GPRS security is given by the ETSI specifications [41]–[42]. The GPRS terminal identification procedure involves terminal identity check procedures already defined in GSM with the distinction that the procedures are executed from the SGSN. The identity check procedure is shown in Figure 3.17 and can be explained as follows:

- The SGSN sends an identity request (identity type) to the MS. The MS responds with an identity response (mobile identity).
- If the SGSN decides to check the IMEI against the EIR, it sends check IMEI (IMEI) to EIR; the EIR responds with check IMEI ack (IMEI).

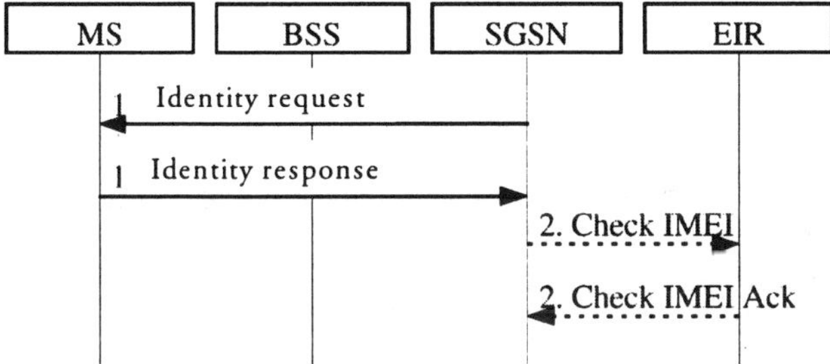

Figure 3.17 Terminal identity check procedure.

3.2.3.2 Phase 2+ Terminal Security Enhancements

In the existing GSM specifications, the IMEI protection at MS side is not sufficient to enable networks to take full advantage of the introduction of EIRs. There were three broad categories of solutions:

- Enhancement of local protection of the ME;
- Use of certificate combined with detection of duplicate IMEIs;
- IMEI authentication.

The merits of various solutions were compared, and it was found that use of a certificate combined with the detection of duplicate IMEIs was the best solution. This was because it provided a countermeasure against most residual threats related to IMEIs and was not too complex. Reprogramming a stolen ME with a valid IMEI (threat 1) and programming a nontype approved ME with a valid IMEI (threat 2) were two kinds of threats.

Threat 1 was not sufficiently covered then and threat 2 was not covered at all. In order to improve the protection against fraud, IMEI was appended with a 32-bit-long certificate (denoted by CERT). CERT was derived from the TAC, FAC, and SNR values of the IMEI and a general secret key MK (stored within the ME at the manufacturer). No special physical protection of CERT was required in the ME, but it was recommended not to display the CERT value on the ME or provide the owner of a ME with this value. CERT used to be transmitted to the network over the radio path at each IMEI checking. EIRs were equipped with security modules containing MK capable of checking the consistency between a received IMEI value and the appended CERT value.

Generation and Installation of Certificates

Each ME manufacturer was provided by its type approval authority (TAA) with a manufacturer's key (K_{MAN}) derived from the general master key (MK), which enabled the manufacturer to generate the certificate value of each ME it would produce. The role of the various entities involved in the certificate generation and installation process (memorandum of understanding [MoU], type approval authorities [TAAs], manufacturers) is depicted in Figure 3.18.

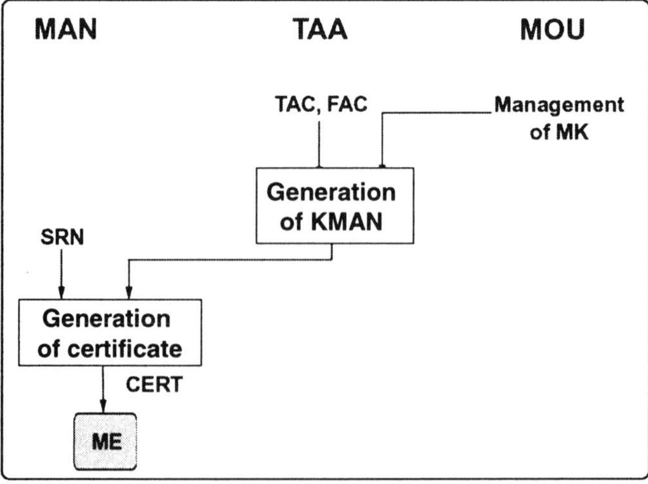

Figure 3.18 Certificate generation.

The general MK is generated and managed under the responsibility of MoU. TAAs are entitled to procure K_{MAN} generation modules containing the MK, capable of deriving the manufacturer's keys K_{MAN} from the MK and the TTA's and manufacturer's codes TAC and FAC, using a diversification algorithm (DIV).

$$K_{MAN} = DIV(MK, TAC, FAC)$$

Each manufacturer is entitled to procure a certificate generation module containing the manufacturer's secret key K_{MAN}, capable of deriving the certificate value CERT of an ME from K_{MAN} and the ME's serial number SRN, using a message authentication algorithm denoted by MAC.

$$CERT = MAC(K_{MAN}, SRN)$$

Certificate Checking

For the purpose of certificate checking, each operator is entitled by MoU to procure physically protected certificates checking modules containing MK, capable of performing the DIV and MAC computations for any TAC, FAC, SRN entry. Certificates checking modules are located in the EIRs. When provided with an (IMEI, CERT) input, these modules perform the cryptological computations required to check the consistency between IMEI and CERT and output a "yes" or a "no" answer.

3.2.3.3 Linkages With Other ACTS Projects

One of the objectives of the ACTS Program was to look for synergy and shared benefits in collaborative research and technical development. The project USECA, because of its scope, was necessarily linked to some other ACTS projects. This section describes different aspects of the collaborative work.

The investigation on PKI, necessary for secure billing using digital signatures, required that USECA joined efforts with the project MOMENTS focusing on the technical feasibility and business viability of the wireless media highway for the distribution of multimedia products. MOMENTS had investigated electronic payment techniques allowing automatic charging for information downloaded by the user.

In order to demonstrate terminal and personal mobility and to realize IN functionality to support UMTS requirements and enhancements to support mobility management USECA relied on the results of the project EXODUS.

Certain selected security mechanisms proposed by the project ASPeCT were in their turn implemented on the UMTS demonstration platform provided by EXODUS. The performance measurements in the above trial were used to assess the resource required by the proposed security mechanism in terms of bandwidth, delay, and storage. The project EXODUS ensured through its work knowledge about the integration of security protocols within the UMTS architecture, and especially about the integration of security messages with the signaling protocols and authentication centers with the SCP.

A major project in the network area of the mobile domain, the project RAINBOW demonstrated the feasibility and evaluated the complexity of the UMTS access infrastructure in its abilities to cope with different UMTS innovative radio access techniques and providing at the same time a solution for migration from 2G mobile system to UMTS.

USECA studied closely the network architecture proposed by RAINBOW. The project needed to base its UMTS security architecture on a specific UMTS network architecture and in the process relied on the proposed by the RAINBOW GRAN approach.

To achieve secure authentication, USECA found an interesting basis in the results of the project UMPTIDUMPTI, which was concentrating on meeting specific requirements for the disabled. Support of security considerations were needed in both projects.

Of course, a strong link existed between the project FRAMES and USECA. Both projects were focusing on development of UMTS with security nonetheless an important issue.

The goal of the project USECA was to develop complete and viable security architecture and to present this architecture to ETSI for facilitating standardization. The success of the project can thus be measured in terms of its impact on the UMTS security specification. The achievements of USECA can be summarized as follows:

- Establishment of a complete list of security requirements and features for UMTS and the incorporation of this list into the relevant ETSI documents;
- Definition of a set of security mechanisms for the new UMTS services and concepts;
- Definition of a complete and viable security architecture for UMTS;
- Definition of a PKI to support UMTS;
- Statement of the role of the USIM within UMTS and determination of an effective interface between the USIM and the terminal.

The results impacted the UMTS security standardization process through the raising of identified security issues, the definition of a series of relevant mechanisms including a PKI, and the provision of a viable and complete UMTS security architecture.

3.3 WIRELESS BROADBAND SYSTEMS

Wireless is a fast-developing technology for data and voice communications. Simultaneously, the requirements for service quality, reliability, and efficiency are increasing. The Internet growth, especially, has been very impressive [43]. The rate at which the Internet has been adopted has surpassed all other technologies, preceding it, including radio, television, and the personal computer. At the same time, there is huge potential for growth in basic mobile voice and data services in areas in and outside the European Union, North America, Asia Pacific, and the rest of the world [38]. A prognosis for the growth of world mobile subscribers is presented in Figure 3.19. Demand in all sectors is determined by the need for access to different types of information and increasing use of video in communications. The changing nature of business processes and the development of electronic information and entertainment services will lead to a far greater use of electronic data transfer. A prognosis for the number of mobile multimedia subscribers is presented in Figure 3.20. Already 3G systems are evolving towards all-IP-based networks in the release 2000 onwards of 3GPP. Commonality among the air interfaces for all operating environments is a goal; however, there is not a single scheme that provides the optimal solution from picocell to satellite cell environments in different user contexts for all possible traffic patterns [44]. Therefore, systems beyond 3G will mainly be characterized by a horizontal communication model, where different access technologies, such as cellular, cordless, WLAN type systems, broadcast systems, short-range connectivity, and wired systems will be combined on a common platform to complement each other in an optimum way for different service and radio requirements [4].

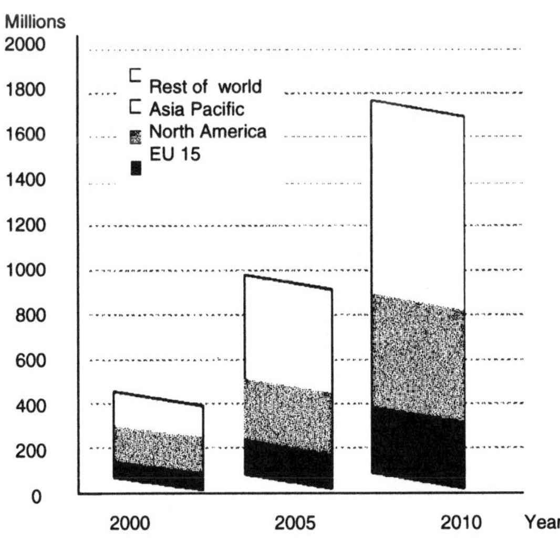

Figure 3.19 Growth rate of world cellular communications up to 2010 [38].

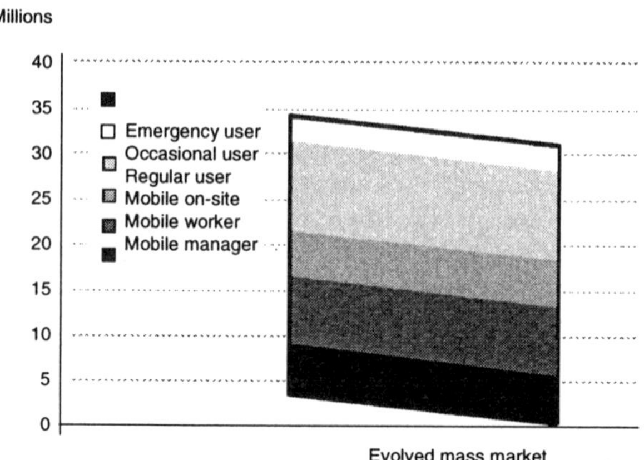

Figure 3.20 European multimedia subscribers for the year 2005 [38].

The next sections address the challenging design aspects on the road to fourth-generation mobile systems, also known as MBSs. A link is made to parallel activities in the field of interactive broadcasting. In that context results are presented of the projects The Magic Wand, MEDIAN, AWACS, and SAMBA of the ACTS program. The project ACCORD developed some joint trials and demonstrations, and it is also included in this section. Figure 3.21 illustrates the range of action of the ACTS WATM projects.

3.3.1 Wireless ATM Network Demonstrator—The Magic WAND Project

A promising technology available today to implement the wireless broadband networks is wireless ATM. It aims at providing sufficient QoS support for real-time and nonreal-time services in high-speed, wireless multimedia networks. Many discussions have been carried out on the definitions of wireless ATM requirements, the most important issue being the MAC protocol [45]. It specifies the method of access to the wireless channel among multiple users while satisfying the basic requirements of ATM. For example, the MAC protocol designed for the project The Magic WAND and called the mobile access scheme based on contention and reservation for ATM (MASCARA) was meant to support all or at least a useful subset of ATM services and to guarantee a QoS for every connection [46]. The project covered the whole range of functionality from basic data transmission to shared multimedia applications. The demonstrator targeted the 5-

GHz band, providing a 20-Mbps data rate, as well as higher bit rate operation (>50 Mbps) in the 17-GHz band.

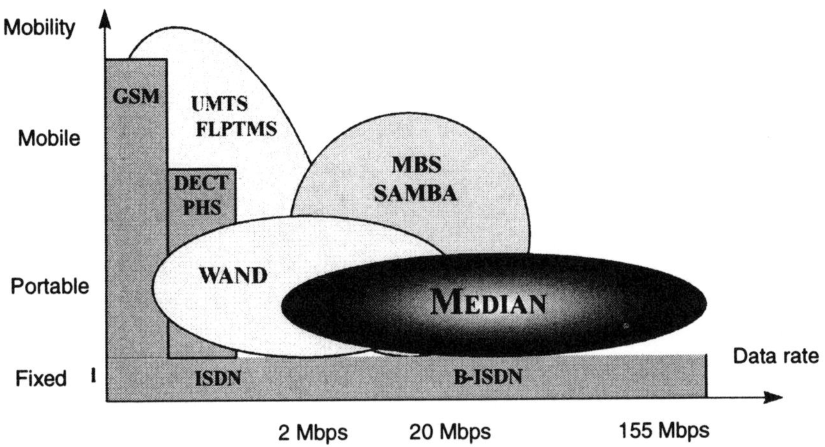

Figure 3.21 Range of WATM projects.

3.3.1.1 High-Rate Wireless LAN Standards

WLANs for the 900-MHz and 2-, 4-, and 5-GHz ISM bands have been available since the 1990s based on a range of proprietary products [47]. The IEEE 802.11 standard, approved in June 1997, specified both MAC procedures and three different physical layers (PHYs). There are two radio-based PHYs using the 2.4-GHz band and the third PHY uses infrared light. All PHYs support a data rate of 1 Mbps and optionally 2 Mbps.

User demand for higher bit rates and international availability of the 2.4-GHz band influenced the development of a higher speed extension to the 802.11 standard. One proposal selected for standardization describes a PHY that provides a basic rate of 11 Mbps and a fallback rate of 5.5 Mbps. It is to be used in conjunction with the already standardized MAC. Practical products, however, will support the high-speed 11- and 5.5-Mbps rate modes as well as the 1- and 2-Mbps modes. Another PHY option being standardized offers higher bit rates in the 5.2-GHz band. The 5-GHz spectrum band is one of the most interesting bands because a significant amount of unlicensed spectrum exists both in Europe and in the United States. As a basis for the new 5-GHz standard to target a range of data rates from 6 to 54 Mbps, OFDM was selected. Following the decision of IEEE 802.11,

OFDM was adopted by HIPERLAN type 2 and multimedia access communication (MMAC) for their PHY standards. Table 3.5 lists some specifications regarding the available frequency bands and the restrictions to devices, which use this band for communications.

Table 3.5

2.4- and 5-GHz bands [47]

Location	*Regulatory Range*	*Maximum Output Power*
North America	2.400–2.4835 GHz	1,000 mW
Europe	2.400–2.4835 GHz	100 mW
Japan	2.471–2.4970 GHz	10 mW
United States	5.150–5.8250 GHz	50–1,000 mW

The Magic WAND project developed the PHY and MAC of a prototype wireless LAN. The project consortium members implemented a prototype wireless ATM network based on OFDM modulation. Since this prototype had a large impact on the current standardization activities in the 5-GHz band, a little will be explained about OFDM for the new 5-GHz wireless standards.

The basic principle of OFDM is to split a high bit-rate data stream into a number of lower bit-rate streams, which are transmitted simultaneously over a number of subcarriers [48]. The symbol duration increases for lower data rate parallel subcarriers; therefore the relative amount of time dispersion caused by multipath delay spread is decreased. The introduction of a guard time in every OFDM symbol eliminates almost completely the intersymbol interference (ISI). In practice, the most efficient way to generate the sum of a large number of subcarriers is by using inverse fast Fourier transform (IFFT). At the receiver side, fast Fourier transform (FFT) can be used to demodulate all subcarriers. Figure 3.22 illustrates how all subcarriers differ by an integer number of cycles within the FFT integration time. This ensures the orthogonality between the different subcarriers, and it is maintained also in the presence of multipath delay spread, due to which the receiver sees a summation of time-shifted replicas of each OFDM symbol. As long as the delay spread is smaller than the guard time, ISI or intercarrier interference (ICI) within the FFT interval of an OFDM symbol can be avoided. The only remaining effect of multipath is the random phase and amplitude of each subcarrier, which has to be estimated in order to do coherent detection. To deal with weak subcarriers in deep fades, forward error correction (FEC) is applied across the subcarriers.

Figure 3.23 shows the basic block diagram of the baseband processing of an OFDM transceiver. In the transmitter path, binary input data is encoded by a standard rate 1/2 convolutional encoder. The rate may be increased to 2/3 or 3/4 by puncturing the coded output bits. After interleaving, the binary values are converted into quadrature amplitude modulation (QAM) values. To facilitate coherent reception, four pilot values are added to each 48 data values, thus a total of 52 QAM values/OFDM symbol is obtained.

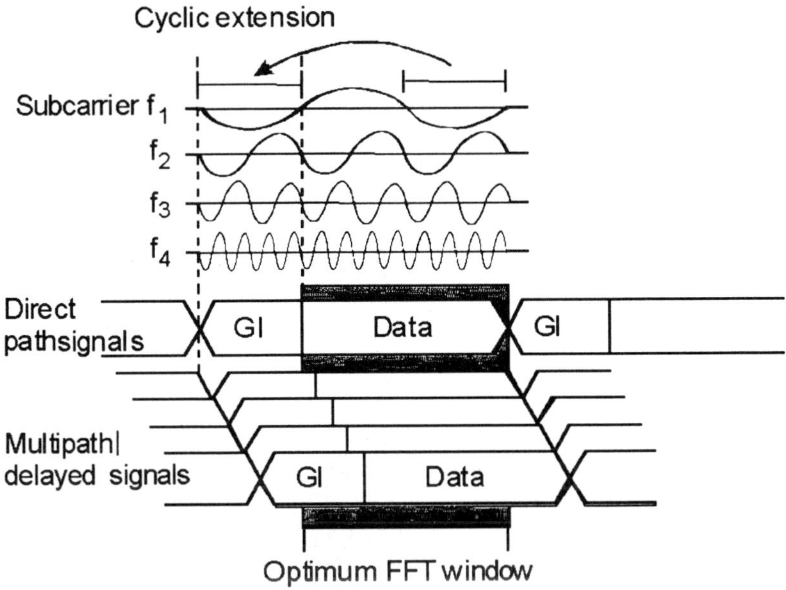

Figure 3.22 An OFDM symbol with cyclic extension.

The OFDM receiver has to estimate frequency offset and symbol timing using special training symbols in the preamble. Then it can do an FFT for every symbol to recover the 52 QAM values of all subcarriers. The training symbols and pilot subcarriers are used to correct for the channel response as well as remaining phase drift. The QAM values are then demapped into binary values, after which a Viterbi decoder can decode the information bits. For more detailed information on OFDM one can refer to [48].

The prototype wireless ATM network based on OFDM and implemented by the Magic WAND project had a great impact on the current standardization activities in the 5-GHz band. The project employed OFDM-based modems, thus helping OFDM to gain acceptance as a viable modulation type for high-rate wireless communications. Also, the wireless ATM-based approach of Magic WAND formed the basis for the current standardization of the HIPERLAN type 2 data link layer.

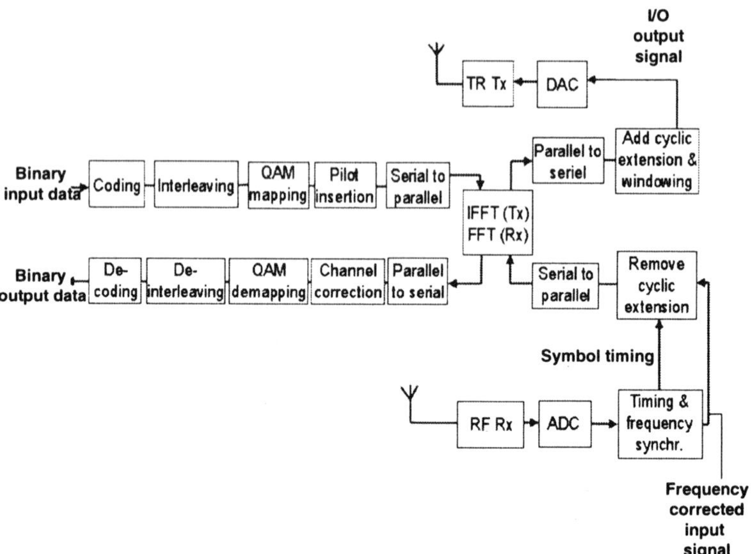

Figure 3.23 Basic block diagram of an OFDM transceiver.

3.3.1.2 The Magic WAND Physical Layer

The WAND modulation scheme is based on 16 subcarriers, each with eight-pulse-shift keying (PSK) modulation. A special coding scheme is employed, providing both error control and peak power control [49]. These codes are known as complementary codes and they reduce the peak power of the transmitted signal to just four times the average power. It is worth mentioning that the eight-length complementary codes used in the 2.4-GHz CCK standard are a subset of the eight-PSK codes used in the Magic WAND. The code set has a minimum distance of four symbols, which means that three out of eight values may be erased without causing errors. There are two-length eight-code words in each OFDM symbol, so it is possible for the receiver to erase up to six of the sixteen subcarriers if these are subject to fading. The number of sixteen subcarriers serves to facilitate implementation and provides a delay spread tolerance of about 50 ns. It should be noted that while this is sufficient for most office buildings and the WAND trial site, a real product would require more delay spread robustness to cover large office buildings and factory halls also.

The OFDM symbol period is 1.2 μs, which includes a guard time of 400 ns. There are 16 subcarriers, which are eight-PSK modulated. At a symbol rate of 13.3 Msymbols/s, this gives a raw bit rate of 40 Mbps. The rate 1/2 complementary coding reduces the data rate to 20 Mbps. The subcarrier spacing is 1.25 MHz, which gives a total bandwidth (3 dB) of 20 MHz. The PHY payload holds an odd number of halfslots. Each halfslot consists of 27 bytes. This number was chosen to allow a full slot of 54 bytes to hold an ATM cell of 52 bytes, and it is a multiple of 3 bytes, which is imposed by the PHY's modulation scheme.

To support packet-based communication effectively, each data unit must be transferred independently of other data units. The implication is that a PHY data unit must be self-contained. All overheads for channel estimation, automatic gain control, timing recovery, and others, must be included in every burst. The packet preamble is 8.4 μs in duration and consists of one OFDM symbol repeated seven times. This preamble is used for frame detection, automatic gain control, frequency estimation, timing, and channel estimation.

The WAND modem used in the trials of the Magic WAND project consisted of five units [50]:

- Transmitter RF unit;
- Transmitter baseband unit;
- Receiver RF unit;
- *Two* receiver baseband units.

The target for implementation was to find a generic platform suitable for the access point and mobile terminal to allow integration of all Magic WAND components in one box. This included the MAC function and the PHY components.

3.3.1.3 The Magic WAND MAC Layer Protocol—MASCARA

The role of the MAC protocol was to provide wired-like services to ATM connections and control the access to the radio medium. MASCARA was specially designed to combine the MAC and DLC layers of the WAND system. The concept was based on a set of ideas and requirements [46]–[47]:

- QoS-aware, reservation-based, demand-assigned protocol;
- Multiple virtual connections per terminal;
- Power efficiency;
- Cell transfer service for optimal interworking with ATM;
- Amortization of PHY overhead over multiple ATM cells;

- Contention-mode request and control channel;
- Automatic repeat request (ARQ)-based cell-loss recovery.

Adverse radio propagation conditions affect the cell loss ratio (CLR) QoS [49]. It is difficult for a radio network to provide the same QoS consistently. To sustain a negotiated CLR QoS as long as possible, the DLC layer adds ARQ capability to the system. ARQ can help retain the agreed on CLR, at the cost of an increase in delay and data overhead. The second important measure of QoS, besides delay, is the connection error performance over the wireless ATM link. In WAND, the radio PHY provided a constant service that could meet typical real-time service requirements (e.g., for voice service). TCP/IP-based data connections cannot tolerate high cell loss ($>10^{-2}$) and are protected by ARQ. Furthermore, services such as video require real-time service with low error rates. The solution applied in WAND was ARQ with a limited number of retransmissions.

The MASCARA protocol was built around the concept of a MAC time frame. It was based on TDMA, to subdivide it further into timeslots. To provide seamless interworking with the ATM backbone and transparency to the user, packets must carry ATM cells. Timeslot duration is equal to the time needed to transmit the ATM cell payload (i.e., 48 bytes) plus the radio and MAC specific header. In the WAND system, the first half of the overhead slot was used for the preamble, the other for the MAC header. To increase efficiency, the size of the packet could be longer than just one cell.

Restrictions on the QoS in terms of cell delay, however, require efficiency of 6 mspcell in the case of 64-Kbps speech traffic. This required a limited packet length of about one or two cells. The conclusion was that packets of variable length with a payload of an integer number of cells should be allowed, depending on the connection delay requirements. The concept of the Magic WAND MAC frame was essentially as follows. Each time frame was divided into three variable length periods: the downlink period, the uplink period, and the uplink contention period.

The components of MASCARA are the data flow and the control information flow. They are depicted in Figure 3.24 by solid and dashed lines, respectively.

Within the project a wireless ATM access network demonstration system was developed to show that wireless access to ATM, capable of providing real multimedia services to mobile users, was technically feasible [51]. The developed applications and the WAND hardware were evaluated in different user trials: medical field trials and office field trials. For the medical trials, a standard ATM testbed was used to evaluate the usefulness and stability of the developed applications. For the mobile computing trials, the WAND terminals were used to retrieve Web pages, enhanced by various multimedia applets using wireless ATM. Different performance evaluation runs were carried out to reflect more deeply on the multiple parts of the software [52].

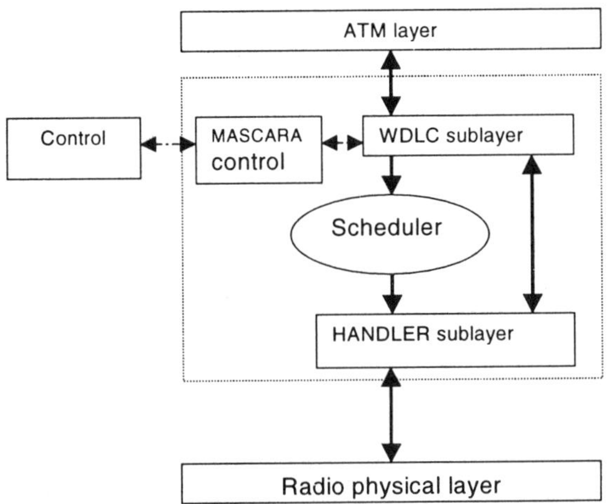

Figure 3.24 MASCARA components.

The objective of the wireless user trials was to evaluate the QoS of the wireless connection and the functionality of the WAND radio system. Also, the trials aimed at assessing the differences between the performances of a standard ATM. Users accessed the applications both through the wireless access and the standard ATM connection. A usability expert observed the trials and guided the users in trial assignments when necessary. A questionnaire to tolerance of the WAND radio system was evaluated outside the user trials by using a hardware radio-channel simulator.

The wireless trials were carried out with a full radio subsystem and evaluation focused on wireless aspects. The main goal was to demonstrate and test the feasibility of a radio-based ATM access system in an office environment. The following aspects were studied:

- Quality of transmitted data;
- Delays in the connection;
- Stability of the connection;
- Radio functionality.

The users were asked to assess the selected attributes on various types of video connections:

- Real-time video connection (wireless radio connection versus a standard ATM connection);

- Video playback (wireless radio connection versus a standard ATM connection).

The trials were conducted with eight office workers who were not familiar with the WAND system, videoconferencing software [53], or application-sharing systems [54]. In addition to the user feedback the actual radio performance was measured and recorded during the trials. The latter were carried out in four sessions. Each session involved two users, one of which used the mobile terminal and the other a wireless local terminal. The users accessed the applications both through the wireless access and the standard ATM connection. A usability expert observed the trials and guided the users in trial assignments when necessary. A questionnaire was designed in cooperation with the USINACTS project, a horizontal ACTS project that supported other ACTS projects in usability issues. Figure 3.25 depicts the radio setup of the wireless user trials.

Quantative measurements were performed with special measurement equipment. The setup of these consisted of two complete WAND modems including DLC statistics monitoring, up- and down-converters to connect the 5-GHz radios to the channel simulator at 1.5 GHz and the radio channel simulator hardware that provided both RF and baseband connections.

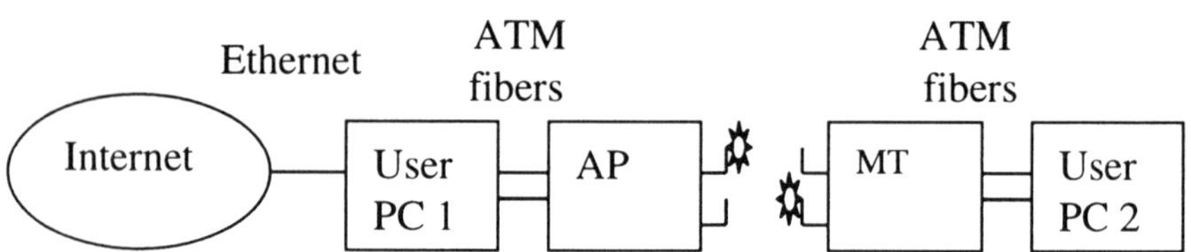

Figure 3.25 Radio setup of the wireless user trials.

In the RF connections the signal quality was set to a high enough level (> 60) to ensure that the residual BER level would be purely caused by multipath propagation. Unfortunately, in the baseband connections the signal quality could not be set. It could be concluded from the measurements that a low signal quality caused increased errors with high, on the order of millisecond, delay spread values. The frame CRC header error rate started to increase significantly after a 40–50-ns RMS delay spread, which was in line with the intended performance of the WAND radio subsystem [55].

Performance of the radio with RF connection was very similar to performance of the radio with the direct baseband-to-baseband connection. However, imperfections in the transmitter and receiver chains, not present in the direct baseband-to-baseband connection, caused the somewhat higher frame header error rates with small rms delay-spread values. Generally, the signal quality proved to be much better as well as the error performance at high rms delay spreads.

The evaluation of QoS was done by several attributes on each quality aspect. All attributes were compared between the radio connection and the standard ATM connection. Table 3.6 summarizes the users' opinions on the acceptability of the audiovisual quality.

The users assessed the acceptability of the overall audiovisual quality with a simple "yes/no" answer based on their subjective opinions. With a stable radio connection the users F1–F4 experienced no differences in the quality of the remote playback video between the radio connection and the standard ATM connection. The opinions about the quality of real-time video differ, but almost universally it was agreed that the overall quality was unacceptable on either connection.

Table 3.6

Users' Feedback Results

Type of Video	*W1*	*W2*	*W3*	*W4*	*F1*	*F2*	*F3*	*F4*
Playback/radio	no	no	no	no	yes	yes	yes	yes
Playback/ATM	yes	yes	yes	no	yes	yes	yes	yes
Video/ radio	yes	no	no	no	no	no	no	no
Video/ ATM	yes	yes	no	no	yes	no	no	no

Another of the trials was meant to investigate the potential for using the WAND wireless ATM system in an outdoor environment [56]. The motivation for doing this arose from the need to do the following:

- Interconnect LANs in separate buildings located in the same area;
- Offer high-bandwidth services (i.e., video-on-demand or Internet access) in the local loop;
- Control equipment or monitor activity in a large industrial installation.

Standardization bodies also showed interest in the potential for synergy between WLANs and the future wireless local loop systems. The WAND system, as an early proposal for HIPERLAN type 2, had to prove to what extent it could fulfill the HIPERACCESS role.

There remained many unanswered technical questions, relevant to HiperACCESS [57]. For example, the choice of antenna systems and PHY technology is still unclear. More complex and powerful antenna systems add to installation costs and may cause problems with mounting and planning permission. High-gain antennas also reduce the potential for system enhancements such as path diversity, forwarding, and listen-before-talk etiquettes. Simpler antennas imply higher path losses, more fading, and higher time dispersion, which will act to reduce range and increase the necessary complexity of the digital circuitry. It is of vital importance for the success of any broadband local-loop product that a solution to this problem is found. The new propagation measurements made for the WAND system in an outdoor environment with medium-gain antennas addressed the following issues:

- Effect of multipath delay spread at 5 GHz, with ranges of 1–10 km and antennas of only moderate gain;
- Ability of path losses to follow closely the free-space propagation law in these conditions;
- Degradation of the channel characteristics by light to moderate blockage of LOS;
- Multipath fading.

The used measurement system is presented in detail in [57]. It consisted of a transmitter and a receiver. Figure 3.26 illustrates the block diagram of the measurement system. The transmitter sends pulses of RF energy, 10 ns long and separated by 40 μs. The receiver has a fast logarithmic amplifier that provides received signal-strength indication (RSSI). An oscilloscope is triggered on the RSSI signal, which is exactly the power envelope of the received pulses. The signals stored in the digital oscilloscope can be downloaded to a PC for later interpretation.

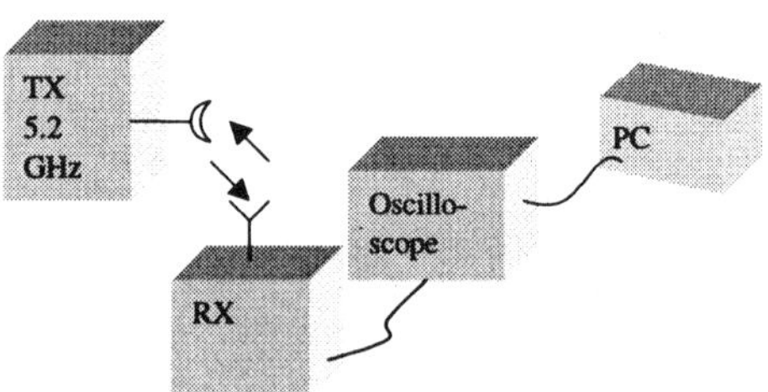

Figure 3.26 Block diagram of the measurement system.

Detailed description of the performed measurements and outcome graphs can be obtained from [56]. Generally, the following could be concluded:

- Delay spread is not strongly linked to range but is more sensitive to pointing offset or antenna type.
- Delay spreads increase as pointing offsets increase.

- The WAND physical layer would have no problems dealing with the measured delay spreads.

To summarize, the project aimed at the following main objectives:

- Specification of a wireless, customer premises access system for ATM networks to maintain the service characteristics and benefits of the ATM networks to the mobile user;
- Promotion of the standardization of wireless ATM access as proposed in the project;
- Demonstration and completion of user trials with a selected user group and testing the feasibility of a radio-based ATM access system.

The technical approach used to achieve the objectives was to cover the whole range of functionality from basic data transmission to shared multimedia applications. The primary goal of the project was to demonstrate that wireless access to ATM capable of providing real multimedia services to mobile users is technically feasible. The project chose to use the 5-GHz frequency band for the demonstrator and performed studies on higher bit-rate operation in the 17-GHz frequency band.

The Magic WAND radio subsystem successfully demonstrated a radio-based ATM access system in the difficult indoor radio propagation environment to show that in good propagation conditions, real-time applications (including video) can be supported. The demonstrated file-transfer speeds of up to 3 Mbps were almost twice the file-transfer speed to be achieved by using the 2-Mbps WLAN products [47].

The main result of the project was the wireless ATM access network demonstration system, which serve as proof of concept for the developed technology and helps the wireless standardization work in the relevant area. The project work was carried out in a close relationship with the standardization research. The Magic WAND project demonstrated the viability of OFDM use in packet-based wireless communication and contributed to standardization efforts going on in Europe, the United States, and Japan, towards an OFDM physical layer standard for the 5-GHz band.

3.3.2 Wireless Broadband CPN for Professional and Residential Multimedia Applications—The Project MEDIAN

To identify a suitably reliable system and associated air interface for the functioning of a high-speed (up to 155-Mbps) local data communication link, substantial research is required. Some of the important issues included are the following:

- Frequency allocation and selection;
- Choice of bandwidth;
- Efficient coding schemes;
- Specification of medium access procedures;
- Definition of link control protocols;
- Connectivity aspects related to connection to other wired or wireless communications networks.

The main objective of the project MEDIAN was to evaluate and implement high-speed wireless customer premises LAN (WCPN/WLAN) pilot system for multimedia applications and to demonstrate it in real-user trials. The pilot system at 60 GHz relied on a multicarrier modulation scheme (OFDM, see 3.3.1.1), with an adaptive access scheme to tailor the data rate to the user requirements and on wireless ATM network extension [58]. ATM-compatibility was achieved by transparent transmission of ATM cells. The demonstrator system provided multimedia, voice, and video applications.

The key issue of the project was to extend the wireless means of existing standards like GSM, DECT, HIPERLAN, and the 3G systems by filling the gap between existing wired video applications, wireless speech, and low-speed data transmission [58]. Figure 3.27 illustrates the office and home operating environments of MEDIAN that together represent the potential applications. The scenario aspects include the following:

- Environment;
- User;
- Applications;
- Specific applications;
- Network service;
- Infrastructure;
- User equipment.

The MEDIAN project identified common topics and problems with respect to the other WATM-oriented projects in the system-platform area of the ACTS mobile domain, interweaving the most with the Magic WAND project.

Some major features of the MEDIAN system [59] are listed as follows.

- Use of an OFDM carrier with a TDD organization and operation up to 150 Mbps;
- Adoption of a simple architecture that consists of a base station (BS). The BS is interfaced with an ATM local exchange (LEX) and with a certain number of ATM portable stations (PS) served by the BS;
- Operation in the indoor environment and use of the 60-GHz band;
- ATM cells that originate at a PS and are carried over the air to the BS and then to another PS or into the fixed ATM network.

The MEDIAN project incorporated research activities in the field of (de-) modulation at 60 GHz [60], power control, medium access strategies, and interworking with the fixed ATM. Additional research focused on designing a model of the ATM-cell streams in the wireless ATM-based LAN to evaluate mobility aspects. It related to functions that inform the network about the PS's location, the capability of the network to locate and keep track of the PS, and the ability to reach them. Furthermore, it ensured continuity of a connection during a change of a physical channel (handover). Handover is the essential mobility issue with respect to cell delay and queue size related to the scheduling principle used in the MAC.

To prove that an adequate handover scheme could be realized in the MAC of the MEDIAN, a mobility-enhanced network model was presented in [61] for the performance simulations. It is worth mentioning that future research focuses on enhancing the model with promising new techniques in the field of broadband communication (e.g., retransmission techniques for real-time services, and traffic shaping).

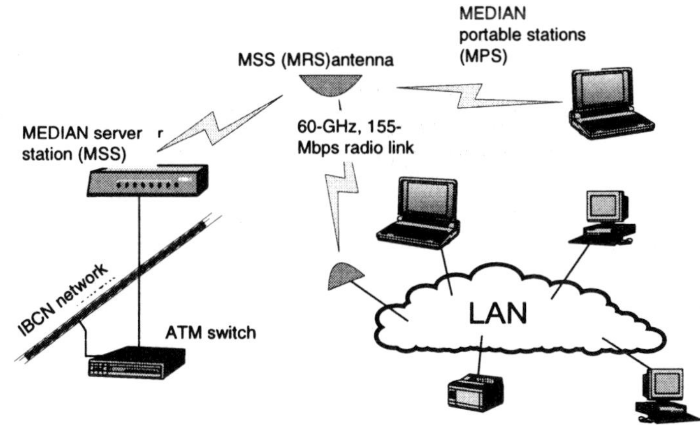

Figure 3.27 MEDIAN architecture and generic scenarios.

3.3.2.1 The MEDIAN Demonstrator

In the MEDIAN demonstrator architecture each PS consisted of an ATM part, called user terminal equipment, that could operate as a standard ATM terminal [62]. The MEDIAN part of the PS was a portable radio part (PRP). The MAC mechanism enabled sharing of the radio channel by multiple PSs, each potentially carrying multiple ATM virtual channel connections (VCC) simultaneously. The basic information unit carried by the system on the air interface was referred to as the physical protocol data unit (PDU). A physical PDU consisted of a MAC PDU (440 bits) and the bits necessary for Reed-Solomon coding (128 bits). Generally, a MAC PDU can contain an ATM cell (424 bits) and a MAC field (16 bits). The schematic diagram of the demonstrator system is illustrated in Figure 3.28.

The transmission takes place by means of an OFDM carrier with a TDD organization on the air interface. Each slot of the TDD frame carries an OFDM block, while each TDD frame consists of two consecutive subframes for the downlink and uplink, respectively. Uplink and downlink frame durations can be varied, on a frame-by-frame basis, to fit the present uplink/downlink traffic mix. The most convenient frame duration and appropriate slots assignment to the calls for the uplink and downlink is decided by the BS. This is the reason why a 16-bit MAC field is added to the 424-bit ATM cells; it informs the BS about the present load of the PS buffers. Hence, the MAC fields are fundamental inputs for the MAC algorithm.

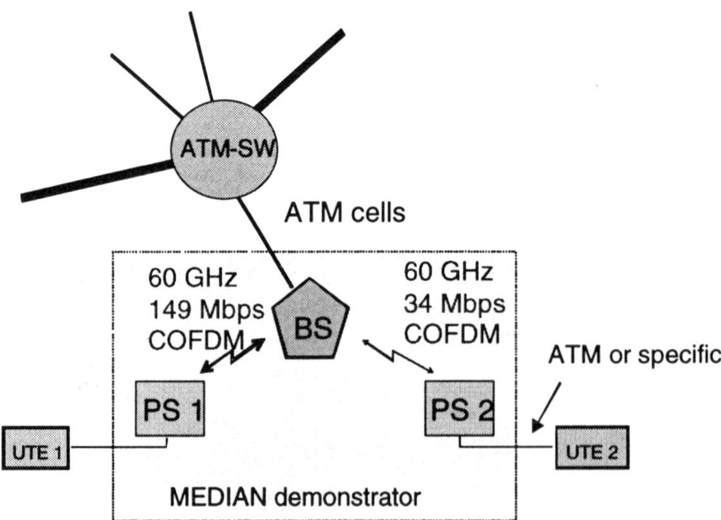

Figure 3.28 Schematic diagram of the MEDIAN demonstrator

The objective of the demonstrator was to show the viability of the chosen concept, rather than the realization of the full range of possible features and services of a WLAN [63]. This was done during field trials that were carried out bearing in mind the following:

- Definition of user groups and user profiles;
- Definition of multimedia applications and services;
- Assessment of user behavior of wireless multimedia applications in the MEDIAN testbed environment;
- Evaluation of the field trials on user acceptance and usability;
- Promotion of public awareness in MEDIAN applications;
- Demonstration and promotion of the MEDIAN system.

The user experiences and the technology performance of the MEDIAN demonstrator were reported and evaluated in order to test the acceptance and usability of the MEDIAN system. The trial environment included a "wired" network component and "wired" user-terminal equipment in addition to the MEDIAN BS and PS. The MEDIAN demonstrator operated as a wireless extension to the telecommunications network environment at the field trial site. Multimedia, voice, and video applications were all tested over the MEDIAN high-speed wireless customer-premises LAN demonstrator. The main user trial activities included the following.

- Development of user scenarios and evaluation criteria;
- Setup of the test site, including infrastructure;
- Identification of applications to be run on the demonstrator;
- Selection and organization of user groups;
- Installation of application terminals;
- Field tests;
- Electronic brainstorm sessions between test users;
- Provision of technical support to the participating user groups during trials;
- Interpretation and summarizing of the field trial results;
- Implications for standardization process.

For testing the usability of the MEDIAN system by potential users, office and home user environments were simulated at the field trial site for testing applications that required more or less generic services such as videoconferencing, fast file/data transfer, real-time audio/video, and file sharing. Table 3.7 gives the parameters of the MEDIAN system [64].

Table 3.7

Parameters of the MEDIAN system

Frame duration (64 timeslots)	170.667 μs
Timeslot	2.667 μs
Cell-error rate	Between 10^{-5} and 10^{-2}
Baseband processing delay	9 timeslots
Negative acknowledgement (NACK) processing delay	± 2 timeslots
Scheduling of new frame	12 timeslots before start of new frame

3.3.2.2 The MEDIAN Physical Layer

Wireless communication utilizing ATM as a transmission protocol requires that the performance of the wireless link can ensure reliable transfer of ATM traffic. In the MEDIAN system the quality of the transfer over the wireless link was improved by adding FEC [64] at the PHY. Thus the link cell loss probability (CLR) was improved up to 10^{-6} after FEC [65], using a (55,71) Reed-Solomon code. The idea was further upgraded in [21], where an ARQ retransmission technique was proposed to be used on top of FEC. ARQ requires smaller extra bandwidth, and it is precisely bandwidth that is expected to be more expensive for future applications because of limited available spectrum.

The proposed ARQ scheme is situated in the DLC sublayer and is especially designed for real-time services that usually have to meet strict delay requirements. The frame construction and frame duration of the MEDIAN system allow for the use of an ARQ scheme that is otherwise barely used for real-time services.

To get useful performance results, the following assumptions were made, in addition to the basic parameter values:

- New cells arriving in the access point (downlink cells) can be scheduled in the next broadcast cell (BC);
- New cells arriving in the mobile terminal (uplink cells) are announced to the access point in the next scheduled uplink cell for this connection and are subsequently scheduled in the next BC;

- NACKs can always be sent along with cells of the same connection but in the opposite direction;
- Retransmissions have a higher priority than new cells and capacity for these retransmissions is always available;
- Buffers in the access point and mobile terminals are large enough to store the expected number of outstanding cells.

By using an ARQ protocol the link CRL was improved considerably. It was interesting to examine how many retransmissions were sufficient to reach an adequate result as requested by a connection through the exchange of QoS parameters during connection setup. In Figure 3.29, the resulting CLR for various numbers of retransmissions is plotted as a function of the CLR of the wireless link. It shows that if the number of allowed retransmissions increases, the resulting CLR improves.

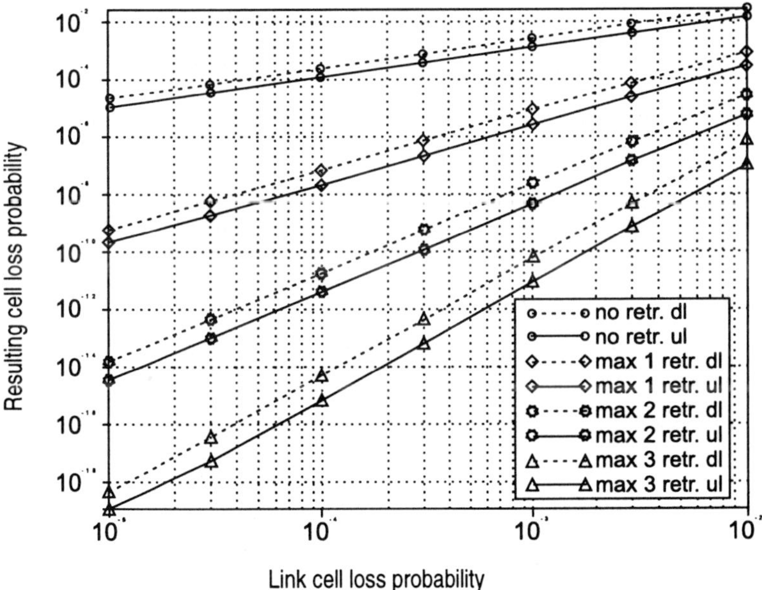

Figure 3.29 Probability of cell loss for downlink and uplink scenarios with different numbers of retransmissions.

The maximum delay of a cell for different numbers of retransmissions was independent from the link CLR. The average cell delay is presented for several values of the link CLR in Table 3.8. The maximum number of allowed retransmissions is four.

Table 3.8

Average Cell Delay for Links with Various CLRs and Cell Rates

	Cell Rate	*CLR 10^{-2}*	*CLR 10^{-3}*	*CLR 10^{-4}*	*CLR 10^{-5}*
Downlink	1 Mbps	0.372 ms	0.359 ms	0.357 ms	0.357 ms
	2.5 Mbps	0.395 ms	0.361 ms	0.358 ms	0.357 ms
	10 Mbps	0.509 ms	0.371 ms	0.359 ms	0.357 ms
	15 Mbps	0.584 ms	0.378 ms	0.359 ms	0.358 ms
Uplink	1 Mbps	0.572 ms	0.568 ms	0.568 ms	0.568 ms
	2.5 Mbps	0.579 ms	0.569 ms	0.568 ms	0.568 ms
	10 Mbps	0.612 ms	0.572 ms	0.568 ms	0.568 ms
	15 Mbps	0.634 ms	0.574 ms	0.569 ms	0.568 ms

3.3.2.3 Conclusions

The MEDIAN system offered fast access (especially to the Internet); interconnection capabilities with other infrastructures like fiber, cable, and GSM/UMTS; reliable connections; mobility; fast file/data sharing; high-resolution imaging and some other services in the office and home environment. Within the project, studies were performed in order to identify WCPN applications and service requirements [66]–[67]. Apart from the office and home environment, the following applications, which should be interpreted as potential areas for MEDIAN [66], were identified (see Figure 3.30):

- *Security:* Warehouses, shopping malls, and construction sites;
- *Health care:* Home care via wireless cameras and portable computers;
- *Education:* Videoconferencing;
- *Traffic congestion problem:* Workplace with extended videoconferencing;
- *Museum/entertainment:* Mobile-guided tour;
- *In-store marketing:* Wireless in-house integrated network linking.

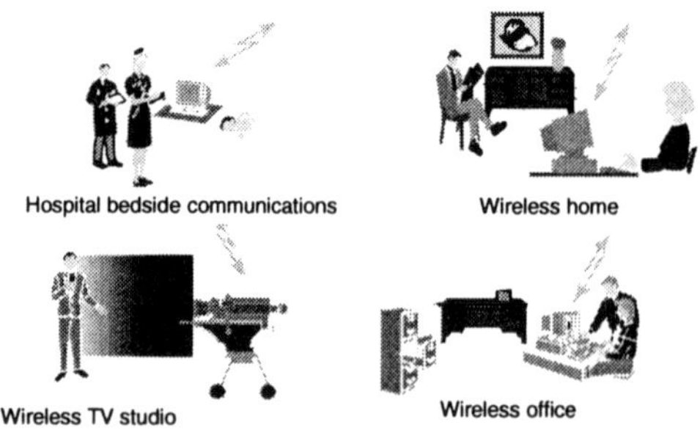

Figure 3.30 Potential application areas for MEDIAN.

A more detailed description of potential applications for wireless use was made in the user requirement study [67]. Already there, the major determinants for the use of a specific application had been translated into characteristics of use and presented in a so-called user requirement map. This set of parameters, closely related to services and system specification parameters, provided a tool for further structured assessment of user behavior and evaluation of user experiences during the trial. Table 3.9 denotes the parameters of the MEDIAN user-requirement map [63]. The requirements can be divided into user requirements and system requirements. The user requirements are the characteristics of the system that are important to the user, and they may include speed, mobility, functionality. The system requirements are the important characteristics of the system itself and as such they are of great importance to the technical specifications of the system but of no value to the user. Therefore, the user requirements should be found and translated into system requirements.

Table 3.9

MEDIAN User Requirement Map

Traffic	*Interconnection options*
Volume per user	Required additional services
Multisharing	Mobility
Delay	Installability
Reliability	Antenna position
Security	Environment
Range	Required services
Resolution	Added value of wireless

The user requirements were an important part of the evaluation data used to draw conclusions about the relevance of the specific system characteristics. In that way, realization of complete MEDIAN networks is pushed into the right direction by inputs from potential future users. The research carried out in the MEDIAN project also contributed to investigations in the area of resource allocation efficiency [68].

3.3.3 ATM Wireless Access Communication System—The Project AWACS

In the design and standardization of new systems [69] it is important to understand the role of present networks and possible evolution paths to operation. Initial broadband coverage is to be provided in 5 GHz (the Magic WAND). As more capacity is required, operation is moved into the higher frequencies; 17–19 GHz in AWACS, 40–60 GHz in MBS/SAMBA, and 60 GHz in MEDIAN. Figure 3.31 summarizes the standardization and ACTS activity schedules in the world towards advanced wireless broadband access systems.

Within this context, the project AWACS contributed to the development of HIPERLAN type 4 with relation to the PHY. AWACS's main objective was to develop a system concept and demonstrate through a testbed wireless public access to B-ISDN.

The operation area regarded the 19-GHz band with full-duplex user bit rates up to 34 Mbps and radio transmission ranges of 50–100 m. The key project component was the ATM wireless access (AWA) equipment made available by NTT from Japan. The availability of this hardware presented opportunities for the contribution of AWACS to HIPERLAN type 2 as well because of its local wireless access conception [70].

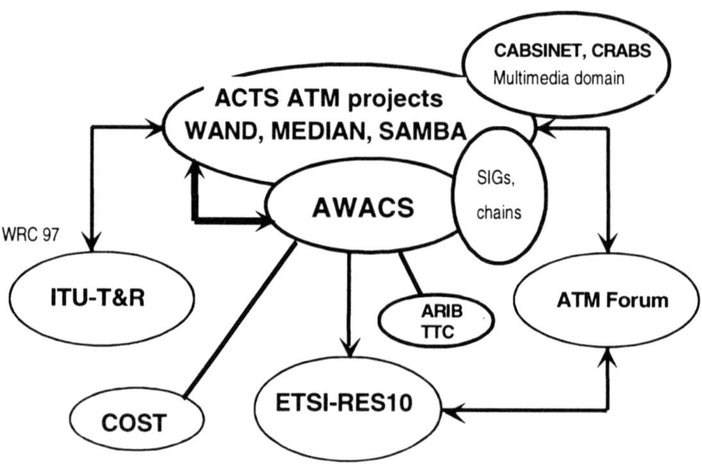

Figure 3.31 Standardization activities for wireless broadband systems through ACTS projects.

Table 3.10 lists HIPERLAN types 2, 3, and 4. These interwork via standard ATM level functions and services to give a wireless architecture that fulfills the demanding needs of a wide population.

Table 3.10

ATM Level Interworking Among HIPERLAN 2, 3, and 4

ATM	*Level*	*Interworking*
Wireless ATM	Wireless ATM	Wireless ATM
Local access	Remote access	Interconnect
(HIPERLAN type 2)	(HIPERLAN type 3)	(HIPERLAN type 4)

Originally the project tried to study and demonstrate a wireless ATM communication system [71]. The AWA equipment utilized TDMA/TDD technique, while directional antennas were used to cope with the multipath and intersymbol interference. In that way use of equalization and multicarrier techniques was avoided. Figure 3.32 illustrates the above-mentioned approach.

The used TDMA/TDD frame structure aimed at allowing both symmetric and asymmetric transparent ATM cell transmission [72]. The instantaneous air rate was about 70 Mbps. The core specifications of the system are presented in Table 3.11.

The frame format consisted of 32 timeslots and a frame duration of 2 ms. In the AWACS system each slot can be used in uplink or downlink in order to allow asymmetric data transmission. The channel coding is based on two Reed-Solomon codes used to protect the ATM cell headers and payload, respectively.

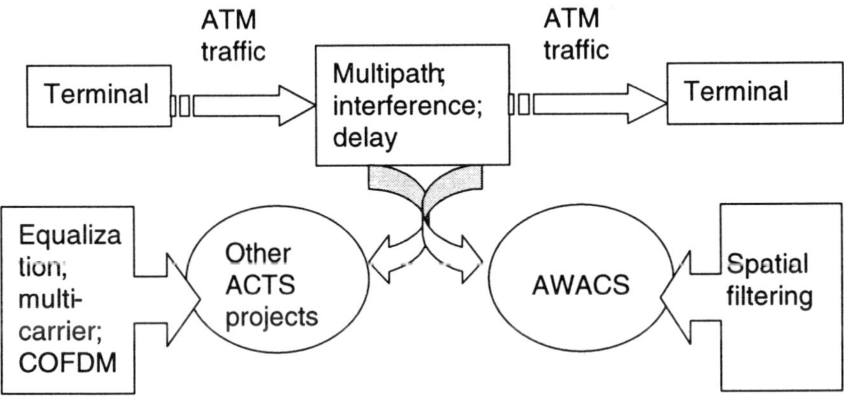

Figure 3.32 Combating multipath for wireless ATM success.

Table 3.11

Major Specifications of the AWACS System

Radio Parameters	*AWACS*	*Radio Parameters*	*AWACS*
Frequency	19 GHz	Mo-Demodulation	OQPSK/Coherent detection
Radio access	TDMA/TDD	Radio cell size	50–100m
Air bit rate	70 Mbps	Environment	Indoor/outdoor

The project addressed the feasibility of directional-antenna technology as a viable alternative to equalization, and multicarrier techniques for the provision of ATM-compatible bit rates over a wireless bearer. In addition to the wideband propagation analysis and transceiver trials, it addressed alternative modulation and coding schemes, MAC protocols, and the potential use of 40-GHz carrier frequencies, as well as the mobility aspects of wireless ATM service provision.

The AWACS measurement trial commenced with a comprehensive set of wideband propagation measurements to provide radio propagation data related to different WLAN operational environments and scenarios [73]. Overall, a total of four indoor environments and nine different propagation circumstances were considered. To characterize the indoor propagation channel as a function of the gain and the aperture of the transmitting and receiving antennas, four different types of antennas were employed during the trials: disc-cone omnidirectional, 60-deg and 30-deg beamwidth patch, and a 15-deg beamwidth horn. The results of these measurements can be found in [73]–[74].

For this purpose, a testbed was established in Europe. The 19-GHz hardware was initially configured to characterize various radio environments at 19 GHz and the BER performance of the modem prior to ATM service provision. Both a stand-alone and a user trial of the ATM wireless technology were planned during the project. The field trial equipment suite consisted of a single mobile platform, one base station, and a fully utilized ATM to user and switch. The equipment was tested in numerous operational environments, and the performance assessed using a set of customized assessment tools. A parallel program of work considered various enhancements to the prototype system including the use of a different source and channel coding system, intelligent antennas, mobility, and handover.

In order to achieve compatibility with HIPERLINK operation, the modem must be able to support data rates of 70 Mbps or more, assuming TDD operation and a bidirectional user rate of 34 Mbps [72]. If the classical Rayleigh model had been assumed for multipath activity, then a single-carrier, QPSK-based nonequalized modem could only operate where the rms delay spread was less than 2.9 ns for a bit-error rate threshold less than 0.1%. However, modems employing directional antennas are known to offer data rates well in excess of 70 Mbps when operating in environments with rms delay spreads far higher than this value. Thus, the use of the Rician K-factor (the ratio between power of LOS and multipath components) was considered to further characterize the environments, as this measure is ideally suited to channel conditions where a stable component of LOS exists. For digital transmission over LOS channels the K-factor is a more important parameter than the rms delay spread. For example, when the K-factor is high, the received "eye diagram" remains open at the sampling points irrespective of the time-delay spread. For K-factors greater than 8–10 dB, even if the random multipaths add up coherently, the LOS component will still remain greater than the random component, and hence, the "eye diagram" will remain open. Using this argument, at high K-factors the rms delay spread is not important and the supported bit rate will be determined by the SNR. This implied that, with sufficient antenna gain and transmit power, for K-factors greater than 8–10 dB very high bit rates of about 155 Mbps could be supported by the AWACS architecture.

Following the channel measurement, performance measurements were conducted using the ATM testbed [75]. To enable the accurate analysis of the relationship between the system performance and the radio-channel parameters these measurements were taken simultaneously with channel parameters. Antenna configurations were carefully selected for a certain system performance level in order to obtain the optimum antenna crate analysis of the relationship between the system performance and the radio-channel parameters; these measurements were taken simultaneously with channel parameters. Antenna configurations were carefully selected for a certain system performance level in order to obtain the optimum antenna configuration for the final real system design.

Link-level simulations to study the performance and some enhancements of the basic AWACS system (i.e., antenna diversity and multilevel modulation schemes) were also carried out by using both the measured channels and a statistical model based on the ETSI RES10 format [76]–[77].

3.3.3.1 The AWACS MAC and LLC Layers

The AWACS concept incorporated an additional functionality in the radio-protocol stack that provided the capabilities for a broadband cellular communication system [78]. The MAC used a central assignment process in the base station to arbitrate the traffic requirements of the individual mobile terminals. This assignment process made use of static weights derived from the long-term requirements of the service and of dynamic priorities derived from the short-term requirements (waiting time of an ATM cell in the queue). For the uplink these dynamic priorities were calculated on the basis of dynamic parameters transferred from the mobile terminals to the base station. The approaches used to improve the link quality were the insertion of FEC capabilities and the introduction of an LLC layer to allow the repletion of erroneous or lost cells.

In the AWACS system, the implementation of LLC functionality was based on the application of ARQ schemes. The focus was on three types of repeat mechanisms:

- Go-back-N (GBN);
- Selective repeat (SR);
- Combinations of GBN and SR.

The MAC provided the following control mechanisms:

- Acknowledgements of slots;
- Reservation of slots;

- Announcement of slots;
- Specific control bursts;
- Control overhead as part of downlink bursts.

The downlink burst can be longer than the uplink burst because the guard period is not needed on the downlink. Figure 3.33 depicts the basic MAC structure. Each mobile terminal has at least one MAC entity.

The corresponding MAC entity at the base station contains the central MAC process. The MAC protocol is a peer-to-peer protocol between a MAC entity in some mobile terminal and the MAC process in the base station.

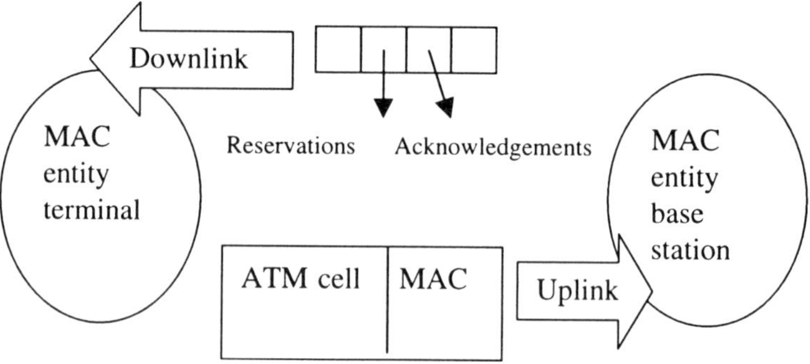

Figure 3.33 Basic MAC functionality.

The MAC protocol is asymmetric in nature. It acts on the air interface comparable to an ATM multiplexer. The cell streams arriving from the various mobile stations served by the base station are merged into one resulting cell stream. In the downlink, the cell stream is demultiplexed into individual cell streams heading for the appropriate terminals. The MAC protocol exchanges control information in the uplink as well as the downlink direction. In order to inform the central control process in the base station, the mobile station inserts into the MAC overhead of each of its uplink bursts a set of dynamic parameters characterizing its load situation. In the AWACS system concept the three parameters listed in the following are transferred in the MAC overhead of each uplink burst:

- Number of ATM cells;
- Minimum residual lifetime of all cells of one mobile terminal;
- Maximum waiting time of all cells of one mobile terminal.

With regard to the requirements and QoS scenarios, important parameters to evaluate were the transfer-delay time and cell-loss rate. The results showed that, as long as the channel was good, the transfer time of the selective-repeat scheme was higher than that of the schemes implementing some GBN mechanisms.

3.3.3.2 Conclusions

The project AWACS was a direct response mobility in personal communications. In particular, it addressed the following objectives:

- Demonstration of future MBS system interactive multimedia applications by means of system trials in distinct environments;
- Demonstration of the interworking of MBS with a cableless LAN and compatibility with B-ISDN and ATM;
- Validation of QoS parameters corresponding to the applications that are tested (especially for low-mobility applications);
- Consideration of the support of mobility management functions, especially for handover;
- Study of signal processing techniques for countermeasures against propagation conditions like adaptive equalizers, intelligent antennas, source, and channel coders;
- Access of possible use of high-frequency (40-GHz) technology for ATM wireless applications.

In order to collect real-life information on the usability and performance of the AWACS trial system, field trials were performed. In the trial the components, the designer, and the production site environments were developed for better usability, and a usability study was made. The main conclusion of the AWACS field trial was that a successful two-way WATM-based multimedia link between designer and production sites of a company was created. This novel application impressed one with a face-to-face telepresence feeling.

The AWACS measurement field trials showed that for a given threshold on the performance requirement of a system, it was possible to specify a bound on the channel parameters in order to ensure good performance. As an example of a BER limit of 10^{-4}, it was shown that K-factors greater than 8 dB and rms delay spreads of less than 12 ns would ensure that the errors would not pass the given threshold for more than 90% of the time.

The project also considered the link-level and system-level simulations in order to assess characteristics and limits of a system concept based on 19-GHz air interface. Regarding the MAC and LLC activities of the project, a QoS framework was established. The need for easy adaptable MAC/LLC protocols was identified, and these were specified accordingly providing a tradeoff between delay error and BER. Finally, the 19-GHz AWACS virtual office field trial system consisting of one ATM-interfaced broadband multimedia application was successfully demonstrated.

3.3.4 System for Advanced Mobile Broadband Applications—The Project SAMBA

SAMBA was a project that developed a broadband digital cellular radio network to deal with high mobility, medium coverage by cellular infrastructure, and symmetric transmission capacity in indoor and outdoor environments [3]. The project was the continuation of the work towards an MBS system, initiated by the MBS projects of the RACE program [79]–[80]. The primary objective of SAMBA was to promote the development of a broadband cellular radio extension to the fixed broadband network, thus allowing the use of broadband fully interactive multimedia services by mobile users [81]. The project focused on a trial platform that implemented full mobility including handover, dynamic resource management, and full-duplex broadband operation over the radio interface. The trial platform depicted a basic MBS configuration with two base stations, two mobile terminals (portable and vehicle-mounted), and ATM interconnection, as shown in Figure 3.34 [80].

The mobile terminal comprised the millimeter-wave transceiver (MWT) with antennas, the baseband processing unit (BBPU), the control unit (CU), and a standard ATM user-network interface (UNI) to connect the user terminals. Two base-station transceivers (BSTs) were connected to one base-station controller (BSC), which is connected to an ATM switch, enhanced with an ATM mobility server (AMS). The trial platform can be connected to an ATM network.

Two cell types were designed, a street cell with dimensions 6 m x 200 m, and a wide cell with 60 m x 100 m. The air interface operated at 40 GHz allowing full-duplex transmission of ATM cells up to 34 Mbps and seamless radio handover; the maximum velocity for the mobile terminal is 50 km/h. A control and monitoring system for system configuration and measurement data acquisition supplemented this basic MBS.

The cellular design of a mobile microcellular communications system operating at the millimeter-wave band in an outdoor environment and a propagation model suitable for the purpose of estimation of coverage and interference were proposed in [82]. Also there is an analysis of cell shapes and cluster size, compared in terms of cochannel interference and capacity efficiency was given.

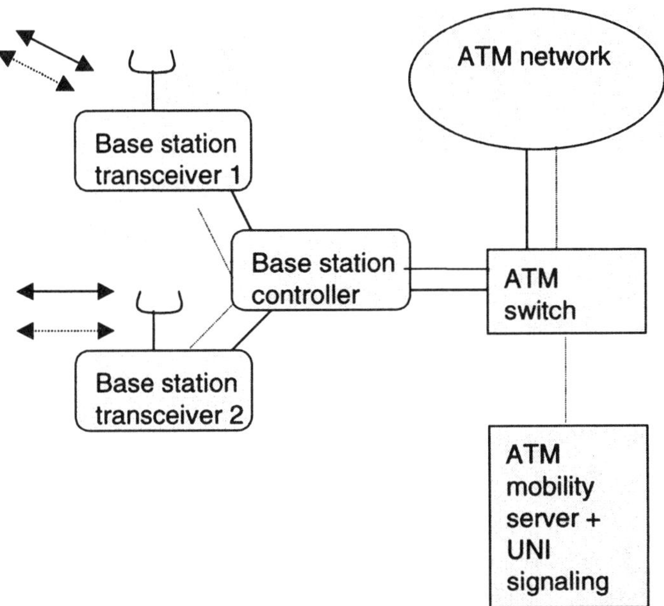

Figure 3.34 The SAMBA trial platform.

At millimeter-wave frequencies, the signal is very attenuated by obstacles and the cell shapes are generally determined by the scenario geometry of the propagation environment [83]. The broadband radio link is very demanding in terms of power budget and channel impulse response characteristics to support high data rates, once the required transmitted power is directly proportional to the carrier bit rate, and this strongly depends on the channel time dispersion. The problem of cell coverage was approached in the field trial scenarios of the project SAMBA by using the antennas designed for the SAMBA trial platform [84]. Since there was a direct relation between the sliding delay window (SDW) and the achievable carrier bit rate for a certain equalization depth, this parameter was chosen to represent the channel time dispersion, while for the budget link the normalized received power (NRP) parameter was used. The main objective of this approach was to get a fairly uniform power distribution over the cell region, while keeping the SDW below a certain level defined by the system parameters. The antenna's role is critical for millimeter-wave systems. The right antenna combination at the BS and MT may contribute significantly to improve the system performance, by providing an adequate signal level over the cell, and favor some multipath discrimination. Two different antenna sets, corresponding to different cell geometry, were designed for the SAMBA trials: the elongated cell and the wide cell [85].

The future MBS system is to be used in urban areas and indoors. Therefore, results for a wide city street or avenue, both with and without obstacles representing trees, cars, bus, trucks, etc., and a pavilion with large dimensions were presented [84]. Two main parameters were used to evaluate the cell coverage:

- NRP defined as the ratio of the received to the transmitted power, respectively, at the antenna input/output terminals;
- SDW defined as the smaller time interval containing 90% of the energy of the channel impulse response as a measure of the channel time dispersion.

The experiments showed that obstacles in the scenario could cause significant long-delayed back reflections that increase drastically the channel time dispersion and shadow regions. Because of that the digital radio link can break down, either due to a high level of ISI caused by high-time dispersion or lack of power. Several simulations were performed to study the effect of the most common obstacles within the scenario on the signal propagation [84]. The results showed that streets with 300-m length could be covered using a single cell and that tilting downwards the BS antenna in order to reduce the power incident on the walls, and thus keeping the time dispersion levels to a controlled level, could efficiently cover large indoor scenarios. The problem of signal attenuation, shadow fading, and significant long delayed back reflections caused by obstacles could be solved by using multiple coverage and sectored antennas. A sample of the first set of measurements carried out with the SAMBA trial platform in two scenarios is presented in [81].

3.3.4.1 Concept of Spatial Division Multiple Access (SDMA) in Wireless Broadband Systems

One of the drawbacks of modern broadband wireless networks is the large amount of required spectrum [86]. The application of SDMA seems to be a possible solution to increase spectral efficiency, because the considered frequency band, in the range of several gigahertz, allows a compact construction of array antennas. The trial platform of the SAMBA project had a generic dynamic slot-assignment (DSA++) MAC protocol applied to it in its FDD version [86]–[88]. Based on the DSA++ MAC protocol, the use of SDMA for the different phases of the MAC protocol was investigated and solutions were described for the concurrent access of more than one wireless terminal to the same time and frequency slot [86]. Because protocol performance incorporating SDMA strongly depends on the physical layer properties, both issues were considered carefully.

The task of the MAC protocol was to coordinate the wireless-terminal access to the shared radio channel. The reservation-based DSA++ MAC protocol grants permission capacity to the terminals per slot to satisfy their current transmission capacity demands. The base station that operates a scheduler centrally controls the protocol, which groups the various control and data channels into MAC frames of flexible structure and length. Therefore, the wireless terminals have to transmit their capacity requests via the uplink to the base station. A complete MAC frame consists of a broadcast phase, a downlink phase, an uplink phase, and a random access phase. According to the current work assumptions of ETSI-BRAN, the protocol differentiates the following channels:

- *Broadcast control channel (BCCH):* Conveys signaling information concerning the whole radio cell;
- *Frame control channel (FCCH):* Transmits information about the structure of the MAC frame and contains announcements and reservations of control and data channels that the individual wireless terminals have to transmit to receive;
- *Data link control channel (DLCH):* Associated with the UDCHs; conveys signaling messages, such as ACKs and resource requests; comprised by the downlink phase and the uplink phase;
- *User data channel (UDCH):* Transmits user data via uplink and downlink;
- *Random access channel-partitioned (RACH-P):* A varying subset of associated wireless terminals is allowed to access this channel;
- *Random access channel-unpartitioned (RACH-UP):* An uplink-connection channel to be accessed by any wireless terminal.

To meet the QoS requirements especially of delay-sensitive real-time services, fast and on-time signaling of capacity requests to the BS was necessary. The DSA++ protocol provided two methods for transmission of capacity requests to the BS via the uplink:

- *Explicit signaling using the DLCH.* The BS-scheduler reserves a slot for the DLCH to receive a capacity-request message of a terminal.
- *Random access using the RACH.* The terminals can randomly access RACH slots to transmit their capacity request messages.

At the end of each MAC frame, the scheduler updated the information database about the capacity demands of the terminal by evaluating the received request messages. This information, together with the downlink transmission demands, provided the input for the scheduling algorithm to determine the structure of the subsequent MAC frame. The BS broadcasts this structure via the FCCH to inform all the terminals about their individual transmission and reception schedule. Figure 3.35 gives a complete system overview for uplink and downlink transmission and signaling channels.

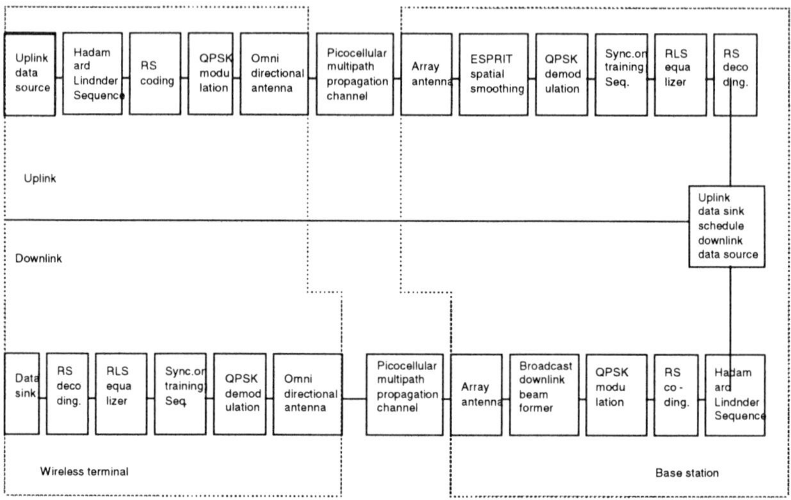

Figure 3.35 Employing SDMA in MBS.

The simulation scenario included the following:

- *Stochastic channel model:* To simulate a picocellular propagation environment;
- *Signals and coding:* Each terminal assigned a training sequence composed of a Lindner sequence for synchronization and a Hadamard sequence for identification with different Reed-Solomon codes applied for FEC of information and signaling messages;
- *Modulation:* QPSK applied to transform the binary source signals to continuous signals;
- *Synchronization:* All signals delayed during their propagation through the channel, whereby the LOS path had the smallest delay and could be expected to be the strongest path;
- *Equalization:* The synchronizer was only able to find the strongest received signal. In multipath environments, this signal was composed of several

delayed versions of the source signal, which overlaid at the receiving unit, creating ISI. Employing training sequences and a recursive least square (RLS) equalizer with a tapped-delay line structure and Ntaps could help notably diminish this ISI.

The application of SDMA technique to broadband wireless systems and the introduction of an uplink pilot-tone phase assured the parameter-estimation accuracy, which was essential for successful downlink beamforming. Concurrent access of several terminals during uplink transmission and random access was enabled by spatial filtering. SDMA can offer increased flexibility for MBS and can help to make efficient use of available spectrum. The frequency band in the range of several GHz allows for compact construction of array antennas and opens up the possible deployment of antenna arrays at terminal side.

3.3.4.2 Internet Protocols in MBS

MBS is designed to extend fixed broadband services to mobile users and is based on an application-transparent wireless-ATM radio interface [89]. Therefore, solutions to support Internet traffic over ATM were considered for MBS and several concepts were presented for serving IP traffic over an ATM-based radio interface. In general, the variable-length IP packets have to be segmented at the transmitting side into the fixed-size packets supported by the DLC layer and are reassembled at the receiver. Depending on the type of IP traffic, the DLC has to make the most efficient use of the available bandwidth.

The approaches for IP over ATM can be categorized in two groups:

- *Overlay concepts:* The classical IP [90] and the LAN emulation [91] concepts that make use of B-ISDN signaling specified for fixed ATM networks in order to establish end-to-end connections [92];
- *Nonoverlay concepts:* Avoid mapping of the IP-session parameters onto B-ISDN connection parameters. Instead of using a static end-to-end connection, IP packets are routed hop-by-hop.

For fixed ATM networks (ATM transmission without B-ISDN signaling), the tag switching and the IP switching have been specified [93]–[94]. The network-specific protocol has to be adapted for the radio-access system in MBS according to the radio protocols used to establish virtual connections serving IP flows. Contrary to the overlay concepts, the QoS parameters characterizing an IP flow can be directly assigned to the radio-specific QoS management located in the DLC layer. This is not possible in overlay concepts where mobility is not supported in B-ISDN protocols. Nonoverlaying concepts simplify the introduction of IP traffic in MBS and allow for efficient use of QoS negotiation; therefore, studies had gone out to the IP-switching concept and QoS mechanisms based on it [89]. The research work was partially funded by the project SAMBA; therefore, some special attention is given to this work in this section.

To perform IP switching in fixed networks, a special IP switch was introduced. An IP switch is a standard ATM switch but without the B-ISDN signaling software, which has been replaced by standard routing software and a flow classifier. At system start-up, virtual channels were established between these IP switches and were used for the hop-by-hop connectionless forwarding of IP packets, which had been segmented in ATM cells.

The flow management protocol was specified in [94]. It is a protocol that allows a network node to instruct a neighboring node to attach a layer label to a specified IP flow. Figure 3.36 gives an illustration of the different steps that an IP switch controller performs. Each IP switch maintains a refresh timer, which initiates periodically verification of each flow. If a flow has not received a refresh message when the timer elapses, it is removed and will be treated as conventional IP traffic.

To facilitate IP switching at the MBS radio interface, the variable-length IP packets had to be segmented into fixed-length packets of ATM-cell length size. For the efficient adaptation of these packets to the radio-DLC service data units (SDUs), it was proposed to define packet payloads of the same size, for DLC SDU and for the switched packets. The convergence layer for IP switching was derived from the concept studied in ETSI BRAN [95]. In this way, segmentation and reassembly needed to be applied only at the sender and the receiver of the IP flow [89]. All intermediate hops were passed transparently, when applying IP switching. The best-effort IP traffic, which was not dedicated to a flow, could be treated hop-by-hop by the IP protocols as depicted in Figure 3.36. That means that best-effort traffic will be segmented and reassembled at every IP switching node, and hence, at the radio interface. Figure 3.37 illustrates the proposed system for the IP convergence layer to adapt IP traffic to the radio-DLC layer.

To support QoS for Internet applications two functions are necessary:

- Network elements along the path from a sender to a receiver have to support mechanisms to control the QoS;
- Signaling communication for QoS negotiation between the applications requirements and the network elements along the path from the receiver to the sender have to be provided and a logical connection has to be established in order to guarantee a certain QoS for a flow.

The first function is provided by IP QoS control services specifying the IP parameters [96]–[97]. The second function may be provided by a resource reservation protocol (RSVP) [98]. As the RSVP specification does not define the format of its protocol fields, either controlled load or guaranteed service may be used with RSVP.

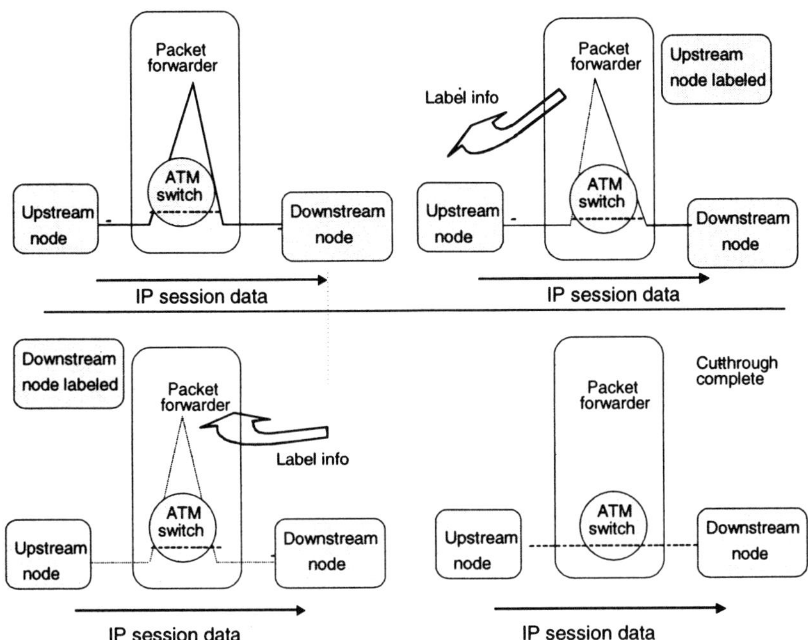

Figure 3.36 IP flow management.

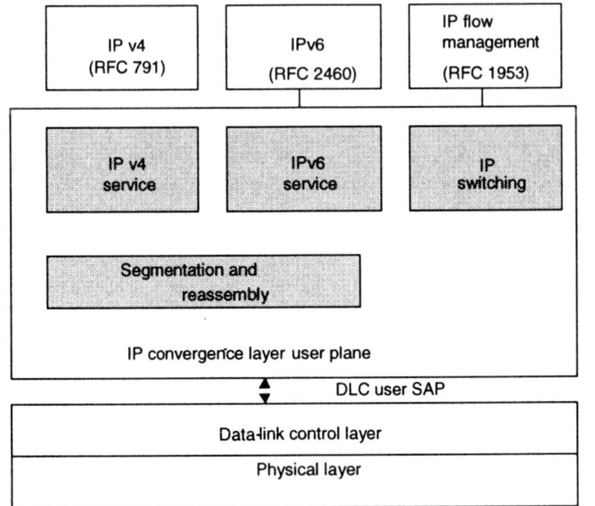

Figure 3.37 Protocol stack at the radio interface of the proposed system.

Using RSVP in the MBS radio-access system allows for adjusting the QoS of an IP session [99]. With RSVP the radio-access system can indicate its QoS capabilities to the receiver, which align its requirements. The capability to allow readjustment of a traffic contract is especially useful in an MBS, where available resources and QoS may change because of degradation of the radio link and handovers. This allows for adapting dynamically the amount of limited radio resources to the needs of an application.

3.3.5 ACTS Broadband Communications Joint Trial and Demonstration—The Project ACCORD

The project ACCORD [100]–[101] integrated four different access systems and a core network, developed in the frames of different ACTS projects, namely the satellite access system of SECOMS (see Chapter 4), the broadband LAN access system of MEDIAN ATM, broadband cellular access system of SAMBA, and the multimedia DECT access system and mobile broadband core network of the project EXODUS (see Chapter 5). The project addressed the study of a target network by means of operative trials through the following:

- Demonstration of the level of functionality possible over the ACCORD trial network in supporting both personal and terminal mobility over a wide set of different complementary radio access;
- Gathering as much possible significant information about the parameters involved in intersegment handover (ISHO) and dual mode terminal roaming (DTMR) and specifying these.

Three trial scenarios, namely SECOMS/DECT, DECT/MEDIAN, and SECOMS/SAMBA, were selected. These trials were representative of the scenarios conceivable with the available access systems. The approach followed in the trial strategy definition was to investigate, with focused tests, the overall ACCORD trial platform capabilities to meet the two-fold project objective as described above.

The trials and related trial strategy were based on a step-by-step approach. The user planes were tested because this was the basis for any integration of any ACCORD platform either experimental or future. From the very beginning, the leading objectives of the test involved:

- Reference to IP based applications and the support of TCP/IP over multimode/multi-access systems;
- Reference to study for ISHO and DMTR behavior.

Both objectives were in line with the indications obtained in the first ACCORD audit. The tests were performed on the EXODUS core network and signaling (IN) encompassing all the network elements (FBT, MB-SSP, MB-SCP).

The project ACCORD investigated the possibility for creating a global mobile broadband (GBM) communication system, which was able to support a wide range of multimedia services (ATM-native and IP-oriented applications). A GBM communication system, so designed, should consist of multiple (satellite/terrestrial) system components (or segments). The project ACCORD addressed this issue by integrating the ATM-based core network developed in the framework of the project EXODUS, with the four different access networks, which were complementary in terms of supported services, mobility facility, and coverage areas.

The project ACCORD aimed at validating a truly global broadband mobile communication system consisting of multiple (satellite/terrestrial) system components (or segments) with different complementary characteristics from both a theoretical and an experimental point of view. It was kept into consideration that for the users roaming outside their office or home network, any visited network should be transparent to the call procedures to which the roaming user is accustomed. A desirable outcome of the project included the appropriate measurements at various levels, namely the physical and IP levels, in order to study the multicoverage in ISHO and DTMR environments.

The following issues were investigated and demonstrated:

- Appropriate level of integration/interworking between all the access networks and the EXODUS core network;
- Design of the required signaling for supporting personal and terminal mobility between SECOMS and EXODUS;
- Design of the required signaling procedures for supporting personal mobility between SECOMS and MEDIAN via the EXODUS core network;
- Design of the required signaling procedures for partially supporting terminal mobility between SECOMS and SAMBA via the EXODUS core network
- Design and implementation of the modular multimode terminals (MMTs);
- ISHO (simulation activities based on measurement data).

The demonstration phase, which was the focus of the project, validated a critical functionality of the target system.

The ACTS projects SECOMS, MEDIAN, SAMBA, and EXODUS complemented each other in terms of the following.

- *Coverage.* The combination of the segments provided a unique coverage area enabling the users to have seamless access to systems while roaming (see Figure 3.38);

- *Services.* The mutually complemented segments enabled the users to take advantage of having access to a wide range of services (with different data rates from 32 Kbps to 150 Mbps). The access to various services was according to the application used and its availability in rural open areas, dense urban areas, or indoor environments;
- *Mobility.* Personal and (in higher levels) terminal mobility using a modular multimode and multifunctional terminals.

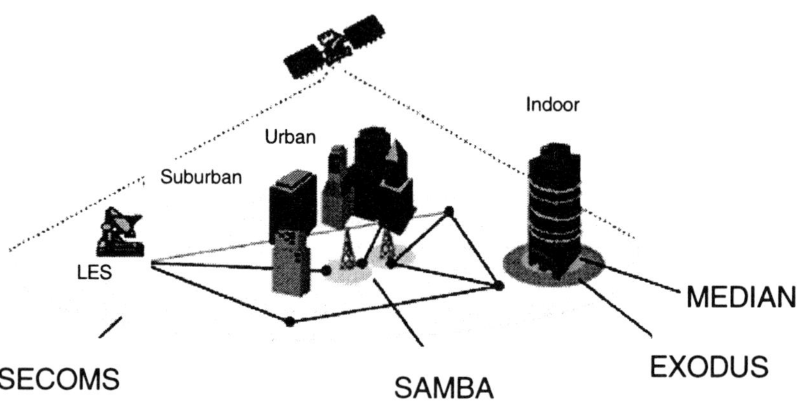

Figure 3.38 The complementary system components.

The study phase identified the mobile broadband integrated network (see Figure 3.39) supporting the SECOMS, EXODUS, SAMBA, and MEDIAN access networks, through the use of an advanced core network. In the target system these networks fully interworked with each other to provide the broadband multimedia services for the users in a seamless manner and with support of personal mobility, terminal mobility, and service portability. To achieve this, access was required to a (only defined but not implemented) MMT, where the terminal actually consisted of a mobile broadband module (MBM) featured with the all-MB signaling needed to communicate with the core network. This was on top of a common host part (CHP) capable of interfacing different access networks via the relevant access modules (AMs).

The ACCORD target system architecture is shown in Figure 3.40. The mentioned four segments can be easily recognized. For instance, the SECOMS segment consists of the SECOMS-specific terminal, the SECOMS-specific air interface, and the SECOMS-specific base station.

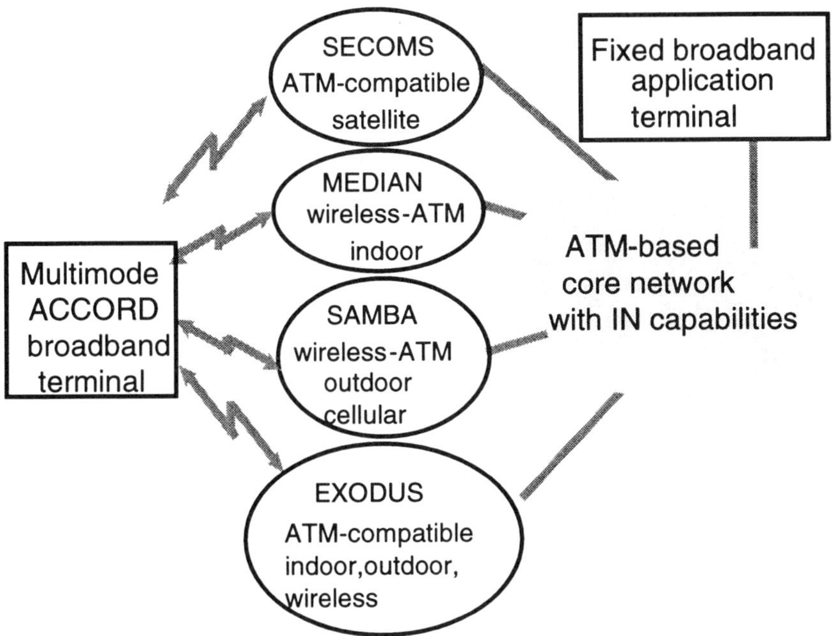

Figure 3.39 Trial network scenario.

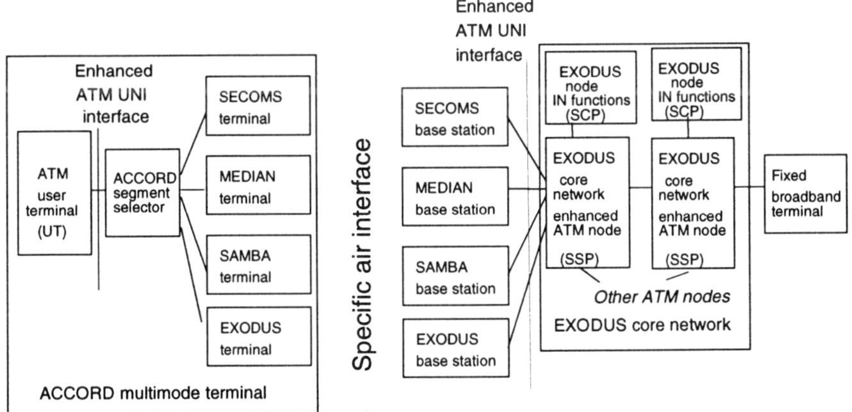

Figure 3.40 ACCORD target system architecture.

The SECOMS, MEDIAN, SAMBA, and EXODUS segments, considered together, can be seen as a single multimode access network of the whole ACCORD target system. As a matter of fact, the cooperation among the four selected segments provided global radio coverage, which guaranteed the generic user, wherever located, access to the ACCORD core network. When more than one segment was available, choosing the segment to use was decided on the basis of fixed priority logic, also taking into account economic considerations. The working of the overall ACCORD system took place at two different strata:

- The *lower stratum* contained all the segment-specific functionalities, which allowed the ATM cells to be carried from the core network to the UT. At this stratum, the segment-specific procedures took place (i.e., the procedures executed inside the different segments). These procedures were completely managed by the segment-specific entities and therefore were invisible to the upper stratum.
- The *upper stratum* was where the ATM traffic and the ATM signaling, and management procedures occurred. In particular, the ACCORD *mobility management procedures* (MMPs) developed in the framework of the project EXODUS and consisted of messages encapsulated in the ATM Q.2932 *facility* message, belonged to this stratum. The ATM signaling and management procedures happened between the UT and the EXODUS ATM-based core network. These procedures were transparent with respect to the specific segments; nevertheless, in order to correctly work in a wireless environment, they needed some information relevant to the segment-specific procedures, and use of the connections setup by the segment-specific systems.

This two-strata approach allowed for as far as possible the decoupling of the segment-specific procedures developed in the framework of the projects MEDIAN, SECOMS, SAMBA, and EXODUS (access network side), from the ATM signaling and management (in particular, the *MMP* procedures) used in the EXODUS core network. As already mentioned, the former procedures belonged to the lower stratum, while the latter procedures belonged to the upper stratum.

The ATM signaling and management procedures were transparently carried over the various segments (i.e., over the lower stratum). In other words, the various segments provided physical support to these procedures. The proper work of the segment-specific entities and procedures required that some information from the ATM signaling and management procedures was collected. Thus, at each segment-specific terminal and at each segment-specific base station, appropriate interworking units (IWUs) existed, able to pick appropriate ATM signaling and management information to be exploited by the segment-specific entities and procedures. These IWUs had already been designed in the framework of the MEDIAN, SECOMS, SAMBA and EXODUS projects.

ACCORD multimode terminal architecture consisted of three different parts:

- ATM user terminal (UT);
- ACCORD segment selector;
- Segment-specific terminals.

The UT was a standard EXODUS ATM-based terminal, which could be personalized by any user through the use of an EXODUS smart card containing that user's profile. The user's profile contained information to his or her identity and the services associated with that user, such as outgoing call setup. The EXODUS smart card allowed the UT to support the MMP procedures.

The ACCORD segment selector retrieved information relevant to the segment-specific procedures (radio attach, radio attach update) from the segment-specific terminals and interacted with the UT. Based on these data exchanges and cooperation with the core network service switching point/service control point (SSP/SCP), the ACCORD segment selector had the fundamental task of selecting the segments to be used by the different connections. The segment-specific terminals, cooperating with the corresponding segment-specific base stations, were in charge of carrying out the segment-specific procedures, aiming to provide the physical support to the ATM traffic and procedures. In addition, the segment-specific terminal contained an IWU, which allowed interaction between the segment-specific procedures and the EXODUS-based ATM signaling and management procedures.

On the core network side, the segment-specific base stations were linked to an SSP, which in turn was linked to an SCP. The segment-specific base stations were the network-side counterpart of the segment-specific terminals. They handled the segment-specific procedures and contained appropriate IWUs to interwork with the ATM signaling and management procedures.

The SCPs were in charge of providing IN functionality that at least for the target system was significantly upgraded in the framework of the project ACCORD. Finally, the SSPs controlled the user and terminal location information required for supporting ATM switched connections between highly mobile end-users.

The trial network is described in Figure 3.41. The ACCORD trial system enabled the execution of relevant trials that involved a couple of access systems each time.

In Section 3.3.5.1, the different access networks and their interconnection within the ACCORD trial system are explained.

3.3.5.1 The Exodus Core Network

This type of network supports terminal and personal mobility at any level. It is suitable to host any type of access system. In EXODUS, only DECT-enhanced and fixed broadband terminal access were implemented and supported. In ACCORD, the same core network was able to support personal mobility and terminal mobility on all the access systems used except for ISHO. An attempt to study this situation was done through the trials.

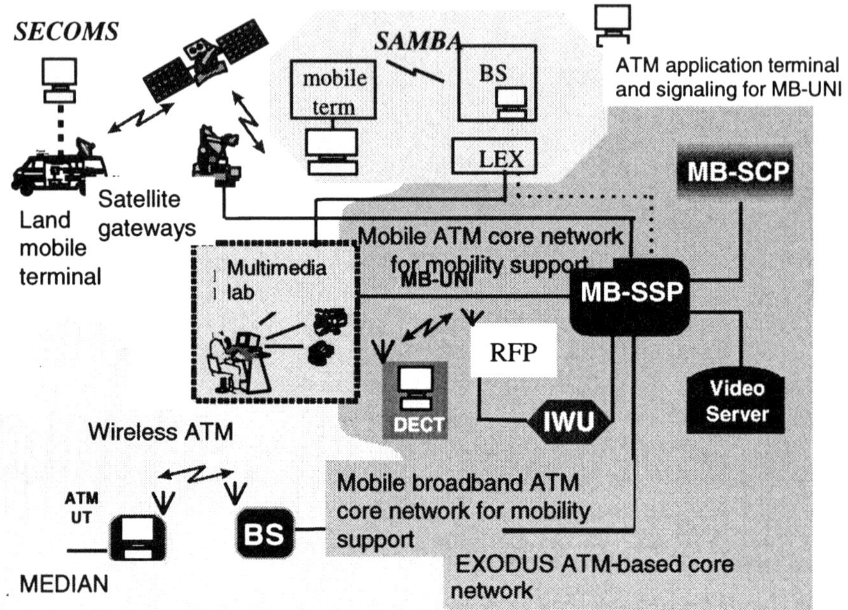

Figure 3.41 ACCORD trial network.

Functional Architecture

The EXODUS functional architecture is shown in Figure 3.42. The various functional entities of this architecture are described as follows:

- *Mobile broadband service control function (MB-SCF).* This corresponds to the SCF defined in the ITU-T Q.12xx series, with enhancements for support of mobile multimedia services in a broadband environment. To support concurrent service and mobility, it was split into an SCF_{sl} for control of user services and an SCF_{mm} for mobility management. The SCF_{sl} functionality included user service support (e.g., call forwarding), subscriber authentication, and control and allocation of resources for multimedia call setup. The SCFmm functionality included location management (e.g., maintenance of the user-terminal relationship—personal mobility, maintenance of the terminal location—terminal mobility, maintenance of temporary identifiers used for air interface confidentiality).

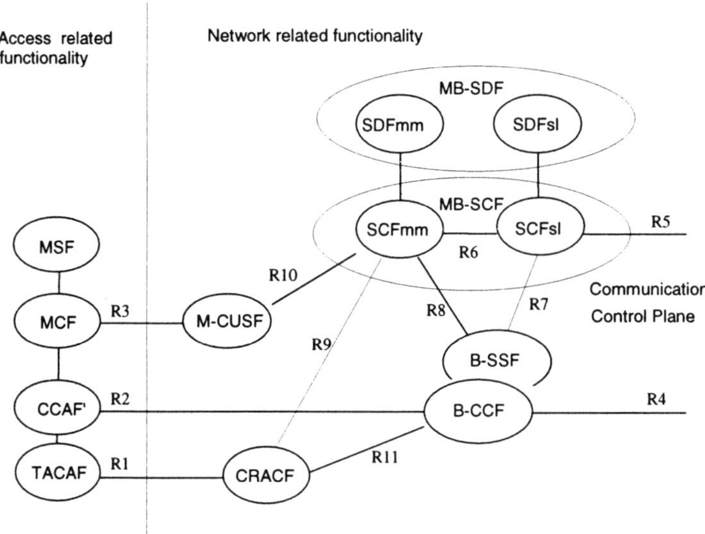

Figure 3.42 EXODUS functional architecture.

- *Mobile broadband service data function (MB-SDF).* This corresponds to the SDF defined in the ITU-T Q.12xx. It was split into two parts, the SDF_{sl} and $SDF_{mm,}$ which represented the database support for the SCF_{sl} and SCF_{mm}, respectively.
- *Broadband call control function (B-CCF).* This functional entity handled B-ISDN calls and all the call-related interaction between the user and the network. It had the capability to detect call and bearer related events relevant for the IN service logic.
- *Broadband service switching function (B-SSF).* In association with the B-CCF, this functional entity provided the set of functions required for interaction between the B-CCF and the MB-SCF. The B-SSF extended the logic of the B-CCF to include recognition and correlation of service control triggers and to interact with the MB-SCF. It managed signaling between the B-CCF and the MB-SCF and modified call/connection processing functions (in the B-CCF) as required to process requests for IN-services under the control of the MB-SCF.
- *Mobile call-unrelated service function (M-CUSF).* This was responsible for the invocation of call-unrelated events in the MB-SCF (e.g., for location

management). It recognized call-unrelated service control triggers and interacted with the MB-SCF; it provided the trigger mechanisms for call-unrelated interaction to access IN-functionality; it modified call-unrelated interaction processing functions as required to process requests for IN-services under the control of the MB-SCF.

- *Call-related access control function (CRACF).* This was responsible for handling call-related events initiated by the SCF_{mm} (e.g., for paging and intraswitch handover of calls and bearers for a call in progress).

The terminal functions are defined as follows:

- *Mobile control function (MCF).* This function represented the service logic and service-related processing in the terminal. It supported all mobile-specific functions (e.g., location management and identity management) and provided local service control.
- *Mobile storage function (MSF).* This represented the data storage function in the terminal for support of the MCF. Data could be stored in the terminal or in a personalized smart card [the user identity module (UIM)]. In addition to subscription- or service-related parameters it stored location information, identity-and-security related parameters, terminal capabilities, and service-related data.
- *Call control agent function (enhanced) (CCAF').* This represented the agent between the user application and the network call control functions. The call control-related interface with the core network interacted with the user to establish, maintain, modify, and/or release a call. It accessed the service-providing capabilities of the broadband-call control function (B-CCF), manipulated and released a call, received indications related to the call or service from the B-CCF, and provided them to the user.
- *Terminal access control agent function (TACAF).* This function was responsible for the paging detection and response to/from the CRACF.

The EXODUS access network supported the following mobility management procedures:

- Location registration/deregistration;
- User registration/deregistration;
- Location update;
- Remote deregistration;
- User profile modification;
- User profile interrogation.

Whenever registration or call setup was performed, authentication took place. The network checked whether the user was allowed to enter the network and if he or she was offered a specific service. Authentication could be considered as part of other procedures such as call setup and location/user registration. The physical architecture of the project EXODUS is presented in detail in Chapter 5.

3.3.5.2 SECOMS Satellite Access Network

The SECOMS access was intended to provide the "global coverage umbrella" in the ACCORD trials. The interfacing of the SECOMS system with B-ISDN was the case considered in the ACCORD project. As a matter of fact, ACCORD foresaw that the SECOMS satellite system acted as access network of the EXODUS core network, which was an ATM-based network provided with IN functionality. In this case SECOMS had to provide the physical link for the mobile broadband UNI (MB-UNI) procedures, taking place between the ATM user terminal (UT) and the EXODUS SSP/SCP, to transparently evolve. The layout of the SECOMS ground demonstrator is shown in Figure 3.43. It comprised three main elements: the multimedia laboratory (MLab) which was the fixed base-band site, the land mobile terminal (LMT), and the satellite (ITALSAT working in global coverage mode). The MLab configuration had two main sections: an indoor section, hosting the fixed end-user for the trial, and an outdoor section, providing a gateway towards the satellite. This last section was based on the TDS-6/A station providing the RF front-end interfacing ITALSAT satellite. The station was adapted and integrated in MLab environment. The TDS-6/A hosted the modem to access the satellite, and much measurement equipment was used for the measurement.

In the MLab with the fixed user terminal, a set of computers was connected by a LAN. A workstation was used as gateway between the LAN and the network system. A PC host controlled the measurement instrumentation.

SECOMS Upgrading for ACCORD Trial Network Architecture

The chosen topology considered access to the WAN satellite system using a segment adaptation unit (SAU) between the UT and the router at ATM cell streams. This was achieved by inserting routers offering ATM I/F, which were capable of extracting ATM cells from SDH frames. The ATM router needed to have traffic-shaping capabilities to adapt the satellite air interface data rate. The topology was obtained by insertion of an ATM adapter in the UT and two commercially available routers to provide a standard ATM STM-1 optical interface on one side, and an RS-449 WAN interface on the other side.

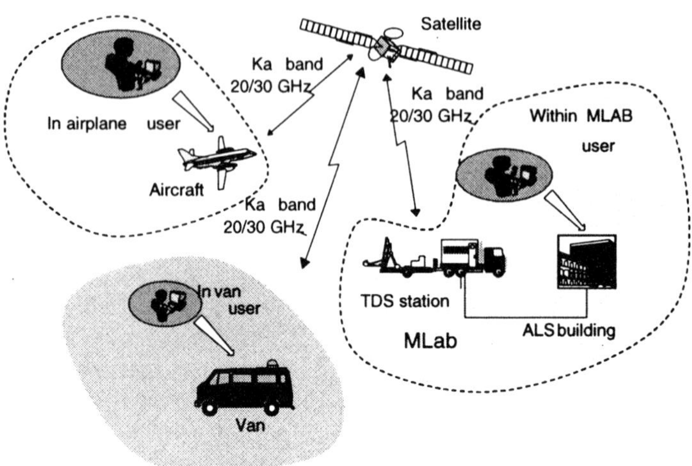

Figure 3.43 SECOMS demonstrator.

A desirable characteristic of these routers was the traffic shaping (buffering) imposed on the ATM traffic received from the STM-1 interface. The protocol architecture is depicted in Figure 3.44. The protocol architecture for the SECOMS/EXODUS interconnection is shown in Figure 3.45 and Figure 3.46 for the control plane and the user plane, respectively.

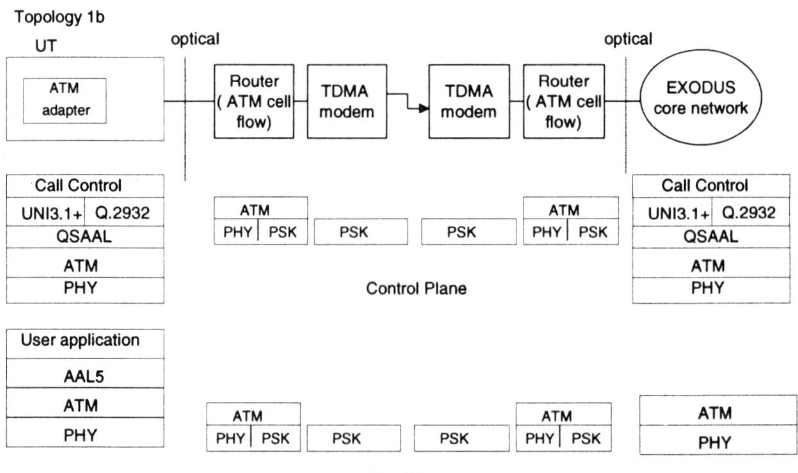

Figure 3.44 SECOMS system protocol architecture in ACCORD.

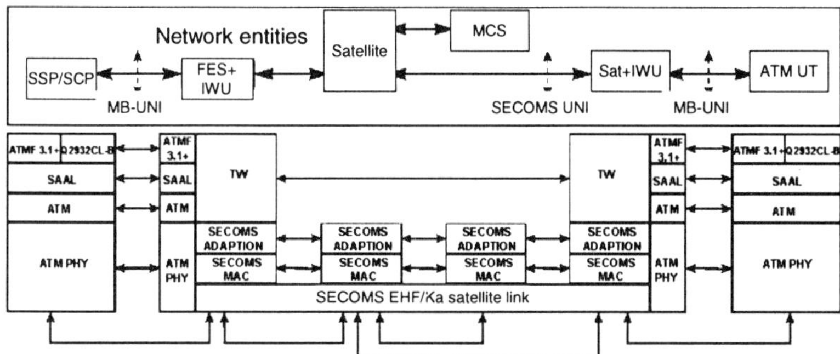

Figure 3.45 Control plane protocols for SECOMS and EXODUS interconnection.

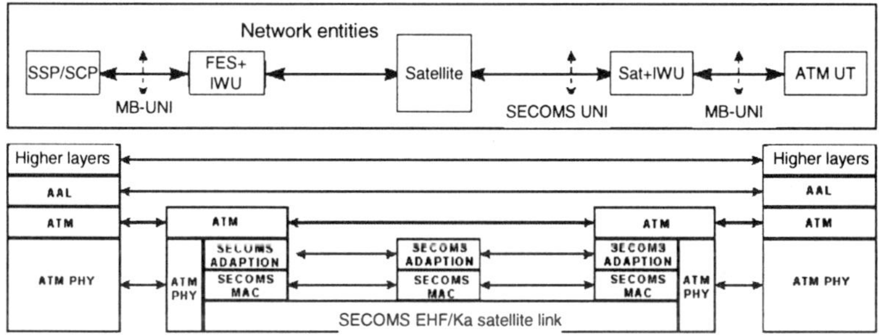

Figure 3.46 User plane protocols for SECOMS and EXODUS interconnection.

The layer structures of the control and user planes clearly show the relationships and interactions between the EXODUS core network and the SECOMS access network. The EXODUS protocols (ATM and higher layers) are superimposed on the top of three SECOMS layers, namely the adaptation, MAC, and EHF/Ka satellite link, which constituted the wireless counterpart of the ATM physical layer and dealt with the SECOMS connections. The interactions between the EXODUS protocols and the SECOMS protocols could happen through IW layer entities that executed the interworking functions.

3.3.5.3 MEDIAN Access Network

The MEDIAN access network in the ACCORD project consisted of one BS and one portable radio part (PRP). A PC-based user terminal was used instead of fixed broadband terminals supplied optionally by the project EXODUS. In a MEDIAN stand-alone configuration, the BS was connected to a commercially available ATM switch, which operated as a LEX.

In ACCORD, however, the BS was connected to an ATM switch that was part of the EXODUS core network. Figure 3.47 shows the MEDIAN protocol stack for the user plane.

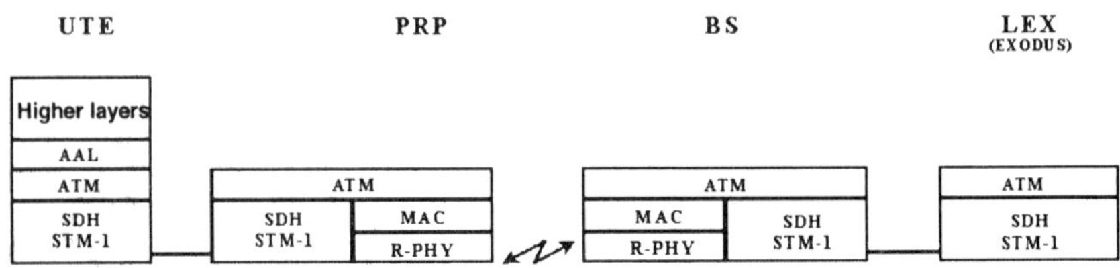

Figure 3.47 MEDIAN protocol stack for the user plane.

The physical interface between the BS and the ATM switch, as well as the physical interface between the PRP and the user terminal, was based on SDH STM-1 framing over multimode optical fibers. The ATM layer functionality in the PRP and BS was restricted to mere forwarding of cells between the "fixed" side PHY interface and the MAC layer interface on the radio side; the VPI and VCI fields in the cell headers were left untouched. Before cells were handed to the MAC layer, however, they were tagged with a radio VCI (RVCI). The RVCI-mechanism allowed the notion of multiple virtual connections with potentially different traffic parameters and QoS requirements to be extended to the MAC layer, which was responsible for scheduling cells into timeslots of the transmission frame structure.

The protocol stack for the control plane is shown in Figure 3.48. The BS intercepted all ATM signaling messages that were exchanged between the user terminal and the switch. The CAC function was part of the MEDIAN connection control (MCC) in the BS. The MCC also took care of RVCI management. RVCI requests, allocations, and revocations were communicated between the BS and the PRP through the exchange of MEDIAN-specific signaling messages. In order to be able to direct ATM signaling messages to the correct user terminal, the BS MCC must learn the ATM addresses associated with each terminal. To this end, the BS must also intercept ILMI messages exchanged between user terminals and the LEX. When an alternative mechanism for ATM address allocation is used between the network and the user terminals, the control plane stack must be adapted accordingly.

The MEDIAN demonstrator also supported the use of permanent virtual channel (PVC) and permanent virtual path (PVP) connections. These could be configured using the control and monitoring terminals connected to the interworking subsystems of both the BS and the PRP. In case of only a single portable station, the MEDIAN demonstrator can be made transparent for all UNI signaling and user channels by setting up a PVP with VPI = 0.

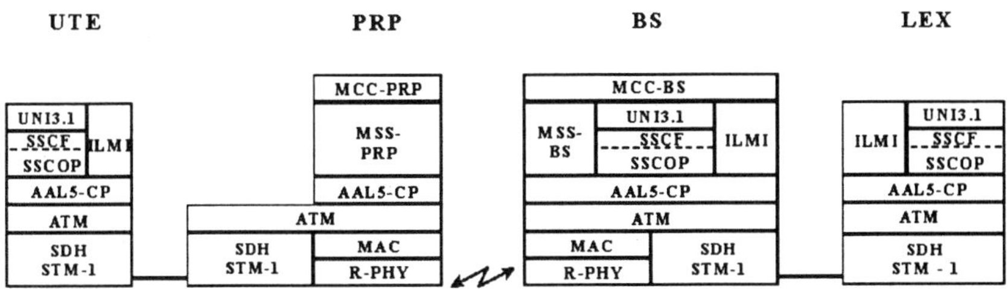

Figure 3.48 MEDIAN protocol stack for the control plane.

3.3.5.4 SAMBA Access Network

The SAMBA trial platform (see Figure 3.49) was designed to provide wireless local access to the public B-ISDN by supporting transparent ATM transmission up to 34 Mbps over a wireless link at 40 GHz.

The trial platform focused on the essential network functions needed for the interconnection of wireless ATM to the fixed ATM network and support of mobile users. The radio access system had two BSTs served by one controller (BSC). The BSC was connected to the ATM network via an ATM switch that was controlled by the ATM mobility server (AMS) responsible for mobility management and call control at the network side. ATM switch and AMS together provided the functions of a mobility-enhanced ATM switch.

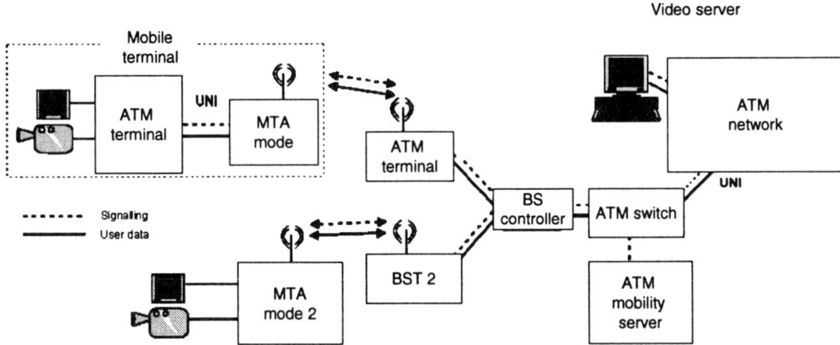

Figure 3.49 SAMBA trial platform.

The two mobile terminals consisted of an ATM terminal connected to a mobile terminal adapter (MTA) for wireless communication. Any standard ATM equipment could be used as mobile terminal. A standard ATM UNI is supported at the mobile side. From the network point of view the trial platform is a conventional LAN, though it was wireless internally. Therefore, standard UNI was also supported as interface to the fixed ATM network.

User data produced by the ATM terminals on either side was conveyed transparently across the network. At the radio interface specific protocols for the MAC and the LLC were used for reliable radio transmission. The signaling protocols of the control plane could be separated into a fixed control plane and a wireless control plane (see Figure 3.50).

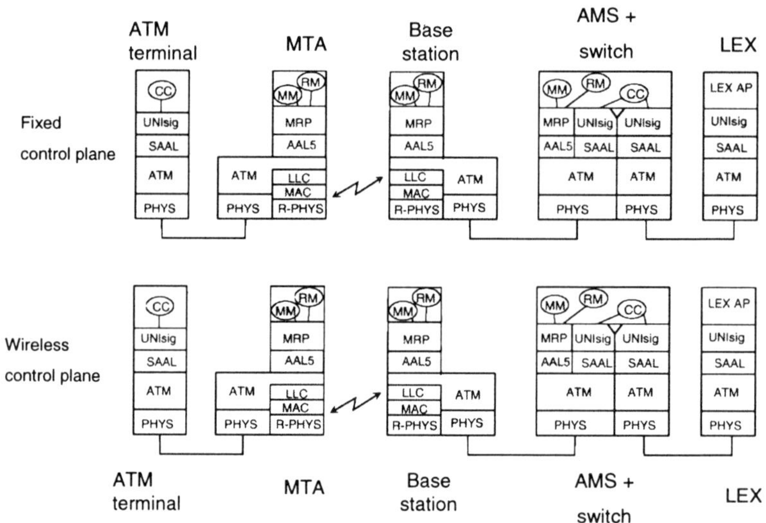

Figure 3.50 Fixed and wireless control planes.

Besides standard ATM signaling protocols, a specific protocol for mobility and resource management (MRP) was used in the wireless control plane. It incorporated procedures for mobile registration, location management, and allocation of radio resources.

3.3.5.5 ACCORD Mobility Management

The four (access) segment-specific systems had to provide the physical support to the ATM traffic and to the EXODUS procedures taking place between the ATM

UT and the EXODUS SSP/SCP. The work of the overall ACCORD system happened at the lower stratum and the upper stratum as was explained earlier.

The segment-specific location areas and the ACCORD MMP location areas within the ACCORD multienvironment provided for coverage. The former were the location area designed in each access system—e.g., a SECOMS-specific location area was the area covered by a spotbeam. The mobility among the segment-specific location area was managed by the segment-specific terminals and base stations, according to the mobility criteria of each access system. Neither the ATM UT nor the ACCORD core network was involved in the segment-specific mobility management. The ACCORD MMP location areas included a certain number (≥ 1) of segment-specific location areas; however, a certain ACCORD MMP location area included segment-specific location areas belonging to the same segments.

Figure 3.51 shows an example of the ACCORD coverage area served by two different access systems (e.g., MEDIAN and SAMBA). For the MEDIAN access system shown, the following can be identified:

- MEDIAN cell (i.e., the area covered by a MEDIAN base station);
- MEDIAN location area grouping n-MEDIAN cells ($n = 4$);
- MEDIAN MMP location area of k-MEDIAN location areas ($k = 3$).

The MMP location areas were associated with a one-to-one correspondence to the EXODUS core network points of access (PoAs), which in turn were associated with proper terminations of an ACCORD SSP. The ACCORD mobility management procedure could be summarized as follows:

- *Segment selection.* The procedure by which the MMT, at its switch-on, selects, among the n segments ($4 \geq n \geq 0$) available from the segment supporting the information exchange with the core network;
- *Segment reselection.* The procedure for a standby of an MMT passing from one active segment to another and changing access networks;
- *User registration.* The procedure for the ACCORD ATM UT and the first user to get registered in the core network and to be reached by the UT terminated calls;
- *User deregistration.* The procedure for the last user to get deregistered from a terminal and the other way around;
- *Location update.* The procedure activated when the MMT is in standby mode and passes from an MMP location area to another MMP location area served by the same access network;
- *ISHO.* The procedure for an in-call MMT to pass from one active segment to another.

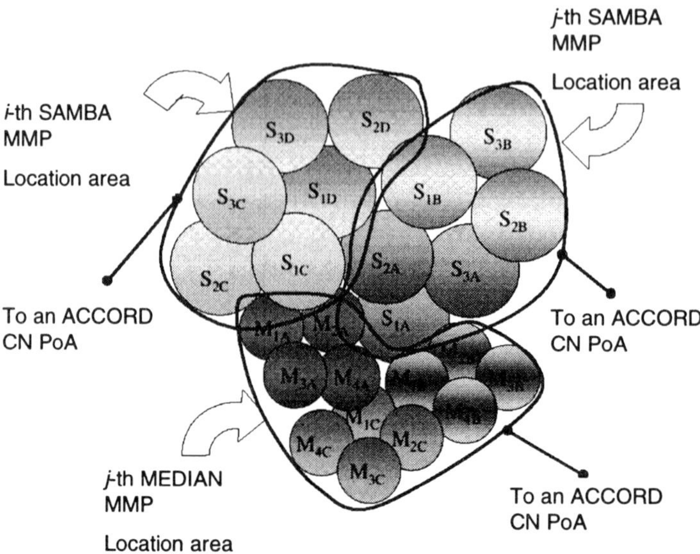

Figure 3.51 Example of ACCORD coverage area.

Figure 3.52 shows the two-strata approach adopted in ACCORD.

The ACCORD procedures utilize both segment-specific and upgraded MMP EXODUS procedures. A selection logic based on a fixed priority rule was adopted for choosing the access segment.

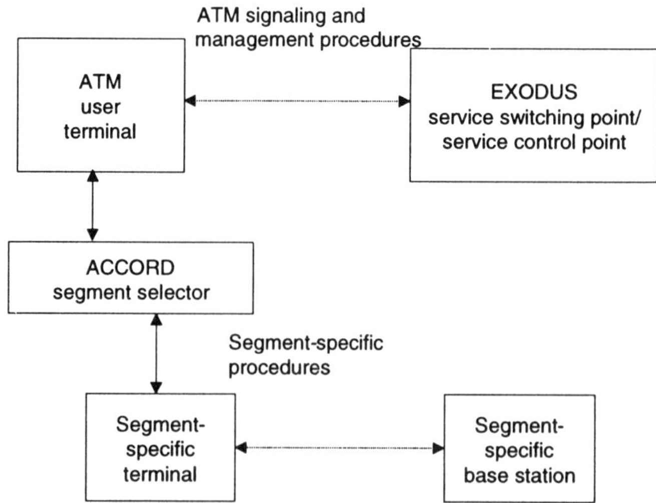

Figure 3.52 ACCORD two-strata procedure approach.

3.3.5.6 System Approach/Trial Design

In order to evaluate the signaling load on the network, appropriate relative weight for each procedure (described earlier) was assigned. For the calculation of the load of the signaling procedures, the following steps were executed:

- Definition of information flows;
- Counting the number of bytes for each message and parameter used in the procedure at application level;
- Counting the framing bytes (overhead) to calculate the actual load on signaling links between network elements;
- Evaluation of transmission time of each message on the signaling link 1 and of the processing time for each procedure. Processing time 2 for a procedure results from the time needed to transmit and process all the messages used in it. This time is a function of protocol involved in the procedure on the network element under consideration. The processing time was calculated as a function of the message length and the number of signaling processes involved.

This sequence of steps enabled the evaluation of the time required for each signaling procedure for transmission and elaboration in the core network. It did not take into account the messages propagation time spent before the arrival to the core network from an MMT to an interworking unit (IWU) (see Figure 3.53).

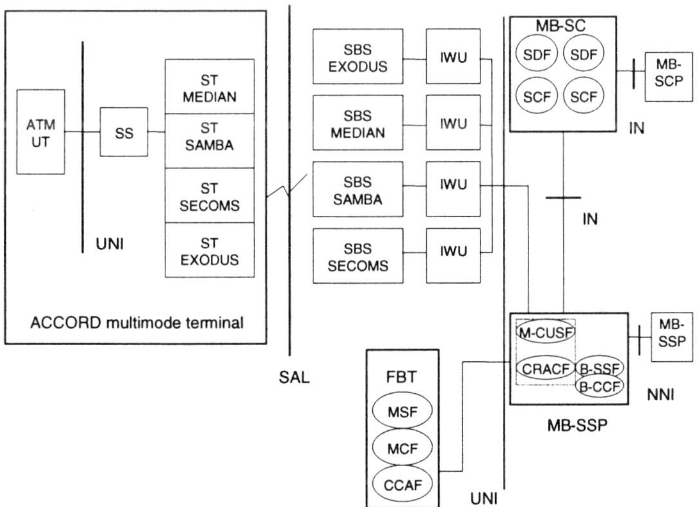

Figure 3.53 ACCORD trial architecture.

The typical propagation delays for the different radio interfaces, measured in milliseconds, are shown in Figure 3.54 for each procedure. The total propagation time depends upon the number of times the considered procedure crosses the SAI interface (see Figure 3.53) and from the used radio segment.

3.3.5.7 Main Technical Development and Summary of Trial

The multidomain target system imposed specific upgrading at the multimode terminal, access networks, and core network levels. To reach this goal, the project investigated the following key issues:

- Characterization of the coverage obtained through complemented segments, through experimental campaigns, based on Ka-band satellite complemented with broadband terrestrial radio coverage;
- Study and experimental verification of a unified approach for the ISHO and roaming;
- Investigation of the achievable grade of service in urban, suburban, and open area environments;
- Identification of relevant access and core network issues and procedures to manage interaccess (or intersegment) mobility.

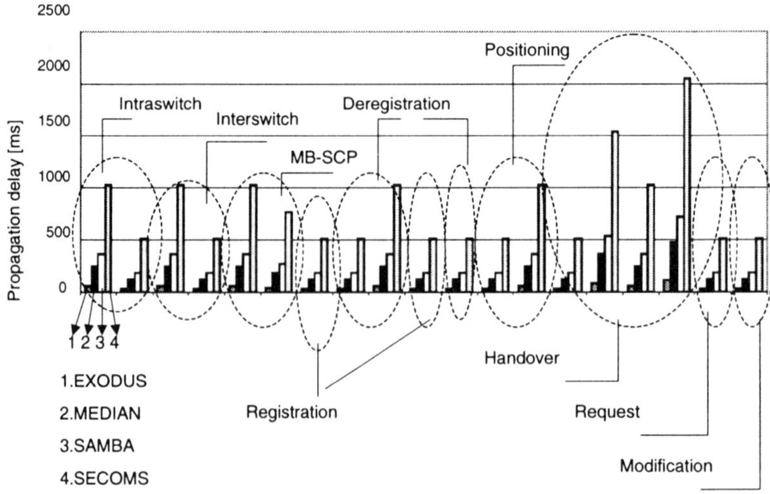

Figure 3.54 Propagation delays at the radio interface for the various segments

Through the series of trials, the ACCORD project aimed at demonstrating how the SECOMS system, which was tailored to a geostationary (GEO) satellite environment could be complemented with a set of terrestrial systems, namely MEDIAN (indoor), EXODUS (picocell outdoor, medium data rate), and SAMBA (picocell outdoor, high data rate), all of which were tailored to particular terrestrial environments. More precisely, through the field trials a selection of the critical interworking/interfacing issues highlighted by the target system studies were investigated. The terrestrial systems considered for trials within the scope of the project (i.e., EXODUS, MEDIAN, and SAMBA) and the satellite component SECOMS, exploited dissimilar air interfaces, and therefore a dissimilar stack of protocols was deployed in the access network. The core network was an ATM-based platform enhanced with an IN capability set and distributed database and processing capabilities to support mobility management, including enhancements of B-ISDN signaling. The EXODUS access network based its control plane on DECT access technology, and SECOMS, MEDIAN, and SAMBA access systems on their own specific protocols. An interworking functionality was required to adapt the access networks protocols to that of the core network. Furthermore, the proposed IN capabilities of the ATM-based core network needed to be further enhanced in order to support multiple levels of mobility and address multimedia service requirements in terms of QoS and mobility service across different systems. Adopting the minimum development approach for the ACCORD demonstrator, the natural configuration had the EXODUS ATM-based core network with IN capabilities as the network behind the ACCORD demonstrator.

The final process of integration allowed implementation of the envisaged set of joint trials for personal and terminal mobility functionality. After the necessary signaling, it was possible for the user to make and receive calls through the new access network. The results of the trials can be summarized as follows:

- *Transport service.* The transport service of the ACCORD network was based on the ATM layer and AAL5, which were inherently supported by the ACCORD user terminal and the core network ATM switch. Alternatively, a point-to-point IP service was also available over the AAL5 through a standard socket interface. The ACCORD network transport layers AAL5/ATM or TCP/IP executed transparently over the various access systems SECOMS/MEDIAN/SAMBA. This offered physical and data link layer services without intervening with the higher layers. In the case of SECOM, a physical and data link layer adaptation was needed as a requirement to interwork the ACCORD network SDH transmission protocol to intermediate protocols and then to a protocol suitable for the satellite access. In the case of MEDIAN and SAMBA, no additional interworking was necessary as both systems used an intermediate user terminal, which offered an SDH interface perfectly fitting the ACCORD network. The physical adaptation to the wireless protocols was done internally within these systems. The DECT wireless terminal was a special case, because the DECT transport protocol was used between the DECT terminal and the DECT base station.

- *Internet applications.* Any proprietary or commercial point-to-point IP application could be deployed in the ACCORD network adopting an IP-over-ATM protocol stack.
- *MPEG-4 testing over the ACCORD platform.* In the ACCORD trials, an MPEG-4 application was developed to test the integration between the SECOMS satellite access system and the EXODUS core network. The aim was to test real-time application data (MPEG-4) over ATM-based satellite and terrestrial segments and to obtain meaningful results concerning the behavior of the MPEG-4 when supported by a satellite link.
- *Measurement tools.* In the framework of the project, the measurement tools used over the ACCORD platform aimed at validating the overall network and its level of reliability in supporting applications and services. A wide set of tools was adopted to test the ACCORD platform and a proper subset of tools was selected for use during the trial sessions. The tools were conceived to have a log of parameters at the TCP/UDP, and possibly, the IP and ATM levels.

ACCORD expedited the transition to MBS deployment by tackling fundamental issues in the field of broadband mobile service supported by a B-ISDN with an IN network. The results of the MMT investigations contributed to precompetitive R&D activities for mutually complemented (satellite/terrestrial) mobile terminals for vehicular and portable terminals.

3.3.5.8 Conclusions

The rationale of the ACCORD project was the integration of preexisting networks. Therefore all the proposed solutions for the system architectures took into account the need to limit as much as possible the modifications to the already working systems. The deliverable focused on the main aspects relevant to the definition of the ACCORD system network architecture. The approach followed in the system design was based on the attempt to develop a possible evolution path to guarantee a gradual transition from the trial system architecture towards the final target system one.

3.4 WIRELESS ACCESS SYSTEMS

Recently, considerable interest has focused on the extension of broadband wired ATM into the wireless medium [99]. Various portable multimedia applications have gained importance by the ability to acquire global access to any other application independent of time or location, thus making such an extension very useful. One of the main design issues of such a system is considered to be the fair rule for the different users to access the radio channel. These, in the context of the wireless (cellular) communications network, are active users within the cell, handovers from neighboring cells, and new users requesting access within the cell.

Different strategies already used for wired or wireless communications networks could be adapted and developed for future systems. In such a wireless system there is a need for an uplink channel used by the mobile terminals for sending requests and data packets and a downlink channel used by the base station for sending signaling messages and data packets. FDD or TDD can be employed for these two channels. Also, a few MAC protocols have been proposed for such a future wireless network [102]. These protocols employ different strategies to fairly allocate their resources to different users. One popular technique is a dynamic reservation of a slot inside a defined, fixed, or variable-length frame, for transmission from the mobile terminals to the base station. However, even with this dynamic reservation some delay constraints are difficult to be met because a predefined frame exists and synchronization is necessary.

The decade of the 1990s has seen the classical boundaries between telecommunications, consumers, and computers merge into a vision of convergence [103]. The goal of this emerging technology is to create a single platform that will allow the consumer to access the Internet, to play DVD discs, to play realistic 3-D games, and receive the latest digital TV broadcasts through one central system. This increasing demand for broadband services is supported, among others, by two developing technologies:

- TV broadcasting systems supplemented by interactive asymmetrical data including the Internet (according to the DVB standard);
- Broadband WLANs, supporting broadcasting of MPEG-2 video/audio programs and the Internet (according to IEEE802.XX LAN/WAN standards).

This section describes the activities of ACTS projects related to wireless access research; CABSINET, developing a flexible cellular broadband architecture; CRABS, which dealt with some problems regarding cellular access to broadband services and interactive television; AMASE, which focused on agent-based mobile access to information services, complemented by the project M3A, which developed mobile multimedia access concepts using intelligent agents.

3.4.1 Two-Layer Network Architecture for Fully Wireless Interactive Broadband Service Access—The Project CABSINET

The market potential for multichannel interactive digital TV and broadband services has been identified and is rising. However, the availability of low-cost, easily installed "last-mile" end-user access is a serious barrier to exploitation. The project CABSINET aimed at designing, building, and demonstrating a working interactive cellular TV architecture at 40- and 5.8-GHz that was capable of providing both broadcast and multimedia services. A low-power cellular wireless, coupled with an in-band interactive return link, offered a potentially attractive solution [103].

The work plan of the project comprised research activities connected to the identification of requirements, architecture design, implementation of the system, and the performance of two trials. The project had the following technical objectives according to its intention to clarify the technical and commercial uncertainties surrounding the technology area of DVB/MPEG-2 and DAVIC return path:

- To compare the performance of the designed system with identified market requirements;
- To derive an exploitation plan for such services in the short/medium term;
- To determine the potential for migration (e.g., to near-symmetric interactive broadband services).

The project's guidelines are described as follows:

- A novel cellular two-layer network (TLN) using macrocells to serve 5-km cells in the 40-GHz frequency band and microcells to serve 50-m cells in the 5.8-GHz frequency band;
- An innovative 40-GHz in-band wireless return link (using both TDMA and CDMA spread spectrum) to maximize interference immunity and to be able to migrate from asymmetric to near-symmetric broadband interactivity;
- Flexible terminal support, including "fixed" units and nomadic wireless units with limited mobility and service delivery to terminals inside buildings;
- Cost-effective consumer transceivers in the 40- and 5.8-GHz bands.

The overall system concept is depicted in Figure 3.55 [104]. The development focused on the optimization of modulation, coding, and DSP methods to maximize performance in difficult propagation conditions, and DVB and DAVIC standards requirements were followed whenever possible.

The trials involved macrocell and microcell deployment, connected to the main platform via a trans-European ATM link. Real interactive cellular TV applications, together with ATM-derived services, were run to determine performance and user acceptability.

Both LOS connections to "fixed" terminals and in-building delivery to nomadic terminals were shown. Wireless access in-building offers several advantages, such as portability of terminals, a possibility for users to add more terminals and to avoid problems with protected properties. These are also supported by the microcell approach. It must be noted that microcell is a tough transmission environment (with non-LOS transmission, shadowing, interference, and so forth).

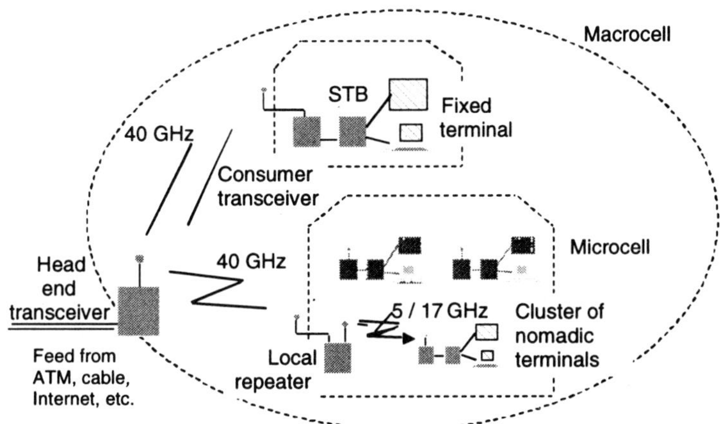

Figure 3.55 Overall system concept of the CABSINET architecture.

However, for propagation indoors, 5.8-GHz penetrates better than 40-GHz, while an OFDM downlink combined with a TDMA/CDMA return link allows for robust transmission and cheap consumer equipment based on DVB-T techniques. The idea of the microcell nomadic terminals is depicted in Figure 3.56.

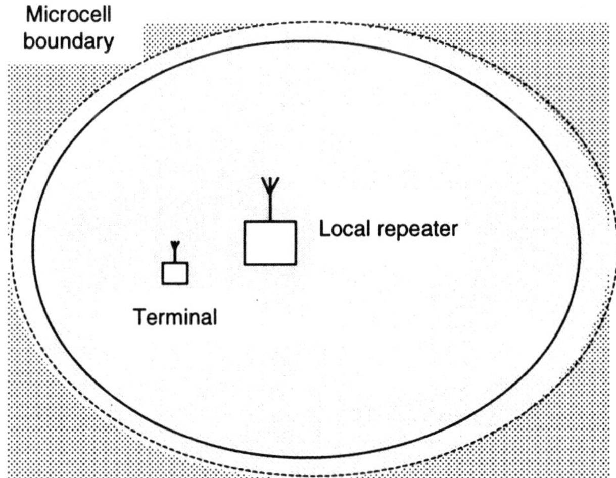

Figure 3.56 Microcell nomadic terminals.

Microcell supports “nomadic” terminals and interfaces to other wireless networks—e.g., HIPERLAN for wireless LANs and portable computing and MBS (see Figure 3.57).

User surveys were conducted with two types of demonstrations:

- *Interactive TV*—Aimed at presenting at least four programs with different aspect ratios, surround sound, TV guide, and pay TV;
- *Interactive multimedia services*—Included video-on-demand (VoD), news-on-demand, and other ATM-derived services.

The innovation of the approach was the nomadic nature of the terminal, which could be moved within the local area without interrupting communication. The demonstration used COFDM at both 40 and 5.8 GHz. Figure 3.58 depicts the block scheme of the CABSINET demonstration scheme.

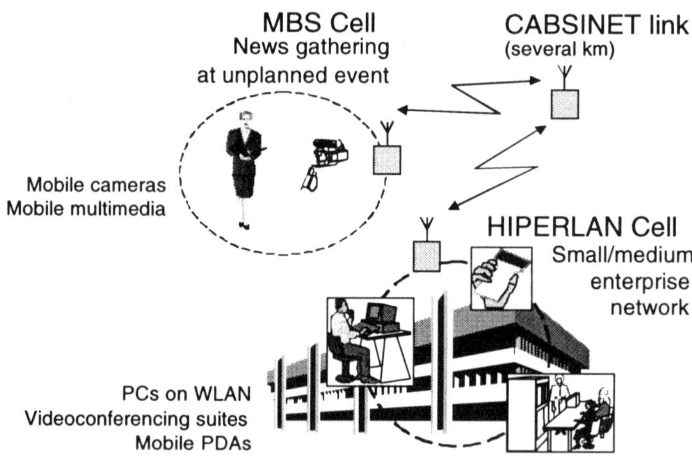

Figure 3.57 Mobility support offered by CABSINET for other networks.

The project activities have influenced both industry and public by doing the following:

- Stimulating new multimedia services and creating benefits for all parties involved in the value chain;

- Delivering new services (e.g., entertainment, healthcare, and education) to users currently unable to receive fiber in the home, or for whom such provision is uneconomical;
- Accelerating the production of affordable equipment and facilitation of the competitive supply of such equipment.

The CABSINET project utilized the results from some other projects and had common research activities with the CRABS project.

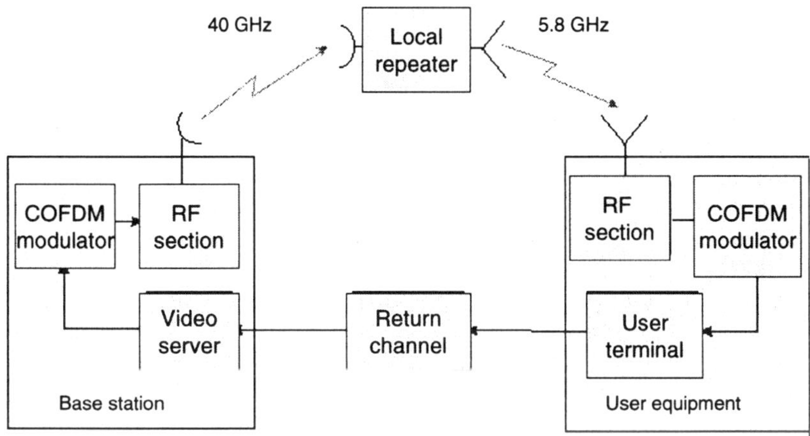

Figure 3.58 Block scheme of the CABSINET demonstration system [103].

3.4.2 Cellular Radio Access to Interactive Television and Broadband Services—The Project CRABS

CRABS concentrated on service definitions and their impact on protocols [103]. Several demonstration platforms were built using available 42-GHz hardware (Oslo, London, Amalgi, Prague, and Athens). In the Oslo trial, asymmetrical provision of data and the Internet was achieved. The Internet services were embedded in the TV bouquet and could reach the users directly through the 42-GHz link at a rate of 5 Mbps, while the reverse path used ISDN lines.

In the Athens trial, the 42-GHz link was used as a backbone for the provision of digital TV and access to the Internet to a number of users at a rate of 2 Mbps [105]. The Internet signal was distributed to the users through a broadband wireless network operating in the ISM-frequency band after it had undergone frequency-division multiplexing. Users equipped with a 42-GHz receiver/down converter and an integrated receiver/decoder were able to watch digital TV. The users could also access the Internet through wireless networking equipment operating in the ISM-frequency band. The Athens trial adopted a symmetric configuration and users accessed the service provider with 2 Mbps in both the forward and reverse path.

The trial located in England examined coverage aspects in a hilly and large-tree areas with populations living in scattered villages and small towns. Cable coverage was expected to be expensive, and a radio solution supplementing it could be promising.

The CRABS project developed a broadband wireless network and through the work of the project, the feasibility of providing an MPEG-2 video/audio program to both portable and mobile users was demonstrated. The idea was based on the platform developed in the Athens trial. The network used the TCP/IP protocol and operated in the ISM frequency band (2,400–2,483.5), with an available throughput of 1.7 Mbps [105]. The quality of the provided video service was evaluated for various MPEG-2 encoding rates and under portable/mobile outdoor conditions.

Originally, the 40-GHz frequency band was considered for pure broadband distribution services and as a supplement to cable systems in areas where installations of such systems would be very expensive or impractical [106]. During the planning period of the project it was observed that digital television represented by the MPEG-2 standardization and the work of DVB, in combination with new interactive services, such as the Internet, would change this picture entirely. Also, broadband services to individuals became imperative.

It was concluded at the time that such services might be delivered through different networks, including the following:

- Satellite connections with terrestrial or satellite return;
- Cable connections making use of existing cable installations;
- Fiber connections to the home;
- Use of existing copper wire connections, ADSL;
- Broadband radio connections.

The project believed that the broadband radio solution could offer several advantages, in particular in the local domain. One of the project's anticipations was a considerable growth in interactive services and their use on a regular basis by a large percentage of the population within a radius of 3–5 km.

The basic architecture of the overall cellular radio access system is depicted in Figure 3.59, also illustrating some of the problems that had to be solved:

- A distribution system connecting the individual cells was necessary.

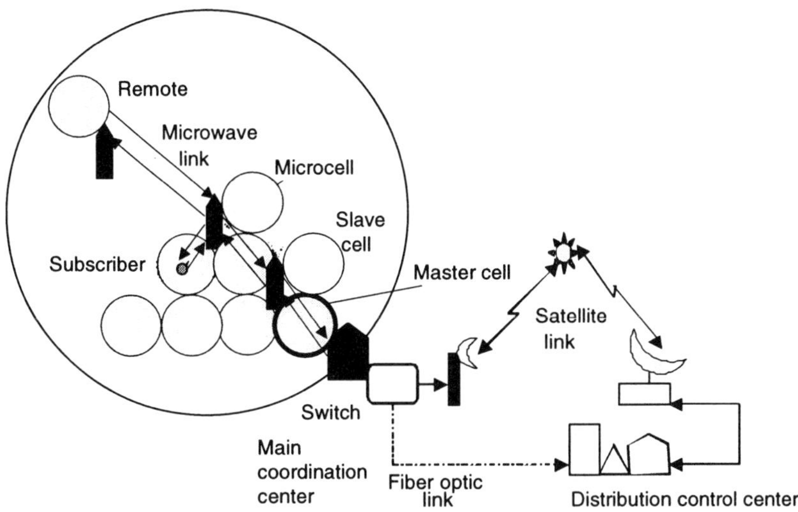

Figure 3.59 High-level block diagram for the overall system.

- Within the individual cells there might be regions that could not be covered (terrain shielding) without the establishment of microcells.
- The complete coverage would require an overlap between neighboring cells.
- The coverage and QoS would be influenced by the local climatical and topographical conditions, buildings and trees, while planning tools allowed for local variations.
- A means for connecting remote cells should be available.
- A two-level control would be required—a main coordination center covering the system and local control centers for each cell taking care of traffic and customers within each cell.

The main coordination center needed to provide as connections broadcast systems based on DVB/MPEG-2 standards (satellites), interactive server systems, ISDN connections, ATM networks, public/private networks for education or telemedicine, and videoconferencing or other multimedia networks.

The project concentrated on the architecture of a total configuration with a cell-based downlink system to contain all required for reliable transmission of the digitally modulated signals from the transmitting site of the cell to the subscriber location. These included the transmitter base station, the front ends for transmitters and receivers, the choice of coding and modulation schemes, frequency use, and reuse planning. A commercial set-top box allowing for connections to both TV and PC was used for reception, and a return path was required for interactive services with path options at least up to 2 Mbps.

During the first phase of the project, the interactive uplink was provided using different types of connections like ISDN, GSM, DECT, and radio-relay connections. For the second phase, the uplink was implemented at millimeter-waves (inband), and different systems were considered satisfying the following requirements:

- Compatibility with ATM;
- Dynamic allocation of return-link capacity in the system;
- Time delay within a cell less than a few tens of μs.

The second phase of the project considered requirements for the uplink capacity. Those were similar to the DECT specifications. For use in the second phase of the CRABS trials, the DECT needed to be upconverted from the output RF of 1,800 MHz to 42 GHz.

It seemed interesting to investigate whether the platform developed in the Athens trial was capable of providing an MPEG-2 TV program using only the wireless networking equipment at the users' premises, because with a system like this, the user would no longer be forced to purchase two individual systems. The idea intended that one purchase would allow the user to access the Internet and to watch an MPEG-2 program. The system also allowed the provision of services to portable and mobile users.

The overall system architecture of the Greek trial consisted of a service provider and a cell main node, located at a distance of 5.3 km away. The cell main node defined a cell with a diameter of approximately 1 km [107]. The communication between the service provider and the cell main node was established in the 42-GHz band using FM modulation, while the communication between a portable/mobile user and the cell main node was established in the ISM frequency band, using the FHSS modulation technique.

The service provider enabled access either to the Internet or to the dedicated server, where MPEG-2 programs were stored. The server ran under the Windows NT operating system and was equipped with an MPEG-2 encoded card, activated by the appropriate software. The analog audio/video source produced a composite PAL signal, which was fed into an MPEG-2 encoder card in the dedicated server. The encoder, operating at level/main profile, captured the incoming composite signal and produced an MPEG-2 data stream (containing the compressed video and audio data). This stream was stored in the hard disc of the dedicated server. The encoder provided also the ability to produce data streams of different bit rates.

The users' requests were received by the service provider, demodulated (baseband 2-Mbps signal), amplified, and then fed into the Rx input of a G703/X.21 converter. The output of the converter was passed to a router that passes the request either to the Internet provider or to the dedicated server. If

Internet services were requested, the Internet provider produced the appropriate reply; otherwise, when MPEG-2 services were requested, the dedicated server produced the reply. In both cases the reply was fed into the router and passed to the converter. The Tx output of the converter was low-pass filtered, FM-modulated at 70 MHz, upconverted to 42 GHz, amplified, and then transmitted to the cell main node.

The cell main node was connected to a portable user (multimedia PC) equipped with an MPEG-2 decoding card. The user generated a request for Internet or multimedia MPEG-2 services using, for example, a common www browser. This request was passed to the Ethernet of the cell main node via the radio LAN access point (2.4 GHz link at 1.7 Mbps). The router received from the Ethernet the users' request and passed it to the X.21/G703 converter. The Tx output of the converter was modulated, upconverted, amplified, and transmitted to the service provider. The corresponding reply was received by the cell main node; downconverted, demodulated (baseband signal at 2 Mbps), amplified, and fed into the Rx input of the converter. The converter output was passed to the router and from there to the user via the radio LAN access point.

A set of tests concerning the trade-off between picture quality and MPEG-2 transport stream characteristics preceded the quality evaluation of digital MPEG-2 programs via the wireless broadband network. An analog PAL signal (of short duration) from a television program was captured, encoded, and saved in the dedicated server's hard disc. This video signal was encoded into MPEG-2 format, twice, using different values for the transport stream encoding parameters. Table 3.12 lists the values of the MPEG-2 transport stream parameters for two different files.

Although the encoding rates below 1.5 Mbps did not comply with the MPEG-2 standard, it was experimentally verified that the picture quality was not significantly degraded. Two questions, however, arose in the provision of digital MPEG-2 programs via a wireless broadband network:

- Did the network throughput affect the quality of the TV program?
- What was the achievable picture quality?

Table 3.12

MPEG-2 Transport Stream Parameters for Two Encoded Programs

MPEG-2 Transport Stream Parameters	*File 1*	*File 2*
Video encoding bit rate	1 Mbps	3.688 Mbps
Audio encoding bit rate	192 Kbps	192 Kbps
Transport stream bit rate	1.2524 Mbps	4 Mbps
GOP size	12	12
I-Frame distance	12	12
Reference distance	3	3
Audio sampling rate	44.1 KHz	44.1 KHz
Audio mode	Stereo	Stereo

An experiment was designed to evaluate the best picture quality to be achieved for a given network throughput of 1.7 Mbps. The testing point was located at a distance of 25 m away from the radio LAN access point under LOS conditions. In the test, the user requested MPEG-2 files from the service provider. Several MPEG-2 files containing the same captured TV program but encoded with different coding rates were stored in the hard disc of the dedicated server and the portable user requested the files during the test. It was experimentally verified that as the transport stream bit rate increased from 1.25 Mbps to 1.7 Mbps, the overall picture quality was the same as the corresponding encoded one and the viewing was uninterrupted. However, increasing the stream bit rate led to the appearance of frame pauses, while the clarity of the received picture remained the same as the corresponding encoded one. Similar sound pauses were observed in the audio signal. The pauses could be explained considering the operation of the MPEG-2 decoder, the video and audio cards in the PC, and the media player software, which were used to reproduce the MPEG-2 file in the user's PC. According to this operation, the network card received the corresponding Ethernet packets from the service provider and passed the appropriate data (MPEG-2 transport stream) to the MPEG-2 decoder. The MPEG-2 decoder decompressed it and produced the appropriate video/audio bit streams, which were passed to the video/audio cards of the PC. The video/audio cards used memory buffers for storing the received video and audio data streams and the media player software enabled the video/audio cards to unload the data from their buffers and produce the appropriate signals on the PC screen and the PC speakers. The procedure implied that the data-loading rate of the buffers depended on the network throughput and that the data-unloading rate depended on transport-stream bit rate. So, a good way to estimate the QoS of the proposed system proved to be a quantitative measurement of the volume of the buffers.

Another experiment was designed to evaluate the performance of the proposed system under portable conditions. For this reason, the user PC was mounted on a car and the transmission antenna was placed at the edge of the rooftop of a 20-m-high building. The measurements also included one with a mobile receiver.

During the car movement the video quality was very good, while an LOS between car and transmitting antenna existed. However, the video quality rapidly degraded when some obstacles intercepted the link, and in some cases there was no service at all.

From the test results it could be concluded that the network throughput dictates the maximum encoding bit rate of the transport stream for pleasant viewing. It was experimentally verified that when the MPEG-2 program-encoding rate exceeded the throughput of the wireless network, picture and sound pauses were observed and that these pauses affected the viewing of the TV program, making it unpleasant when their overall duration exceeded 15% of the actual video duration. It was also shown that an MPEG-2–encoded program could be provided to mobile users, as long as propagation conditions permit it. For the transmitted power used, only LOS conditions permitted an acceptable picture quality. Recommendations for future work include higher transmitted power and multicell configuration.

3.4.3 Agent-Based Mobile Access to Information Services—The Project AMASE

The AMASE project investigated techniques and solutions for enhancing an existing agent platform to support stationary and mobile agents for wireless mobile communications. AMASE was able to provide a high degree of user mobility between various networks, devices, and physical locations. The main focus of the project has been on small and mobile devices, such as a portable PC or PDA. An overview of the scope of AMASE is shown in Figure 3.60.

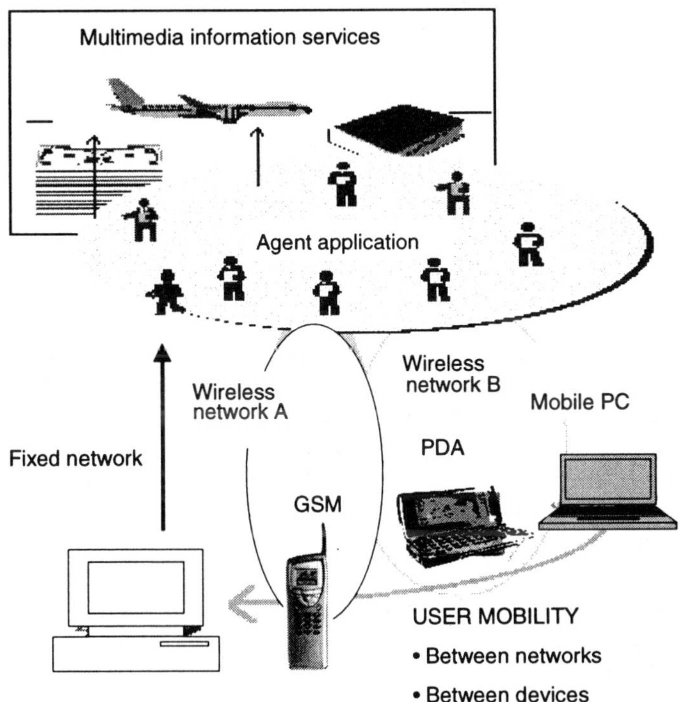

Figure 3.60 Scope of the project AMASE.

An agent environment (see Figure 3.61) could be built using the AMASE system. It consisted of hosts running the agent platform and hosts running supporting services, such as a directory system or a database. Agent platforms could be added and removed dynamically from an agent environment, which allowed for large-scale—even Internet-wide—installations. The AMASE testbed demonstrated the agent environment on European-wide scale. The agent environment was open in the sense that agent platforms were allowed to join or leave the agent environment without intervention from the system administrator.

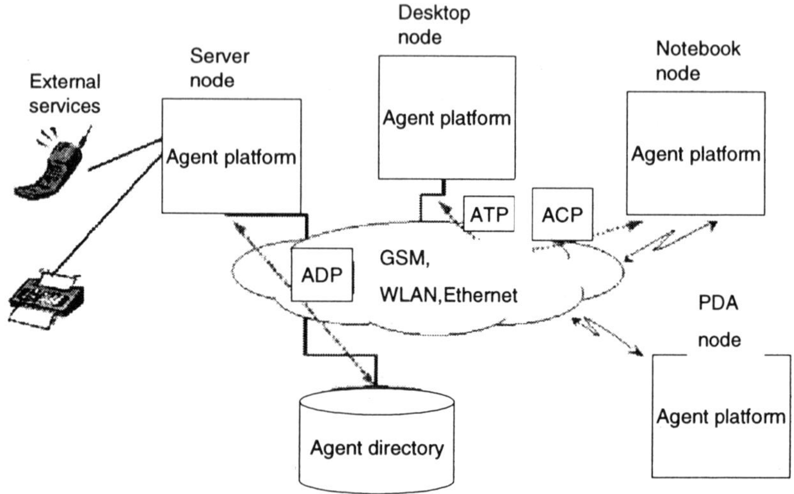

Figure 3.61 AMASE agent environment.

It was possible to install the AMASE agent platform on servers, desktops, and mobile devices running Java or Personal Java. The AMASE system was successfully ported to the EPOC32 simulator for the PSION 5 palmtop. While the base software was the same for servers and desktops, it was different for a mobile device like a PDA because of the device-inherent resource constraints.

The main objectives of AMASE based on available agent technologies were the following:

- Enhancement of an existing agent platform in order to support both stationary and mobile agents for wireless mobile communication environment;
- Design, specification, and implementation of the AMASE agent environment by incorporating features, such as a high degree of user mobility, a high degree of scalability for various device platforms (from desktop computers to handheld devices), secure and efficient agent operations, and cost-effective adaptability to wireless network characteristics and UMTS service capabilities;
- Verification and demonstration, with an agent-based application using the agent environment, of the technical and economic suitability of agent technology for commercial wireless mobile applications;

- Evaluation of the project results for providing recommendations for refinements and enhancements for OMG´s multiagent facilities (MAF).

AMASE also addressed other projects. The main goals of the project AMASE in the overall perspective were in the following:

- To identify and develop efficient and effective techniques for agent-based technology for wireless mobile applications;
- Benchmark stationary and mobile agent techniques in the wireless environment to understand their efficiency in cost and response time under different operating conditions;
- Demonstration of the application of agent technology for customizable and adaptive applications in the wireless mobile environment;
- Development of agent-related standards.

The project distinguished between the agent environment (agent facilities plus communication facilities) and agent-based application. The following steps were significant:

- *Step 1*—Specification of the mobile agent facilities. The differences between the agent systems prevented interpretability and proliferation of agent technology. To promote both interoperability and system diversity, the project reviewed the emerging OMG MAF specification where it was applicable in the AMASE agent facilities. Additionally, the specification of the AMASE communication facilities was based on the AMASE specification of the OnTheMove project.
- *Step 2*—Enhancement of an existing agent platform (see Figure 3.62) for wireless mobile applications: The used system was fully written in JAVA, known within the consortium, and the source code was available. The platform enhancement development supported secure and efficient agent operations, cost-effective adaptability to wireless network characteristics, a high degree of user mobility, a high degree of scalability for various device platforms, and UMTS service capabilities.
- *Step 3*—The wireless mobile application scenario, or the AMASE demonstrator; AMASE applied the developed agent environment in the domain of wireless mobile applications. The application could enable bank customers to access and trigger electronic banking services from their wireless mobile devices.

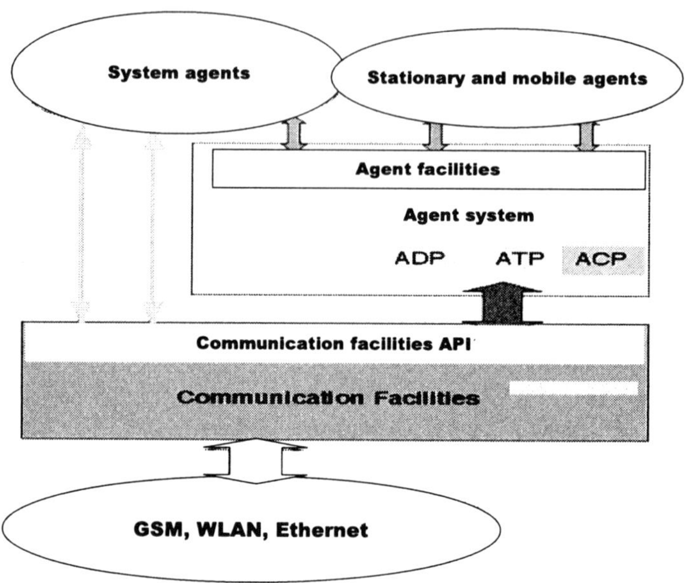

Figure 3.62 The AMASE platform.

- *Step 4*—Evaluation of the AMASE demonstrator; an expert evaluation that conducted qualitative and quantitative evaluation of the performance of the functions of the AMASE agent environment and recommendations for the relevant standard form was performed.

The first objective was to develop a scalable agent platform. A minimal agent platform core was developed to fit various handheld devices to optimize agent platform operation in various device platforms. The memory size, processor speed, power budget, and network access were given special consideration in the development of the platform. The scalable agent platform was meant to have the following capabilities:

- Support of network access from the device platform;
- Support of flexible and modular agent platform software architecture to allow various degrees of functional richness, such as the suitability to support a number of networks and telecommunication services in a device platform;
- A scheme for agent platforms to reliably deal with the potential overloading of visiting agents and events regarding the available resources and capabilities with, for example, refusal and replacement policies.

The network adaptation and mobile operation was designed in view of the following objectives:

- To incorporate QoS adaptation in the agent platform protocols for the available QoS of the accessed network, such as network bandwidth and delay;
- To incorporate wireless adaptation in the agent platform protocols to allow reliable and efficient agent operations over sporadic wireless connections, with atomic operational recovery strategy;
- To incorporate mobility support in the agent platforms protocols to allow agent platform operations to continue seamlessly or recover while a device moves between networks;
- To support geographic location information in the agent platform API to allow agent-based applications to provide location-dependent services.

The approach of AMASE concerning the secure execution of agents was to adapt existing security mechanisms for agent communication to solve the usual security problems of authentication, authorization, confidentiality, integrity, and nonrepudiation. Research on the protection of the mobile agent against the malicious agent system was carried out.

AMASE regarded security especially relevant when an agent migrated and carried security-relevant data or keys and such others with him or her. This is illustrated in Figure 3.63.

Potential approaches for the migration in agent systems are listed in Table 3.13. The AMASE chose to do the following.

- Adapt well-known security mechanism for the AMASE agent platform;
- Investigate the protection of agents against malicious agent systems, such as the one shown in Figure 3.64 through trust domains and enabling the handling of insecure domains on the application level.

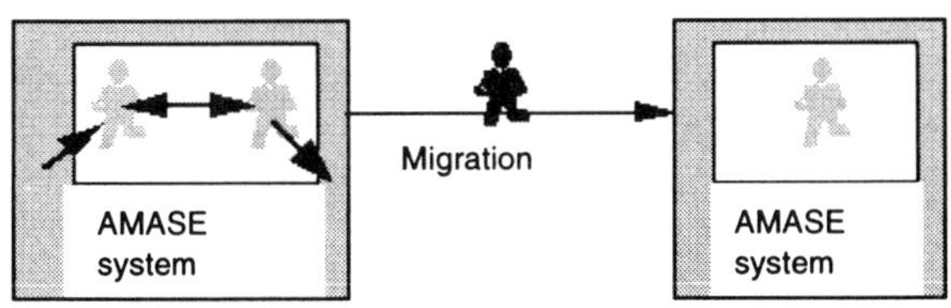

Figure 3.63 AMASE migration system.

Table 3.13

Secure Agent Execution

Malicious Entity	*Attacked Entity*	*Comments*
Agent	System	JAVA Sandbox, known security mechanism
System	System	Well-known security mechanism
System	Agent	Problem: Open research issue
		Known approach
		Like code obfuscating
		Manipulation checks
		Security log-in

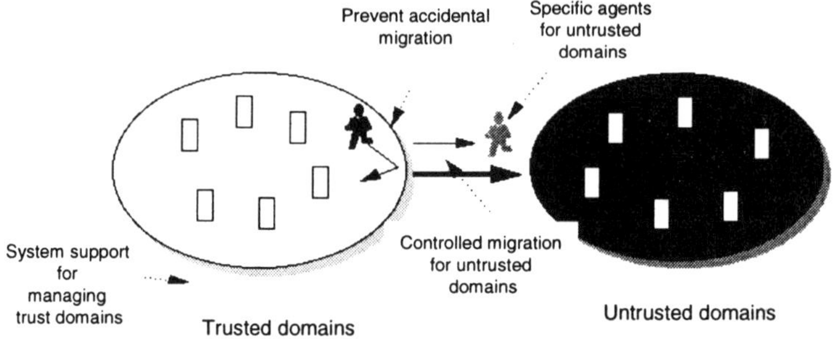

Figure 3.64 System support for managing trust domains.

R&D issues addressed by AMASE were related to customization of agents and involved the following:

- Exploration of the ease of customization via agents to adapt to user needs and application environment;
- Investigation of a basic development tool into an agent-based application to facilitate the composition and integration of agents;
- Investigation of the appropriate agent structure and agent execution model to allow an easy update of the agent code and data;
- Investigation of the appropriate communication language for typical application domains.

Major emphasis in the project AMASE was put on the development of a generic open communication platform based on agent technology. It enabled users with handheld mobile devices to access value-added services (VASs) over wireless networks in the AMASE agent environment. The open communication environment allowed service providers to create application-specific, customizable, and adaptive services. This allowed deployment, upgrade, and service changes to meet new requirements (see Figure 3.61).

An open communication service to the agent facilities was provided via the communication facilities (CF) application programming interface (API), through which the AF could have access, independent of any specific network technology. In particular, the CF TCP/IP networking services supported user mobility in agent operations by means of dynamic switching between networks.

Specifically based on existing networking middleware and the contributions from the OnTheMove ACTS project, AMASE managed to do the following:

- Support access to GSM, WaveLAN, and Ethernet, simultaneously if permitted by the system platform;
- Define telecommunications services abstraction based on UMTS service capabilities as a means of accommodation to current and future telecommunications networks in the CF API;
- Support telecommunications data, fax, and SMS services;
- Develop appropriate adaptation of supported networks to support the CF telecommunications services abstraction;
- Support TCP/IP networking over telecommunications data services;
- Add mobility support in the TCP/IP stack by means of mobile IP or OnTheMove technique;
- Monitor the status of the supported networks;
- Switch dynamically and cost-effectively between supported networks according to user mobility and the specified QoS requirement;
- Provide and simulate, if not readily available from the supported networks, geographic location information.

The agent system (AS) and the agent directories (AD) were the two entities defined in the agent facilities. The situation is depicted in Figure 3.65.

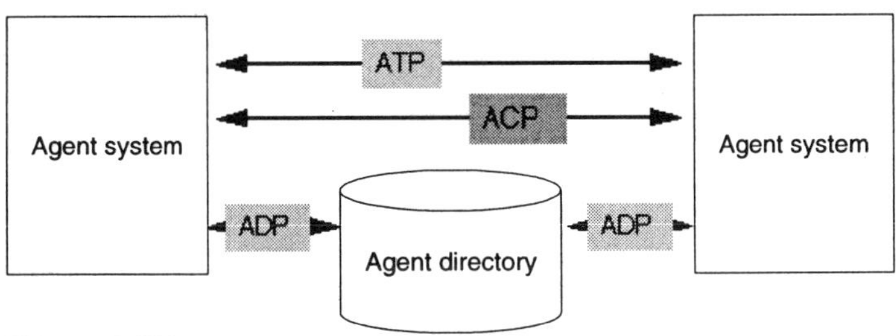

Figure 3.65 Agent facilities.

The AS provided an execution environment for both stationary and mobile agents via the agent facilities application programming interface (AF API). The idea included actions, such as the following:

- Look up, create, execute, and terminate agents;
- Communicate with and transfer agents;
- Access various agent operation security options;
- Obtain geographic location information.

The AF API also provided agents with access to telecommunications services by means of system agents within the AS. The agent directory maintained the directory of agents, so that they were accessible to other agents or applications. The concept of agent system and agent directory could be explored with the most general (and most complicated) configuration. Because of the resource constraints, but without losing the essence, the simplest configuration model (which included multiple agent systems), was adopted, and there was one and only one agent directory (centralized directory function). The agent facilities needed to establish the following protocols:

- Agent directory protocol (ADP) to support the directory update and interrogation;
- Agent communication protocol (ACP) to support the communication between agents;
- Agent transfer protocol (ATP) to support the transfer of an agent from one agent system to another.

The AMASE demonstrator implemented applications in the banking domain using the AMASE agent environment. Writing an agent-based application corresponded to creating a set of agents that together performed a required functionality. Mobile agents were responsible for the active parts of an application, whereas system agents provided access to local services. Both communicated using built-in communication mechanisms. Agent-based applications were inherently not monolithic because they relied on the cooperation of different independent objects. It was common practice to build agent-based applications by reusing existing agents.

The project AMASE introduced the concept of an agent package—a set of agent classes required to perform a specific task. These included classes for both mobile agents and system agents. Thus, an agent package could be thought of as the conceptual equivalent of a programming language module in the agent world.

Mobile agent-based access to real-life value-added banking services was implemented to demonstrate the feasibility of the AMASE approach for commercial applications. The associated AMASE application addressed three fundamental aspects of the service paradigm: service configuration, service access, and service-triggered notification. In the context of the project the Alpha Credit Bank in Greece granted the project members access to the Alphaline home-banking services, which could be accessed via the APIs of a dedicated server. It allowed customized client applications to invoke the basic Alphaline banking services, and the agent paradigm used allowed for rapid introduction of new services to the user.

The architecture of the AMASE application was partitioned in two systems—the external banking system and the AMASE agent environment. There were also two management domains, namely the banking domain and the user domain.

Figure 3.66 depicts the agent-based architecture for the case where a single user interacts with the application.

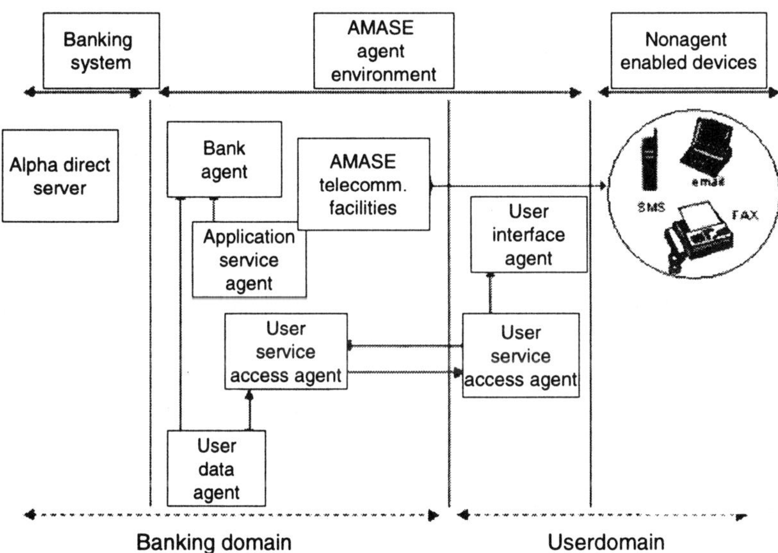

Figure 3.66 Architecture of the AMASE applications.

The demonstrator was designed to give special attention to user mobility, disconnected operation, user reachability, and automated notification. The implemented services were context-aware and used information contained in user profiles and supporting databases to locate currently used terminals and users at their respective geolocation. Before migrating to the user, agents had to specify their migration policy. This included negotiation of the resources needed on the destination platform/device. Furthermore, agents were able to behave according to the currently provided network QoS by using different interaction schemes with the mobile user.

Three agent-based value-added banking services provided realistic evaluation of the service and the technology:

- Funds transfer service allowing the user to transfer fund between bank accounts;
- Account balance update service;
- Share price information service, allowing the user to track a set of shares on the Athens Stock Exchange.

The AMASE approach demonstrated access to real-life banking services, and although no standardized agent system was used, it was expected to have an impact on the standardization activities within OMG MASIF. Furthermore, compliance with and integration of the WAP standard were exploited, and AMASE had evaluated results to provide recommendations for the WAP and OMG MASIF standardization activities.

In summary, AMASE managed to achieve the objectives of user-friendliness, accessibility, convergence, and QoS of the information society for scalable agent-based mobile access to multimedia information services. The key issues investigated by AMASE are listed as follows.

- Scaleable agent platform to investigate, develop, and support the operations in the agent platform according to the system resources;
- A flexible and modular agent platform software architecture to allow various degrees of functional richness;
- A scheme for the agent platform to deal with potential overloading of visiting agents and events;
- A minimal agent platform core, which could fit various handheld devices;
- Network adaptation—the protocols in the agent environment were enhanced to adapt to the available QoS of the accessed network (e.g., bandwidth and delay) and to allow reliable agent operations over sporadic wireless connections;
- Mobile operation to investigate a scheme for operations in the agent environment to continue seamlessly or recover while a device moved between networks and an agent operation failure notification scheme to allow recovery;
- Secure agent execution to investigate the adaptation of existing security mechanisms for agent communication to solve traditional security problems;
- Research on the protection of mobile agents against malicious agent systems;

- Customizable agents to explore and validate the ease of customization via agents to adapt to users' needs and application environments, and to investigate a basic development tool to facilitate the composition and integration of agents into an agent-based application.

3.4.4 Mobile MultiMedia Access Using Intelligent Agents—M3A

M3A was a project within the IT for Mobility Theme, which addressed the near-term technology and integration issues to permit users of information devices to begin to experience the ubiquitous multimedia communications services implied for the UMTS vision. This vision included high-bandwidth communications to low-power mobile terminals (with long battery life) over arbitrary distances. The UMTS vision portrayed a cellular network topology of medium bit-rate service cells interconnected by the broadband infrastructure. The Internet applications already demand high-bandwidth communication for good responsive service of complex, color Web pages and possibly real-time audio or video. For mobile access to the Internet, the problem is that no single technology exists to satisfy the bit rate/distance metric for both in-building (short-range) and wide area communications.

M3A planned to take multiple existing technologies and integrate them into a single mobile host platform. These technologies would be unified on the host system by communications software to permit data delivery over whatever path—if access was available. This multinetwork approach to multimedia access needed significant support from intelligence within the main public network as well as sophisticated data routing capabilities able to deal with multiple networks. The results included data-optimized modular implementations of existing communications technologies (GSM, DAB, and HIPERLAN) in formats suitable for mobile multimedia terminals together with agent intelligence in the network supporting the terminal mobility via filtering and rerouting over the multiple network types.

The M3A work plan consisted of a set of parallel technology tasks, which prepared communications subsystems for integration into the host platform; integration tasks that built the M3A terminal, and any required access devices; and service integration tasks, which tested and evaluated the performance of the M3A model of multimedia communications.

M3A attempted to make available network and device technology to permit true mobile multimedia services to be deployed well in advance of the availability of the first instantiations of the UMTS vision. The experience gained in such early implementations is well used to validate the market requirements currently envisaged for UMTS and hence improve the success of the early-to-be-deployed UMTS systems. A secondary, but equally important impact occurred from the delivery of IN services oriented toward individual users rather than network systems. Finally, the modular technology developed for the M3A terminal permitted a large variety of mobile applications to be more easily realized than was possible at the time.

The M3A consortium consisted of an excellent blend of equipment manufacturers, technology providers, and network operators. All the technology results of the project are to be exploited by the technology and equipment suppliers either as devices or as licensable technology. The network operators gained valuable insights in the implications of UMTS networks and users and practical, marketable services to support both current corporate user demands and demands for future spectrum requirements for mobile applications.

REFERENCES

[1] Pereira, J., et al., "From Wireless Data to Mobile Multimedia: R&D Perspectives in Europe," *Insights Into Mobile Multimedia Communication*, Academic Press, 1999.

[2] ETSI Web site, http://www.etsi.org/.

[3] Prögler, M., and C. Evci, "Air Interface Access Schemes for Broadband Mobile Systems," *IEEE Communications Magazine,* September 1999, Vol. 37, No. 9, pp. 106–114.

[4] Mohr, W., and R. Becher, "Mobile Communications Beyond 3G," *Wireless Personal Communications,* June 2001.

[5] Evci, C., "HIPERLAN: A Viable Solution for the Future Market of High Resolution Multimedia Services," *MBS Conference,* London, March 1997.

[6] Jajszczyk, A., "What Is the Future of Telecommunications Networking?," *IEEE Communications Magazine,* June 1999, Vol. 37, No. 6, pp. 12–20.

[7] UMTS Forum, "Enabling UMTS 3G Services and Applications," *UMTS Forum Report Number 11,* October 2000.

[8] Zeng, M., A. Annamalai, and V. Bhargava, "Recent Advances in Cellular Wireless Communications," *IEEE Communications Magazine,* September 1999, Vol. 37, No. 9, pp. 128–138.

[9] Callendar, M. H., "Future Public Land Mobile Telecommunication Systems," *IEEE Personal Communications*, April 1994, Vol.1, No. 4, pp. 18–22.

[10] Special Issue on IMT-2000, "Standards Efforts of the ITU," *IEEE Personal Communications,* August 1997, Vol. 4, No. 4, pp. 8–40.

[11] http://www.itu.int/imt.

[12] http://www.etsi.fr/ and http://www.tiaonline.org/.

[13] Baier, A., et al., "Design Study for a CDMA-Based 3G Mobile Radio System," *IEEE JSAC,* 1994, Vol. 14, pp. 733–743.

[14] DaSilva, J. S., B. Barani, and B. Arroyo-Fernandez, "European Mobile Communications on the Move," *IEEE Communications Magazine,* February 1996, Vol. 34, No. 2, pp. 60–69.

[15] Koipillai, D., S. Chennakeshu, and B. Gudmundson, "Evolution to 3G Cellular," *Proceedings ISWCM,* India, September 1998, pp. 27–34.

[16] Bull, D., N. Canagarajan, and A. Nix, *Insights into Mobile Multimedia Communication,* Academic Press, 1999.

[17] Ojanperä, T., and R. Prasad, eds., *Wideband CDMA for 3G Mobile Communications,* Norwood, MA: Artech House, 1998.

[18] Anderson, P.-O., et al., "Investigation in the FRAMES Project on the UMTS Radio Interface," *4th ACTS Mobile Communications Summit,* Sorrento, Italy, 1999.

[19] ETSI SMG, *Proposal for a Consensus Decision on UTRA,* ETSI SMG Tdoc 032/98.

[20] Nikula, E., et al., "FRAMES Multiple Access for UMTS and IMT-2000," *IEEE Personal Communications Magazine,* April 1998, pp. 16–24.

[21] Ojanpera, T., et al., "Comparison of Multiple Access Schemes for UMTS," *Proceedings 47th IEEE Vehicular Technology Conference,* Phoenix AZ, May 5–7, 1997, pp. 490–494.

[22] Prasad, R., W. Mohr, and W. Konhauser, *3G Mobile Communication Systems,* Norwood, MA: Artech House, 2000.

[23] Anderson, P. O., et al., "FRAMES Multiple Access: A Concept for the UMTS Radio Interface," *Proc. ACTS Mobile Communications Summit,* Aalborg, Denmark, October 7–10, 1997, pp. 470–475.

[24] Virtanen, T., E. P. Pereira, and J. Vialen, "Frames Radio Protocol Overview," *Proceedings ACTS Mobile Communications Summit,* Aalborg, October 7–10, 1997, pp. 678–683.

[25] Nikula, E., et al., "Flexibility in Providing UMTS Services with FRAMES FMA1," *Proceedings ACTS Mobile Communications Summit,* Aalborg, October 7–10, 1997, pp. 690–696.

[26] Makelainenn, T., et al., "FRAMES Demonstrator Implementation Roadmap," *ACTS Mobile Summit,* 27–29 November, Granada, November 1996.

[27] Richardson, K., et al., "The FRAMES Demonstrator First Implementation Steps," *Proceedings ACTS Mobile Communications Summit,* Aalborg, October 7–10, 1997, pp. 659–666.

[28] "Concept Group Delta Wideband TDMA/CDMA, Evaluation Report," Tdoc SMG2 368/97, Cork, Ireland, December 1997.

[29] Girard, L., et al., "Presentation of the Measurement and Demonstration Campaigns with the FRAMES TDD TD-CDMA Demonstrator," *4th ACTS Mobile Communications Summit,* Sorrento, Italy, 1999.

[30] Eldstahl, J., and A. Nasman, "WCDMA Evaluation System Evaluating the Radio Access Technology of 3G System," *Ericsson Review,* Vol. 76, November 2, 1999, pp. 48–55.

[31] 3GPP Web site, http://www.3gpp.org/.

[32] ETSI, The UMTS Terrestrial Radio Access (UTRA) ITU-R RTT Candidate Submission, ETSI SMG 2, Tdoc 260/98, June 1998.

[33] Chaudhury, P., W. Mohr, and S. Onoe, "The 3GPP Proposal for IMT-2000," *IEEE Communications Magazine,* December 1999, pp. 72–81.

[34] OHG: "Harmonized Global 3G (G3G) Technical Framework for ITU IMT-2000 CDMA Proposal," Operator Harmonization Group, May 1999, IRU-R, Beijing, June 1999.

[35] Irvine, J., "Report on MOSTRAIN Access Technique, Selection for SIG1," *European Commission: Document number AC104/USC/WP7/IN/I/080,* September 1996.

[36] Moody, P., "Final Report," *European Commission: CEC Deliverable Number: AC104/SMI/WP1/PI/460/b1,* February 1998.

[37] Vainikainen, P., et al., "Characteristics of the 2GHz Radio Propagation Channel in HST Environment," *ACTS Mobile Summit,* 27–29 November, Granada, Spain, 1996.

[38] UMTS Forum, "The Future Mobile Market," *Report Number 8,* March 1999.

[39] USECA D05, "Intermediate Report on Terminal Security," Version 1, December 1998.

[40] GSM 02.16, "Digital Cellular Telecommunication System; IMEI," *Version 5.0.0.,* November 1996.

[41] GSM 02.60, "Digital Cellular Telecommunications System, Phase 2+, GPRS Service Description, Stage 1," Version 5.2.0.

[42] GSM 03.60, "Digital Cellular Telecommunications System, Phase 2+, GPRS Service Description, Stage 2," Version 6.0.0, March 1998.

[43] Awduche, D., "MPLS and Traffic Engineering in IP Networks," *IEEE Communications Magazine,* December 1999, Vol. 37, No. 12, pp. 42–47.

[44] Prasad, R., J. S. Da Silva, and B. Arroyo-Fernandez, "Air Interfaces Access Schemes for Wireless Communications," *IEEE Communications Magazine*, December 1999, Vol. 37, No. 12, pp. 70–71.

[45] Ngo, C.Y., et al., "Contention-Free MAC Protocol for Wireless ATM," *ACTS Mobile Communications Summit,* Aalborg, Denmark, 1997, Vol. 2, pp. 728–733.

[46] Paskalis, S., et al., "Performance of Traffic Scheduling for Wireless ATM Networks," *ACTS Mobile Communications Summit,* Aalborg, Denmark, 1997, Vol. 2, pp. 751–756.

[47] Nee, R. van, et al., "New High-Rate Wireless LAN Standards," *IEEE Communications Magazine,* December 1999, Vol. 37, No. 12, pp. 82–88.

[48] Nee, R. van, and R. Prasad, *OFDM for Mobile Multimedia Communications,* Norwood, MA: Artech House, 1999.

[49] Nee, R. van, "OFDM Codes for Peak-to-Average Power Reduction and Error Correction," *IEEE Global Telecommunications Conference,* November 18–22, 1996, pp. 740–744.

[50] Aldis, J., et al., "Magic Into Reality, Building the WAND Modem," *ACTS Mobile Communications Summit,* Aalborg, Denmark, 1997, Vol. 2, pp. 775–780.

[51] Schoettner, M. et al., "WAND User Trials—Evaluation of a Wireless ATM System," *ACTS Mobile Communications Summit,* Sorrento, Italy, 1999, CD-ROM.

[52] 6D5 Deliverable, "User Trials-Magic WAND," *European Commission,* November 1998.

[53] Kassler, A., O. Schirpf, and P. Schulthess, "A Multimedia Transport System for Wireless ATM Networks—Architecture and Performance," *SPIEE Multimedia Systems and Applications,* Boston, MA, 1998.

[54] Schoettner, M., et al., "Application Sharing—Architecture and Performance Aspects," *ACTS Mobile Communications Summit,* Aalborg, Denmark, 1997, Vol. 2, pp. 587–592.

[55] Aldis, J., M. Althoff, and R. van Nee, "Physical Layer Architecture and Performance in the WAND User Trial System," *ACTS Mobile Communications Summit,* Granada, Spain, 1996.

[56] Vaisanen, A., J. Aldis, and U. Lott, "Radio Propagation Measurements at 5 GHz for Broadband Radio Local Loop Systems," *ACTS Mobile Communications Summit,* Sorrento, Italy, 1999, CD-ROM.

[57] ACTS WAND Project Deliverable 2D8, "Evaluation of the WAND System for Outdoor Point-to Multipoint Configurations," *European Commission,* August 1998.

[58] Oelsner, M., "Access Schemes and Modulation Techniques," *AC006/IMST/aaa/DS/P/076/al deliverable,* European Commission, September 1997, pp. 3–28.

[59] European Commission ACTS MEDIAN Project AC006, "Wireless Broadband CPN/LAN for Professional and Residential Multimedia Applications: Final System Design," December 1996.

[60] ATM Forum, ATM User-Network Interface (UNI) Specifications, Prentice Hall, 1995.

[61] Meulenhof, D. v.d., and P. F. M. Smulders, "Modelling Handover in a High-Speed Wireless ATM Based LAN," *ACTS Mobile Communications Summit,* Sorrento, Italy, 1999.

[62] Priscoli, F. D., and R. in't Velt, "Design of Medium Access Control and Logical Link Control Functions for ATM Support in the MEDIAN System," *ACTS Mobile Communications Summit,* Aalborg, Denmark, 1997, Vol. 2, pp. 739–744.

[63] Dam, C. van, et al., "MEDIAN 60 GHz Wireless Field Trials," *ACTS Mobile Communications Summit,* Sorrento, Italy, 1999, CD-ROM.

[64] Berkvens, W. A. H., "Performance of a Selective Repeat ARQ Scheme for Real-Time Services in the MEDIAN System," *ACTS Mobile Communications Summit,* Sorrento, Italy, 1999.

[65] Borowski, J. et al, "Is ARQ Needed in the Demonstrator MAC?—A New Demonstrator Radio Physical Layer Nonlinearity Study," *AC006/TUD/WP2/DN2_ARQ1 deliverable,* Dresden, Germany, April 1997.

[66] Kuiper, J., and W. van Herwijnen, "MEDIAN, Potential Applications for User Trials," *BEAM'IT Report Version 1,* December 1998.

[67] Overduin, R., et al., "MEDIAN User Requirement Study," *TNO Report FEL-97-C041,* February 1997.

[68] Progler, M., and C. Evci, "Air Interface Access Schemes for Broadband Mobile Systems," *IEEE Communications Magazine,* Vol. 37, p. 107.

[69] Lupo, G., et al., "A Model For Cellular Systems With Broadband Users: Effect of Users' Speed on Resource Allocation Efficiency," 4th *ACTS Mobile Communications Summit,* Sorrento, Italy, 1999, CD-ROM.

[70] Hoz de, A., and C. Evci, "Impact of AWACS Project on the Evolution of HIPERLAN," *ACTS Mobile Communications Summit,* Aalborg, Denmark, 1997, Vol. 2, pp. 851–856.

[71] Tolonen, M., and M. Araki, "AWACS Project Field Trial Platform," *2nd ACTS Mobile Communications Summit,* Aalborg, Denmark, 1997, Vol. 2, pp. 369–373.

[72] Evci, C., et al., "AWACS: System Description and Main Project Achievements," *4th ACTS Mobile Communications Summit,* Sorrento, Italy, 1999, CD-ROM.

[73] Hafezi, P., et al., " Indoor Channel Characteristization Measurements with Directional Antennas for Future ATM Wireless Access Systems," 2nd *ACTS Mobile Communications Summit,* Aalborg, Denmark, 1997, Vol. 1, pp. 246–251.

[74] Sun, Y., et al., "19-GHz Channel Characterization Measurements for Future ATM Wireless LAN Systems," *PIMRC Conference Proceedings,* Finland, September 1997, pp. 184–188.

[75] Nix, A., et al., "Overview of the AWACS Testbed," *COST 260,* Dubrovnik, December 1997.

[76] Melis, B., G. Romano, and M. Araki, "Link Level Simulations in AWACS Project," *2nd ACTS Mobile Communications Summit,* Aalborg, Denmark, 1997, Vol. 1, pp. 476–481.

[77] Melis, B., and G. Romano, "Link Level Simulations in AWACS Project II-System Enhancements," 3rd *ACTS Mobile Communications Summit,* Rhodes, Greece, June 1998, pp. 640–646.

[78] Rheinschmitt, R., et al., "AWACS MAC and LLC Functionality," *2nd ACTS Mobile Communications Summit,* Aalborg, Denmark, 1997, Vol. 2, pp. 745–750.

[79] Fujino, T., et al., "Design of Baseband Signal Processing Unit in SAMBA Trial Platform for MBS," *3rd ACTS Mobile Communications Summit,* Rhodes, Greece, June 1998, pp. 854–859.

[80] Fernandes, L., "Developing A System Concept and Technologies for Mobile Broadband Communications," *IEEE Personal Communications Magazine,* February 1995, Vol. 2, No. 1, pp. 54–59.

[81] Dinis, M., et al., "SAMBA: A Step to Bring MBS to the People," *ACTS Mobile Communications Summit,* Aalborg, Denmark, 1997, Vol. 1, pp. 495–501.

[82] Dinis, M., et al., "The SAMBA Trial Platform in the Field," *4th ACTS Mobile Communications Summit,* Sorrento, Italy, 1999, CD-ROM.

[83] Bedini, C. and L. Correia, "Cellular Planning in Urban Regular Structures for Millimetre Waveband Mobile Broadband Systems," *4th ACTS Mobile Communications Summit*, Sorrento, Italy, 1999, CD-ROM.

[84] Velez, F. and L. Correia, "Optimization Criteria for Cellular Planning of Mobile Broadband Systems in Linear and Urban Coverages," *2nd ACTS Mobile Communications Summit,* Aalborg, Denmark, 1997, Vol. 1, pp. 199–205.

[85] Fernandes, J., A. Marques, and J. Garcia, "Cellular Coverage for MBS Using the Millimeter-Waveband," *4th ACTS Mobile Communications Summit,* Sorrento, Italy, 1999, CD-ROM.

[86] Fernandes, J., and C. Fernandes, "Impact of Shaped Lens Antennas on MBS Systems," *PIMRC Conference Proceedings,* Boston, MA, 1998, pp. 744–748.

[87] Vornefeld, U., "Employing SDMA in Mobile Broadband Systems," *4th ACTS Mobile Communications Summit,* Sorrento, Italy, 1999, CD-ROM.

[88] Petras, D., and U. Vornefeld, "Joint Performance Of DSA++ MAC Protocol and SR/D-ARQ Protocol for Wireless ATM Under Realistic Traffic and Channel Models," *1st International Workshop on Wireless Mobile ATM Implementation,* Hangzhou, April 1998.

[89] Petras, D., "Medium Access Control Protocol for Wireless Transparent ATM Access," *IEEE Wireless Communications System Symposium,* Long Island, New York, November 1995, pp. 79–84.

[90] Kadelka, A. and S. Sippel, "The Internet Protocols in Mobile Broadband Systems," *4th ACTS Mobile Communications Summit,* Sorrento, Italy, 1999, CD-ROM.

[91] The Internet Engineering Task Force, "Classical IP and ARP Over ATM," *RFC 1577,* January 1997.

[92] ATM Forum, "LAN Emulation Over ATM Version 2.0," July 1997.

[93] Kadelka, A., et al., "B-ISDN Interconnection of a WATM Demonstrator," *International Zurich Seminar on Broadband Communications,* Zurich, February 17–19, 1998.

[94] The Internet Engineering Task Force, "CISCO Systems' Tag-Switching Architecture Overview," *RFC 2105,* February 1997.

[95] The Internet Engineering Task Force, "Ipsilon Flow Management Protocol Specification for Ipv4 Version 1.0," *RFC 1953,* May 1996.

[96] ETSI BRAN, "Broadband Radio Access Networks (BRAN); IP Convergence Layer for HIPERLAN 2 and HIPERACCESS," *DTS/BRAN-00240004, V.0.0.1,* April 1999.

[97] The Internet Engineering Task Force, "Specification of the Controlled-Load Network Element Service," *RFC 2211,* September 1997.

[98] The Internet Engineering Task Force, "Specification of the Guaranteed Quality of Service," *RFC 2212,* September 1997.

[99] The Internet Engineering Task Force, "Resource Reservation Protocol (RSVP)—Version 1 Functional Specification," *RFC 2205,* September 1997.

[100] ACCORD Annual Project Review Report Part 1, 2, November 1999.

[101] ACCORD Trial Strategy Description, WP3100, AC348-ITL-RS-031-al.

[102] Georganopolous, N., et al., "Asynchronous MAC Protocol with Polling for Wireless Access," 4th *ACTS Mobile Communications Summit,* Sorrento, Italy, 1999, CD-ROM.

[103] Sanchez, J., R. Martinez, and M. W. Marcelin, "A Survey of MAC Protocols Proposed for Wireless ATM," *IEEE Network Magazine,* November–December 1997.

[104] http://www.cpk.auc.dk/asap/cabsinet.htm.

[105] Flint, B., "CABSINET—A Flexible Cellular Broadband Architecture," *Annual Technical Audit, European Commission,* Brussels, Belgium, 1998.

[106] Pallis, E., et al., "Provision of the Internet and Digital MPEG-2 Programmes Via a Wireless Broadband Network," *4th ACTS Mobile Communications Summit,* Sorrento, Italy, 1999, CD-ROM.

[107] Kourtis, A., and C. Mantakas, Demonstrations Hardware Used in CABSINET (Oslo, London, Amalgi, Prague, and Athens).

[108] Kormentzas, G., et al., "A Cellular System for Providing Wireless Services," *3rd ACTS Mobile Communications Summit,* Rhodes, Greece, June 1998, pp. 672–677.

Chapter 4

Satellite Air Interfaces

4.1 INTRODUCTION

The applications foreseen for future broadband satellite communications correspond to the satellite component of an integrated global broadband system [1]. The latter also includes the terrestrial component of UMTS, and the multimegabit per second broadband services, mainly for low-mobility applications as discussed in Chapter 3. Such development has entailed in Europe the emergence of supranational operators, such as Unisource and Global One, aiming to provide customers with one-stop shopping services over large geographical areas [2]. Aside from regulatory considerations, technological progress in the field of satellite communication has reached a level of maturity that allows deployment of global systems with the capability of serving millions of user terminals. The introduction of digital technologies has repositioned space communications as an alternative to be taken into account. It is expected that for the space segment, the deployment of high-capacity global constellations of satellites supported by new technology will provide access to millions of user terminals, while for the ground segment, increasing availability of integrated devices allows for the manufacture of less expensive user terminals. With the evolution of multimedia service requirements, however, the lack of global coverage of a terrestrial infrastructure will constrain their provision, so it is important to identify the role of the satellites in this emerging market.

Availability of standards, many originally developed for terrestrial network applications, provide a favorable ground for the provision of multimedia services by satellite. It is expected that the world market for physical users of mobile and multimedia satellite services will be 11.5 million users by the year 2005, rising to 18 million by 2010 [3]. Satellite systems for digital TV broadcasting have built their success on the MPEG and DVB standards. Some of the novel systems proposed to provide personal communications services with satellites (S-PCS) partly exploit the developments of GSM (e.g., Iridium, Globalstar, and ICO), while other proposed systems (e.g., Skybridge) will offer multimedia services reusing as much as possible standardized solutions of terrestrial broadband networks, such as ATM.

The satellite component of UMTS (S-UMTS) will be positioned as the successor of the many global S-PCS systems, currently proposed to provide narrowband services, such as voice, fax, and low-rate data (4.8–9.6 Kbps). Forecasts for worldwide and European satellite UMTS/IMT-2000 users have been segmented into nonmultimedia users (those requiring only nonmultimedia services) and multimedia users [3]. It is expected that all the forecast mobile satellite services (MSS) multimedia users will be UMTS/IMT-2000–compliant, while only a portion of the MSS nonmultimedia users will be UMTS/IMT–2000 compliant.

For nonmultimedia users, services extend only to basic speech services (but at high quality, 8–16 Kbps) and low-speed data (9.6–16 Kbps). For multimedia users, the requirements are for a variety of different services and applications. The forecasts are summarized in Figure 4.1.

	Worldwide		EU	
Year	**2005**	**2010**	**2005**	**2010**
MSS subscribers (000s)				
Nonmultimedia	4,875	7,500	609	938
Multimedia	6,585	10,975	395	659
	11,460	18,475	1,004	1,596
Average usage per subscriber (kBs per month)				
Nonmultimedia				
Voice	8,709	8,491	8,709	8,491
Low-speed data	6,208	5,587	6,208	5,587
Multimedia				
Voice	1,194	1,561	1,194	1,561
Low-speed data	2,584	3,380	2,584	3,380
Asymmetric	26,154	34,247	26,154	34,247
Interactive	1,781	2,334	1,781	2,334
Total annual traffic (million MBs)				
Nonmultimedia				
Voice	509	764	64	96
Low-speed data	491	736	45	63
Multimedia				
Voice	94	206	6	12
Low-speed data	204	445	12	27
Asymmetric	2,067	4,510	124	271
Interactive	141	307	8	18
Total	3,506	6,968	259	486
Annual traffic (million MBs)—excluding nonUMTS/IMT-2000-compliant traffic				
Nonmultimedia				
Voice	34	123	4	15
Low-speed data	33	119	3	10
Multimedia				
Voice	94	206	6	12
Low-speed data	204	445	12	27
Asymmetric	2,067	4,510	124	271
Interactive	141	307	8	18
Total	2,573	5,710	158	354

Figure 4.1 Forecast for the worldwide satellite market [3].

The implementation of S-UMTS will not be constrained to any particular type of orbit [geostationary (GEO), low-Earth (LEO), medium-Earth (MEO), or elliptical] [4]. Rather, it is likely that no particular orbit in the future will meet singularly the QoS requirements of the users for the broad range of services and applications to be provided. Also important are the increasing bandwidth requirements and the restricted capacity of lower frequency bands [5]. Satellite-based systems providing mobile services are termed MSSs or S-PCNs.

The development of global satellite systems drives technology developments in a number of areas:

- *Space segment:*
    - High-capacity satellites (multibeam, multigigabit per second);
    - Extensive use of advanced antenna systems;
    - Extensive use of onboard digital technologies and components;
    - Highly adaptive systems capable of allocating power to where the traffic is;
    - Integration with terrestrial telecommunication networks and adoption of terrestrial standards, where appropriate.
- *Ground segment/terminal:*
    - Cheap integrated terminals, with low voltage low power integrated DSPs and ASICs;
    - Multimode terminals for dual-use terrestrial-satellite;
    - Software reconfigurability of mobile handsets;
    - Utilization of satellite diversity for fixed terminals used with broadband constellations;
    - Utilization of higher frequency bands (Ka, EHF, and V-band);
    - Maximum standardization above the pure physical layer (waveform specific to each satellite system);
    - Systematic use of robust and adaptive modulation and coding;
    - Packet/cell type transport.

Finally, interworking/interoperability requirements are becoming more and more important as the systems become global and are designed to support user applications seamlessly with terrestrial infrastructure.

In the scope of the ACTS program, a number of projects dealt with S-UMTS development. The projects spanned technical and economic areas because both influence the performance of the UMTS service. Also, terrestrial concepts and standards are evolving quickly and are not fully reflected in the current developments taking place in the satellite communication environment. The following developments can be mentioned:

- TMN concept for network management, not used in the satellite communications world;
- Telecommunications information networking architecture (TINA) and the common object request broker architecture (CORBA), the intelligent-agent developments allowing for customized service implementation;
- The use of the GRAN approach to separate the radio-independent part from the radio-dependent part of the access network;
- Reconfigurable multiservice provider environment;
- Operator versus service roaming.

These concepts, protocols, and environment need more and more to be supported by global satellite systems, which are likely to be connected to a terrestrial infrastructure, operating and being managed according to such standardized approaches. Figure 4.2 illustrates the European position and participation in the R&D process towards global satellite systems, and Table 4.1 outlines some of the key spacecraft technologies and compares the European and U.S. status on some of these.

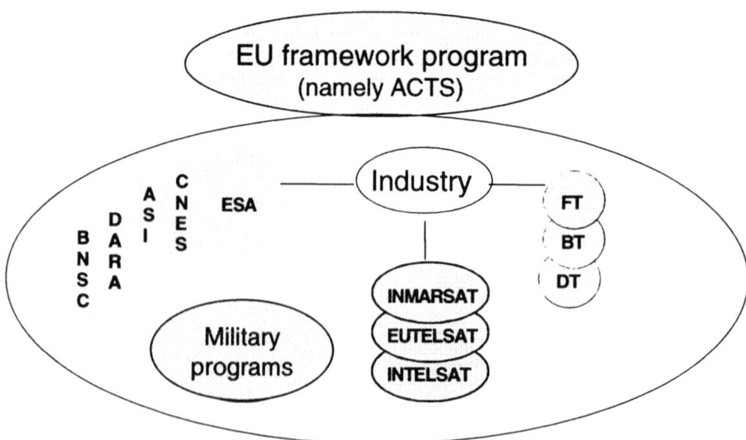

Figure 4.2 Satellite communications in Europe.

Table 4.1

Comparative Status for Some Key Spacecraft Technologies

Technology	*Used for*	*U.S. Status*	*European Status*
Multispot beam antennas	Concentration of power on ground, frequency reuse, and system capacity maximization	Commercially available through Raytheon at L band	Prototypes developed through ESA and CNES
Onboard switch	Connectivity between spot beams	Available, developed by under the "Star Wars-SDI" military program and under the ACTS space program; used in Iridium; planned in major U.S. constellations	Limited prototype developed by Alenia Spazio under ESA contract, proposed to be flown in the Euroskyway system and in the Matra Marconi Space (MMS) wideband European satellite communications (WEST) system for broadband multimedia services
High power platforms	High-power system allowing size reduction of user terminal	16-kW platforms announced	Commercial platforms have capabilities up to 10 kW and plans to develop a 15-kW platform (MMS, Aerospatiale)
Intersatellite links	Connectivity of spacecraft of the constellation	Developed and tested under military SDI program	First generation of optical ISL developed by MMS through ESA contract to be flown in 2000; 2G ISL required for constellations in early R&D stage
Large antennas	Mobile GEO systems needed to generate narrow spot beams at mobile frequencies	Commercially available	40 M ECU R&D-invested through ESA and CNES

This chapter addresses the ACTS projects contributing to the development of satellites. First, the project SINUS is discussed. It aimed at the development of a software simulator to allow the validation of complex specifications, such as the satellite air interface and to investigate the degree of integration with the terrestrial component of UMTS (project FRAMES, see Chapter 3) and existing networks (project RAINBOW, see Chapter 5). The project defined the services and applications for the UMTS space segment, particularly the services to be implemented in the testbed for trials.

The work of SINUS was complemented by the project TOMAS. It used previously developed equipment together with specific satellite facilities, primarily procured through the Italsat and Inmarsat operators, to propose an open platform to host projects that test equipment developed within the ACTS program and using real satellite channels.

Higher frequency bands were also investigated to provide broadband mobile satellite services with low mobility. The project SECOMS/ABATE examined the Ka- (20–30 GHz) and EHF- (40–50 GHz) bands, and the techno-economic feasibility of a new satellite-system generation for the provision of broadband services to fixed and mobile small-size terminals on continental-wide coverage [6]. Finally, the convergence of existing technologies for the provision of future applications and related problems are discussed in the scope of the project COIAS.

4.2 SATELLITE UMTS

Satellites can provide coverage to remote, rural, and maritime areas where it would be too expensive or impossible to set up terrestrial infrastructure. The forecasts for the worldwide satellite market have already been presented in Figure 4.1. This evolution is illustrated in Figure 4.3 for the S-PCS case.

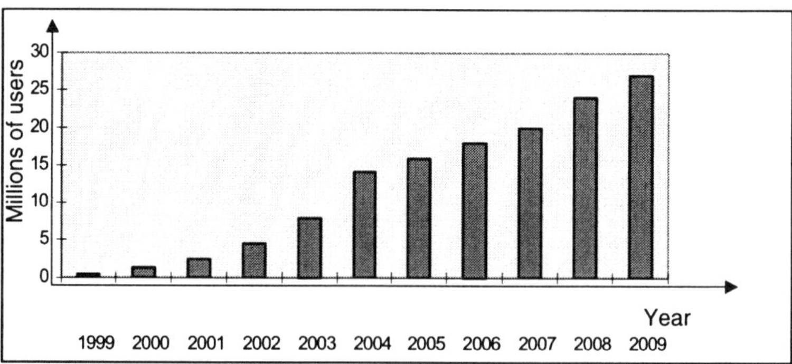

Figure 4.3 Evolution of S-PCS-terminal market.

The evolution of 2G S-PCS systems is expected to materialize through the UMTS system, which, at the satellite level, will provide for data rates up to 144

Kbps; a tenfold improvement relative to first-generation S-PCS systems. The UMTS Forum evaluated the market potential of S-UMTS, and the figures listed in Table 4.2 were derived from an independent market analysis of the satellite working group [2].

Table 4.2

UMTS Forum S-UMTS Market-Potential Evaluation (millions of users)

	2005			2010		
Type	Business	Nonbusiness[b]	Total	Business	Nonbusiness[b]	Total
Handheld voice and low-speed data	2	1	3	5[c]	2.5	7.5
Multimedia	1	0.5	1.5	2	1	3
Total	3	1.5	4.5	7	3.5	10.5

Note: Forecast users do not include potential terrestrial replacement situations (i.e., where satellite is used to create a hub/base station to provide localized terrestrial area). It also excludes provision of temporary service to terrestrial infrastructure.

It is expected that the S-UMTS market will justify the development of several systems, especially if satellite terminals can be developed at marginal cost as an upgraded version of a terrestrial terminal. From an R&D point of view, development of S-UMTS requires availability of key technologies, such as those listed in Table 4.1 and in some areas it will push the technology limit one order of magnitude higher than in the S-PCS case. S-UMTS is intended to complement the terrestrial networks in terms of coverage and capacity [7]. During the initial deployment phase when terrestrial coverage will be limited, satellites will play the key role to provide regional and global services especially in areas where there is no seamless terrestrial coverage. In fact, it needs to be understood that where GSM is available, UMTS will use it as a fallback system. Without the need for a satellite component GSM is almost a global system.

The current standardization of the UMTS air interface is based on WCDMA- and TDMA-access schemes. UMTS-services provision via existing satellites will have to share the transponders with current services. Hence, the access schemes and air interfaces must be compatible with the operational systems in terms of bandwidth and power requirements and must provide interworking capabilities with existing access control systems.

A key point for 3G systems is service flexibility [8]. The definition of service types is done in a new and more generic way relative to 2G systems. 3G services are defined in terms of bit rate, bit error rate, and maximum transfer delay requirements. With the definition of S-UMTS, a common set of satellite specific services integrated into the UMTS network shall be defined allowing to create a system-independent compatible basis for present systems willing to adapt and future satellite systems.

The services envisioned for this common basis can be subdivided into services to be provided by handheld terminals and services to be provided by portable terminals. Table 4.3 is a proposal for S-UMTS bearer services as they could be defined taking two terminal types and a two-step deployment time frame for the handheld terminals into account.

Table 4.3

Proposal for S-UMTS Bearer Services

Bearer Service (Kbps)	*Handheld*	*Terminals*	*Portable Terminals*
	Medium	*Long-Term*	
Low-speed	0.3–4.8	0.3–64	1.2–64
Medium-speed	—	64–144	N x 64 (n = 1, ..., 6) asymmetric
High-speed	—	—	N x 64 (n = 6, ..., 32) asymmetric

The following teleservices compatible with UMTS services can be provided:

- Voice services of various quality;
- Paging, fax, SMS, and positioning;
- Multimedia (e.g., H.32M and MPEG-4);
- Internet/intranet access;
- LAN interconnection.

These services fulfill the need of most of the industrial and private applications and give a perspective for the long term.

The ACTS S-UMTS projects have implemented a wide range of services, which are representative for S-UMTS. SINUS and INSURED concentrated on low-speed services provided by handheld terminals and their integration into UMTS, whereas the TOMAS testbed provided high-speed services. The focus of the latter project was also on the implementation of high-speed services, realization of highly portable or even mobile terminal equipment, and their application in field trials using GEO satellites.

4.2.1 Novel Satellite Mobile Application—The Project SINUS

One of the main objectives of the project SINUS was the investigation of different and novel radio access, modulation, and coding schemes through a simulation testbed. Basis for the research work carried out through the project were the results of previous projects in the scope of the RACE II program. It used the results of, namely the project SAINT to implement S-UMTS specifications [9].

SINUS was a service-oriented project that aimed at defining and demonstrating an end-to-end communications network incorporating a remote-user terminal, satellite and terrestrial channels, and an access network. All elements were studied and defined, including satellite and terrestrial interfaces and network procedures.

SINUS emulated a fully representative satellite channel, removing the risk of physical satellite unavailability. Any satellite constellation and the ability to characterize satellite channels through adaptive simulation could be moduled/emulated. For this reason, the project proposed and implemented a testbed demonstrator of the satellite component. The basic concept of the proposed demonstrator is illustrated in Figure 4.4.

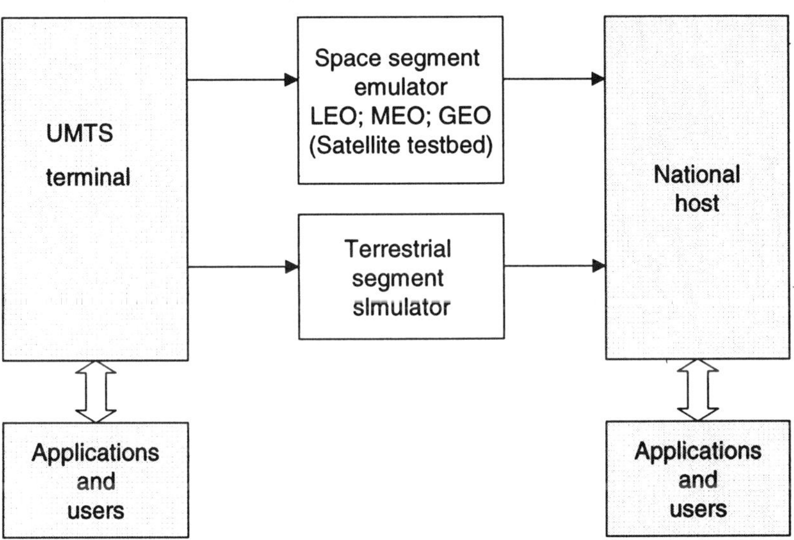

Figure 4.4 Overview of the SINUS system demonstrator.

Another key issue of the SINUS project was the achievement and maintenance of service flexibility. The project approach considered data rates up to 144 Kbps and to provide for such W-CDMA was chosen as the most appropriate radio-access scheme for the satellite segment, adopting a variable spreading rate for the uplink link and a multicode option for the downlink [8]. Several parameters, such as coding and framing, however, were studied and evaluated. For the analysis of the radio access of the return link, a Walsh-Hadamard block code/orthogonal modulation with rate 64/6 as a convolutional encoding, performed by puncturing a code with different rates and constraint length 7 or 9 at the initial phase, was proposed. The convolutional codes were concatenated, with shortened Reed-

Solomon codes, whose rate was varied according to the service. The overall processing gain was obtained by combining the codes, the Walsh-Hadamard code/modulation block and PN sequences with different processing gains. As the access scheme should offer good performance in several environments and with different constellations, a variable interleaving depth and power control with a variable step was proposed.

4.2.1.1 SINUS Satellite Requirements Towards FRAMES Multiple Access

For the necessity of as many as possible common features between the terrestrial- and satellite-radio interfaces, cooperation between the projects SINUS and FRAMES was established. The coordination of the research work of the two projects was set to define common baselines, such as bandwidth and clock for W-CDMA, so that a SINUS terminal could be realized to ensure the interoperability of the terrestrial and satellite UMTS segments.

The starting point of the cooperation was the definition of top-level SINUS requirements [10]:

- Candidate services characteristics:
    - Data rates (up to 64 Kbps with a possible extension to 144 Kbps) and associated BER and delays (see Table 4.4);

Table 4.4

Candidate Service Data Rates

Candidate Service Type	*Typical Source Bit Rate (Kbps)*	*BER*	*Max End-to-End Delay (ms)*
Speech	2.4–64	10^{-3}–10^{-4}	400
Audio	2.4–64	10^{-5}–10^{-6}	400
Data	2.4–144	10^{-5}–10^{-6}	500
Text	20	10^{-5}	500
Image	2.4–144	10^{-5}–10^{-6}	400
Video	8–144	10^{-6}	400

Table 4.5

Link Budget Example for the User Uplink (Handheld-to-Satellite)

	LEO (*h = 780 km*)	*ICO* (*h = 10,385 km*)	*GEO* (*h = 35,863 km*)
Uplink frequency (GHz)	2.01	2.01	2.01
Range (km)	2,464	13,636	40,042
Handset peak power (dBW)	3 (TDMA) -3 (CDMA)	3 (TDMA) -3 (CDMA)	3 (TDMA) -3 (CDMA)
Handset average power (dBW)	TBD (TDMA) -3 (CDMA)	TBD (TDMA) -3 (CDMA)	TBD (TDMA) -3 (CDMA)
Handset antenna gain (dBi)	0	0	0
Handset peak EIRP (dBW)	3 (TDMA) -3 (CDMA)	3 (TDMA) -3 (CDMA)	3 (TDMA) -3 (CDMA)
Antenna-body loss (dB)	3	3	3
Range loss (dB)	166.3	181.2	190.6
Atmospheric + rain + polarization loss (dB)	0.7	0.7	0.7
Satellite system temperature (K)	480	480	480
Noise spectral density (dBW/Hz)	-201.8	-201.8	-201.8
Satellite antenna gain edge (dB)	Gs TBD (ex. 16.8)	Gs TBD (ex. 31.9)	Gs TBD (ex. 38.8)
Received power (dB)	-167.0+Gs (TDMA) -170.0+Gs (CDMA)	-167.0+Gs (TDMA) -170.0+Gs (CDMA)	-167.0+Gs (TDMA) -170.0+Gs (CDMA)
Satellite diversity gain (dB)	TBD	TBD	TBD
Bit rate (Kbps)	TBD > 64 (TDMA) 1 - 64 (CDMA)	TBD > 64 (TDMA) 1 - 64 (CDMA)	TBD > 64 (TDMA) 1 - 64 (CDMA)
Bit rate (dB)	> 48.1 (TDMA) 30 - 48.1 (CDMA)	> 48.1 (TDMA) 30 - 48.1 (CDMA)	> 48.1 (TDMA) 30 - 48.1 (CDMA)
Eb/No (dB)	TBD	TBD	TBD
Margins (dB)	TBD	TBD	TBD

- Provision of VBR services;
- Provision of different levels of BER.
- Link budget: In a satellite environment the path loss depends on the adopted constellation, and the satellite antenna gain depends on the antenna size, which cannot be too large. Some examples of budgets for different constellations are given in Table 4.5.

The following parameters have been identified as key ones for SINUS demonstration and are important for the offer of a right multiple-access scheme:

- Capacity (not defined for the SINUS project);
- Signaling transmission capability;

- Channel bit rate flexibility;
- Satellite diversity;
- Satellite and beam handover;
- Doppler and synchronization acquisition and reacquisition;
- Coding and modulation;
- Power control and MT power limitation;
- Implementation complexity/feasibility.

For the aspects related to radio-resource management, like diversity and satellite/beam handover, several solutions were proposed [10]. One was the introduction of the pseudocell concept (splitting the satellite coverage in several reduced coverage cells, called pseudocells) to limit the differential Doppler and delay inside such a cell. The other was the application of dynamic channel allocation techniques or accurate frequency planning to reduce interference and to cope with overlapping satellite coverage areas. The project provided a preliminary summary of the major requirements towards a TDMA solution.

Another big limitation in satellite communication is the available power. As a consequence, the signaling power has to be minimized to optimize capacity, which is valid for both TDMA and CDMA schemes. In particular, with CDMA it is necessary to reduce the number of physical channels used in each connection to reduce both the interference among different physical channels of the same connection and interference among different users. Under this hypothesis, it was considered inappropriate to adopt a reference signal to allow channel estimation and coherent reception. In this case, it was considered more appropriate to adopt noncoherent reception, although this solution introduced loss in terms of E_b/N_o and with respect to coherent reception. The same minimization for the reference signal power was considered necessary both in the uplink and in the downlink.

To reduce intraspot interference, using synchronous CDMA in the downlink was recommended. The adoption of a RAKE receiver was considered unnecessary, since the propagation channel in the most common environments for S-UMTS usually presents a very low delay spread. Its use, however, was considered for the exploitation of satellite diversity and to make interbeam and intersatellite handover easier. Satellite diversity was regarded as similar to terrestrial macrodiversity since performance could be improved, and it allowed for the reduction of shadowing margins.

Other parameters, described as critical, were related to synchronization, acquisition, and reacquisition. In general, the pseudonoise (PN) sequences to be used have to be small to speed up synchronization, but they must be sufficiently long to allow discrimination of signals coming from different satellites.

4.2.1.2 SINUS Testbed Architecture

To prove the feasibility of the integration of a satellite segment with terrestrial UMTS, a laboratory testbed was implemented [11]. It was composed of a MT, a satellite, and a terrestrial access network. The satellite component was composed of a satellite channel emulator (SCE) and a fixed Earth station (FES), while the terrestrial segment simulator (TSS) realized the terrestrial component. Under the demonstration activity, the TSS contributed to the assessment of the impact of intersegment handover on the QoS and to the derivation of reference models for the ISHO and the mobility/intelligence network services. Figure 4.5 gives a more detailed look at the SINUS testbed architecture [12].

The system architecture of the SINUS testbed was based on the GRAN concept, as proposed by the project RAINBOW (see Chapter 5). The approach was to develop a generic interface between the higher protocol layers of the UMTS-access network and the lower layers, which were radio-technology dependent and provided the ability to connect different radio-access modules (both 2G and 3G systems).

The MT was virtually a dual-mode terminal. It had been totally developed for the satellite mode and partially developed for the terrestrial part (the functionalities implemented concerned only the upper layers in order to demonstrate handover from a satellite spotbeam to a terrestrial cell and from a terrestrial cell to a satellite spotbeam).

The MT supported the following.

- Call establishment using a selection of multimedia applications;
- Handover within the satellite segment.

The FES allowed a connection from the mobile terminal to the network. Radio access and fixed network transport functions used different protocols and consequently, transport-interworking functions in the FES needed to be implemented to allow end-to-end connection. The terrestrial access network was composed of a simulator that contained only functionalities used to execute ISHO. The radio components were not simulated.

The testbed included, also, a core network composed of an ATM switch and two workstations: one containing software to control service and mobility aspects and the other to allow control of the switch and conversion of UMTS call control to Q2931 call control.

As already mentioned, the SINUS architecture was based on the GRAN principle of separating the radio access–dependent parts from the radio access–independent parts. The protocol stack was based on the layering concept of the OSI model. The detailed description of the protocol stack is given in Figure 4.6, and it includes the user plane and the control/management plane.

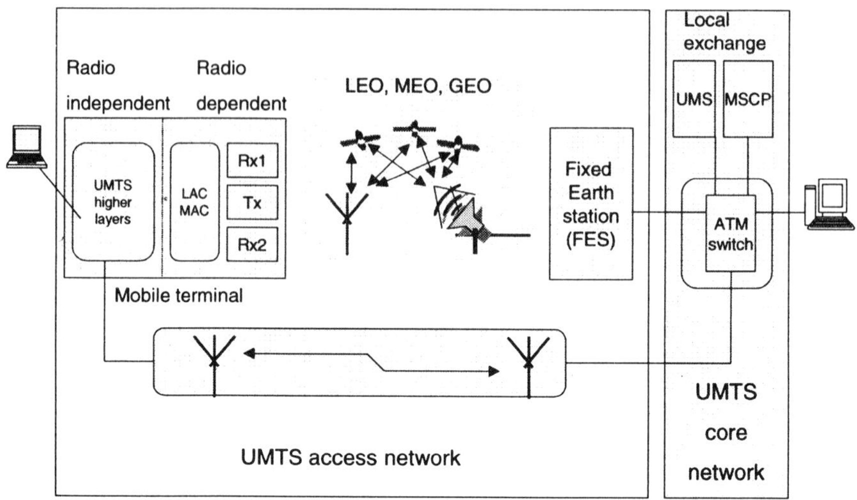

Figure 4.5 SINUS testbed architecture.

The structure used common protocols for radio layers (layer 1 and 2) but was different for signaling (control and management plane) and user data above the radio layers. User data was transported using a network layer and for signaling data the routing function was used in the signaling network layer. As underlined in Figure 4.6 the radio subsystem protocol stack was based on the principle of separating the radio-access–dependent parts from the radio-access–independent parts.

The separation between the dependent and independent parts was made at the data-link layer (layer 2). This layer was split into two sublayers, one radio-independent [link access control (LAC)] and one dependent (MAC). The radio-independent layers had to be defined for all radio interfaces and did not contain radio-dependent functions. Some functions, however, implemented in these layers needed to be configurable in order to be adapted to specific radio interfaces:

- *Application layer.* In addition to the application and services functions of the user plane, this layer contains functions, such as call control, mobility management, and radio-resource management (control and management plane). Call control and mobility management were mainly independent of the radio interface but needed to be adapted to work with the satellite and the terrestrial components. Furthermore, the radio-resource management would be different for the satellite and the terrestrial components; therefore availability of a configurable module seems important.

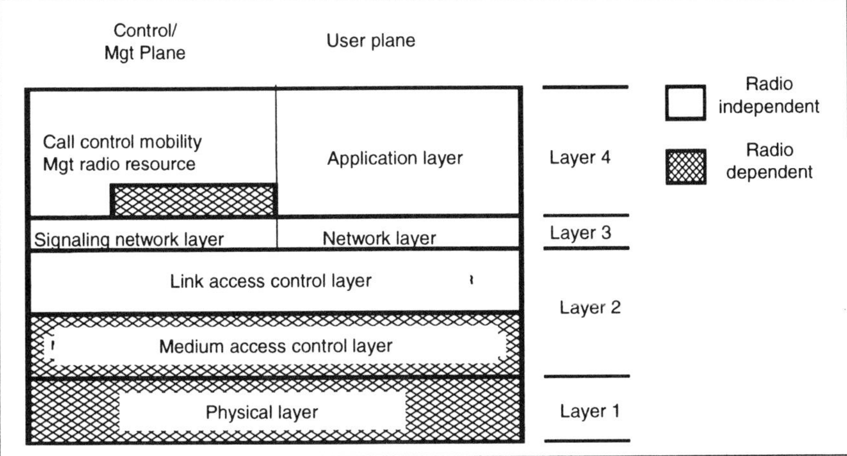

Figure 4.6 Transmission-dependent and -independent parts of the protocol layers.

- *Network layer.* The control and management planes of this layer contain routing functions (signaling network layer). The service offered by the SNL is similar to that of the OSI layer 3. The main role of the SNL is to route the signaling messages between different entities, which were not located in the same node (MT, FES, and so forth). Functions performed by the SNL depend on the type of node the SNL entity resided on (MT, FES, UMS, and so forth). Regarding the user plane, the network layer is the interface towards the user services for the purpose of transporting user data.
- *Link access control layer (LAC).* This layer aims to control and maintain the quality of the connection as required by the supported service (e.g., via flow control, retransmission).

The radio-dependent layers included the following.

- *MAC.* This layer means adaptation of the signaling data, the user data, and the common control information exchanged between the transmitter and the receiver parts to the physical layer format within the imposed logical channel structure.
- *Physical layer.* It provides a radio link over the radio interface(s), characterized by its throughput and data quality.

The TSS was considered a key component of the SINUS testbed because it was required to assess the impact of ISHO on the QoS requirements. The ISHO protocol was implemented in the TSS for both space-to-terrestrial and terrestrial-to-space directions [11]. The functionalities and the associated entities required to implement ISHO were located in the application layer (layer 4). The basic function of the signaling network layer (layer 3) was to provide routing for the messages to/from the TSS or to retransmit messages to/from the mobile terminal and fixed network. A description of the TSS's SNL structure was provided along with the SNL functionalities located in the TSS, whose target was to perform the signaling tasks for the ISHO.

4.2.1.3 Terrestrial Segment Simulator and ISHO

To achieve the project's goal of validating the integration of the satellite component into the terrestrial UMTS network, an important target for the TSS was to perform the signaling tasks for the ISHO. The first step was the investigation of the applicable algorithm for the ISHO, which fulfilled the system requirements and reduced the signaling delay by minimizing the signaling requirements. Within the SINUS project, backward mobile-assisted handover (MAHO) with signaling diversity was derived and chosen as the optimum ISHO algorithm for the integrated UMTS environment. The advantages and constraints of the handover algorithm were described in detail in [13].

The MT and the TSS were developed on the same hardware platform in the signaling implementation terminal (SIT). The idea is depicted in Figure 4.7.

The TSS was connected via an ATM adapter interface to the LEX and via an Ethernet link to the central control unit (CCU). The upper layers were developed by means of appropriate simulation software, while the SNL and application layers contained the ISHO signaling protocols and the application entities used to host the protocol. A PC was chosen as the development platform. Figure 4.8 shows the organization of layer 4 in the TSS.

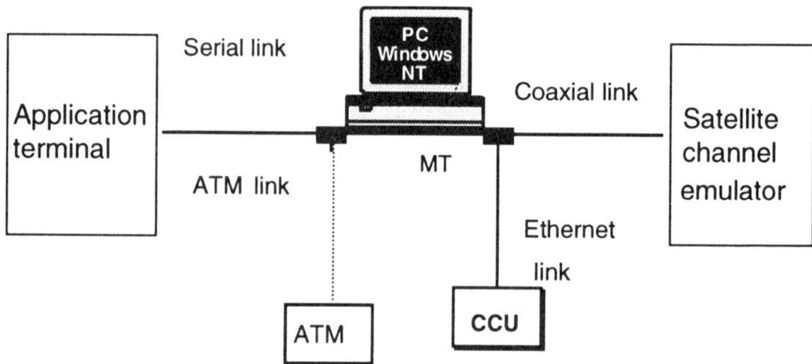

Figure 4.7 Internal architecture of the signaling implementation terminal [12].

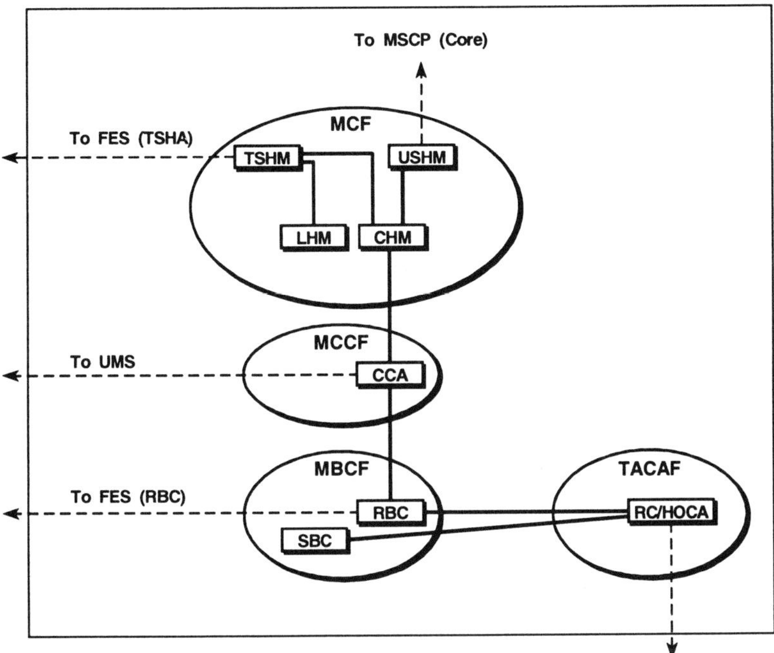

Figure 4.8 Application layer in the TSS.

In the TSS, two functionalities related to the ISHO were identified: radio-bearer control functionality (RBCF), responsible for the control of the radio bearer resources in the TSS, and radio-access control functionality (RACF), which provided the control functionality required in the TSS. For each functional entity, a state diagram was identified, and the required interfaces between functional entities were specified.

Both the RBCF and RACF functionalities included a number of functional entities that interacted with each other in order to perform the required signaling exchanges for the ISHO protocol. The interfaces between the entities were classified as internal or external; the former related to interfaces between entities within the TSS, and the latter referring to entities between interfaces within the TSS and those located in other network elements.

In the project SINUS, the RBCF located in the TSS was only involved in the handover procedures, and more precisely in the ISHO execution phase. It was assumed that the bearer connections had already been set up and that the role of the RBCF in the TSS was to trigger the establishment of the new channel and the release of the old one during an ISHO handover. With the above roles, the RBCF only included the radio-bearer control (RBC) entity. However, because in SINUS there was no radio bearer and radio interface between the MT and the TSS, the RBC did not perform any real functionalities [14].

The RACF provided the control function required in the TSS. Its role was to coordinate the actions to be performed by other functional entities for radio resource allocation/deallocation, radio-bearer connection, handover monitoring, and handover execution. It performed several functions, regarding the handover control [15]:

- Indication of relevant data needed for radio-resource allocation request;
- Processing of data (decision of handover, monitoring);
- Coordination of the bearers control by providing a reference that will be used by the RBCF and BCF to cross-connect the connection element.

In the TSS, the RACF was not fully implemented because of system constraints (no bearer between MT and the BTS) and priority of tests. The following functional entities were implemented in the RACF:

- Target cell and connections network (TCCN);
- Handover decision (HD);
- Resource allocation (RA);
- Handover initiation (HI).

The initiation of the handover in the SINUS testbed was performed with the "push-button" solution; therefore the functionalities of the HI were reduced to a minimum. The functional entities TCCN and RA were implemented as simple databases, and their description was not considered essential.

The project devoted special attention to the description of the HD for both directions of the ISHO: satellite-to-terrestrial (HD_{new}), located in the $MSCP_{new}$ and terrestrial-to-satellite (HD_{old}), located in $MSCP_{old}$. The state diagrams for both entities are depicted in Figure 4.9(a, b). The HD entity considered in the TSS included not only functionalities related to the decision of the handover, but also control functionalities performed during the execution phase of the handover. It is important to mention in particular the SNL layer. Its basic functions are [16] listed as follows.

- Routing of messages between noncolocated functional entities independent of the adopted radio-access technique;
- Extracting the location of the mobile terminal from the application-layer functional entities;

- Extracting of the signaling applications from mobility-related aspects.

The SNL was composed of three distinct network layer entities; the $SNL_{terminal}$, SNL_{fixed}, and SNL_{edge}. A detailed description of these entities can be found in [16]. For the SINUS project the TSS includes the SNL_{edge} entity [14].

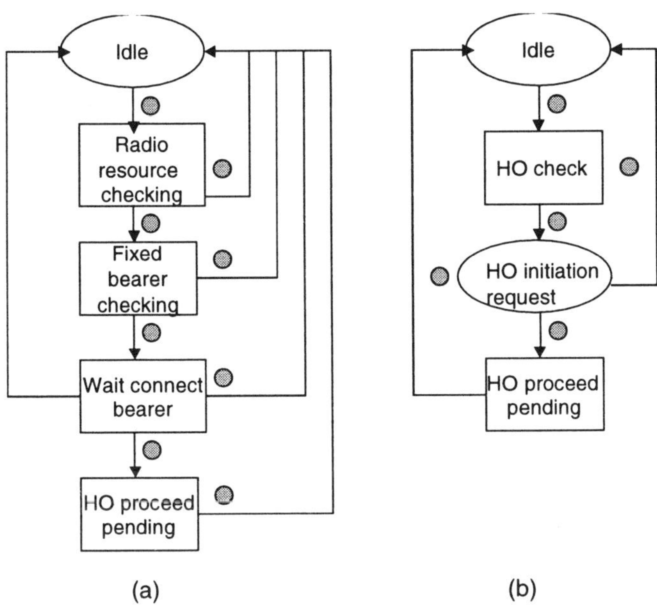

Figure 4.9 (a) HD_{new} state diagram and (b) HD_{old} state diagram.

From a signaling point of view, the SNL in the TSS had to communicate with the SNL peer entities located in other SINUS components (MT, FES, MSCP) and the application layer in the TSS. The constraints of the SINUS testbed defined the following functions of the SNL within the TSS:

- Assumption that the addresses of the fixed nodes were assigned at the initialization of the testbed;
- As only one BTS was considered, the routing update function was not performed;
- As the MT and the TSS were to be implemented in the same workstation, there was no need to establish a signaling connection between these two components.

The state diagram for the SNL in the TSS is presented in Figure 4.10, which describes how the SNL_{edge} entity, located in the TSS, processes the different messages [11].

The SNL message contained in its header a data part that could vary depending on the type of message. Three types of messages were defined:

- *Unit data (UDT).* To carry the application layer PDU and the only message to carry user data. All other network layer messages were generated by the network-layer internal events;
- *Route update (RTU).* To distribute routing information between NL-entities;
- *Identify (IDN).* To identify the MT when a link is established, it could be generated locally by the BTS in response to the setup of a signaling bearer.

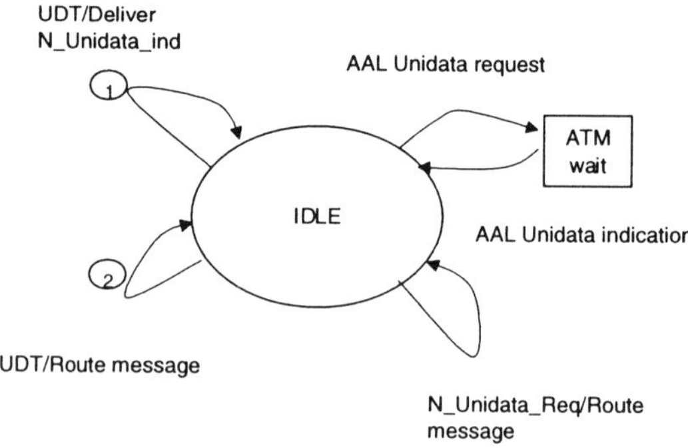

Figure 4.10 State diagram for the SNL in the TSS.

Upon the receipt of a message with UDT type, the SNL can process the message in two different ways:

- If the destination is the BTS, the SNL_{edge} (BTS) delivers N_Unidata_ind;
- If the destination is not the BTS, the SNL_{edge} (BTS) routes the message.

In general, the functions of the SNL in the TSS were limited to the routing of the signaling information independent of the radio technology.

The ISHO requires a high level of cooperation between the MT and the network. The following various handover schemes were identified for the SINUS project [9]:

- Network-controlled handover (NCHO);

- Mobile-controlled handover (MCHO);
- Network-assisted handover (NAHO);
- MAHO.

Backward and forward strategies were considered as connection establishment schemes, and both soft handover (with switched or combined diversity) or hard handover as connection transference schemes were taken into account. The implementation of the different handover schemes followed an approach similar to that of ISHO. Detailed information is provided in [12].

In the SINUS approach, the MT did not initiate the handover [9]. Nevertheless, in order to obtain downlink measurements it was necessary to include link-quality measurement capabilities in the MT. Therefore, as already mentioned the chosen handover strategy for the SINUS project was the MAHO scheme. The handover was as seamless as possible and a significant break in transmission was avoided as the traffic channel was not released until the new link had been established.

The handover phases taken into account within the project were the handover initiation phase and the handover execution phase. In the latter, both radio- and terrestrial-link procedures were introduced within the testbed architecture.

4.2.2 Integrated S-UMTS Real Environment Demonstrator—The Project INSURED

The demonstration of ISHO as an important S-UMTS facility was also one of the main goals of the project INSURED [17]. It aimed at indicating a viable way towards S-UMTS with a basic concern of how to define a smooth and gradual migration path from 2G to 3G, based on the coexistence of the GSM and UMTS networks. Such a migration path was also recommended to ETSI.

For this purpose the project developed and demonstrated a new communication architecture based on an interworking-function (IWF) approach, to allow for integration between satellite systems and terrestrial networks. The IWF approach is a generic system concept and IWFs are appropriate equipment placed at the interface between two networks that have the scope of guaranteeing the interworking. In the IWF approach, no common protocol is imposed on the two networks, which would imply the modification of the way a network usually works. In particular, an IWF placed at the interface between networks A and B is able to emulate both network A and network B protocols. In the project INSURED, the implementation of two specific IWFs helped to realize relatively easily ISHO, without causing any impact on either standard GSM, or the standard Iridium entities and procedures used in the demonstrator architecture.

Based on the INSURED trial architecture, it was recognized that access to services could be provided at several levels [18]. As a consequence, an attempt was undertaken to upgrade the currently available UMTS reference models from ETSI and from other ACTS projects, in particular the RAINBOW project. The enhanced UMTS reference model was to foresee additional interworking functions in both the user's end-system and amongst core networks, because some of the proposals on UMTS reference models suffered from a lack of consideration of the latter.

By combining the RAINBOW, ETSI, and INSURED concepts, target architecture was conceived to mark the first migration step towards S-UMTS [18]–[19]. This integration is illustrated in Figure 4.11. For simplicity reasons, only four radio accesses were considered, namely, GSM, IRIDIUM, and two UMTS modes [18]. Additional radio access schemes (e.g., DECT and HIPERLAN) can be added, including even S-UMTS. In the last case, depending on the onboard functionalities, the UMTS payload can be provided with both the payload-radio-independent layers (when the satellite must cope with mobility issues) and the payload-radio-dependent layers.

The upper part of the figure shows multimode user equipment (UE), to work in any of the four considered radio access types, and in this case, to be consistent with ETSI UE [20]. The UE is equipped with single terminal equipment (TE) belonging to a specific overlaying network (OLN). The considered OLNs were the PSTN, ISDN, IP and X.25 networks, and B-ISDN, whereas the term underlying network (ULN) designates the GSM, Iridium, and UMTS systems. The OLNs were, in a way, superimposed on the ULNs, meaning that at least the application layer protocols of the OLNs were transparently carried on the ULNs, thus providing the physical support to the OLN application protocols.

The OLN TE was linked to the user subscriber identity module (USIM), which supports personal mobility. A certain UE can be personalized by simply plugging in the USIM. The OLN TE was also linked via the OLN terminal inter-ULN IWU with some OLN-ULN terminal adapters (TAs), which provided the adaptation between the OLN and the ULN protocols. In this respect, the GSM-OLN IWF and Iridium-OLN IWF were the network-side peer entities of the terminal side OLN-UMTS TA. OLN-GSM TA, or OLN-Iridium TA.

Still referring to the UE, it should be noted that the USIM was neither OLN-specific, nor ULN-specific. The OLN TE was OLN-specific, but not ULN-specific. The OLN terminal inter-ULN IWU was OLN- and ULN-specific. The OLN-ULN TA were both OLN- and ULN-specific. The ULN-specific equipment (i.e., the UMTS terminal blocks, the GSM MT, and the Iridium terminal) is ULN-specific but not OLN-specific.

Figure 4.11 shows that the proposed UE block diagram, was consistent with the GRAN approach of having a modular UE [18]. A UE designed in the described manner can be provided with the equipment necessary to work in the radio accesses, which are of user interest; or it can be provided with the OLN TE relevant to the OLN.

In the UMTS terminal, the I4 interface separates the terminal UMTS radio-dependent layers from the terminal UMTS radio-independent layers (i.e. it corresponded to the RAINBOW terminal interface) [21] –[23].

Finally, the OLN terminal inter-ULN IWU is OLN- and ULN-dependent equipment. It assured the interworking between the various ULNs in the considered OLN. By following this approach, the OLN terminal inter-ULN IWU could realize, in cooperation with the OLN network inter-ULN IWU, the inter-ULN procedures, which actually was a first form of OLN terminal inter-ULN IWU linking the GSM mobile terminal and the Iridium mobile terminal. Thus, inter-ULN roaming and inter-ULN handover was accomplished between the Iridium and GSM ULNs.

The fixed part of the access network comprised the blocks of the four considered radio accesses. For the UMTS BS, the interface I5 separated the BS-UMTS radio-dependent layers from the BS-UMTS radio-independent layers, and this again corresponded to the RAINBOW terminal interface. For the terminal part applied the same considerations.

From Figure 4.11 it is easy to see how the UMTS access network was separated from the core network by the IU interface. The GSM and Iridium networks were separated from the core network by the GSM A and Gb interfaces, which was consistent with the ETSI architectures [20, 23].

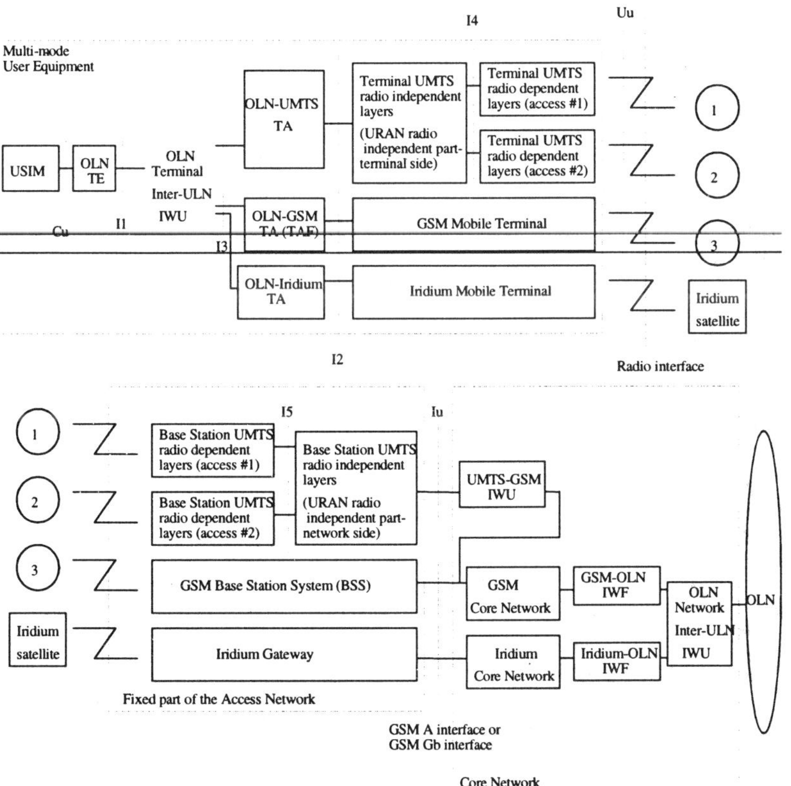

Figure 4.11 RAINBOW, ETSI, and INSURED concepts, combined to realize the first migration step towards S-UMTS.

By following this approach, the GSM network could be seen as one, which in the first migration phase provided the underlying support to the OLNs. Figure 4.12 illustrates the four protocol layers referring to the GSM ULN [18]. Similarly, four protocol layers exist with respect to the Iridium ULN. They can be identified in Figure 4.13.

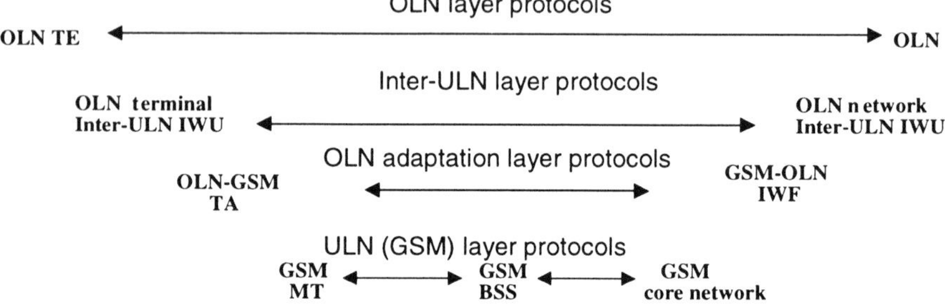

Figure 4.12 Four protocol layers for the GSM ULN (first migration step).

The project INSURED provided also a realization model for the second migration step, when a UMTS core network would be integrated with the OLNs. The situation is depicted in Figure 4.14 and its scenario was consistent with the second migration step identified by ETSI [24]. The same considerations as for the first-migration step situation apply again here for the multimode user equipment, whereas the four protocols relevant to the GSM-UMTS ULN and the ones relevant to the GSM ULN remain the same as in Figure 4.12 and Figure 4.13, respectively.

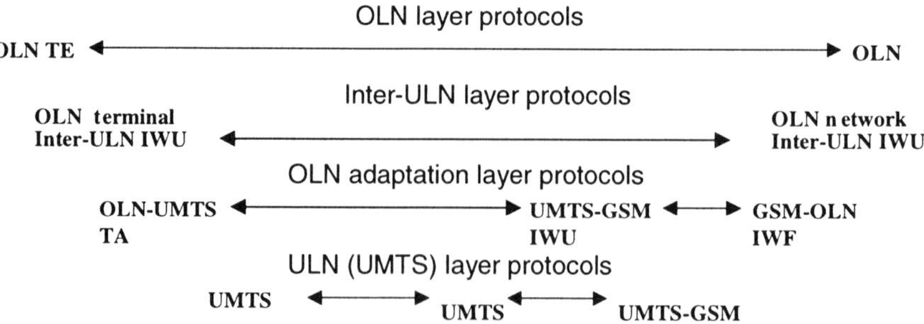

Figure 4.13 Four protocol layers for the UMTS ULN (second migration step).

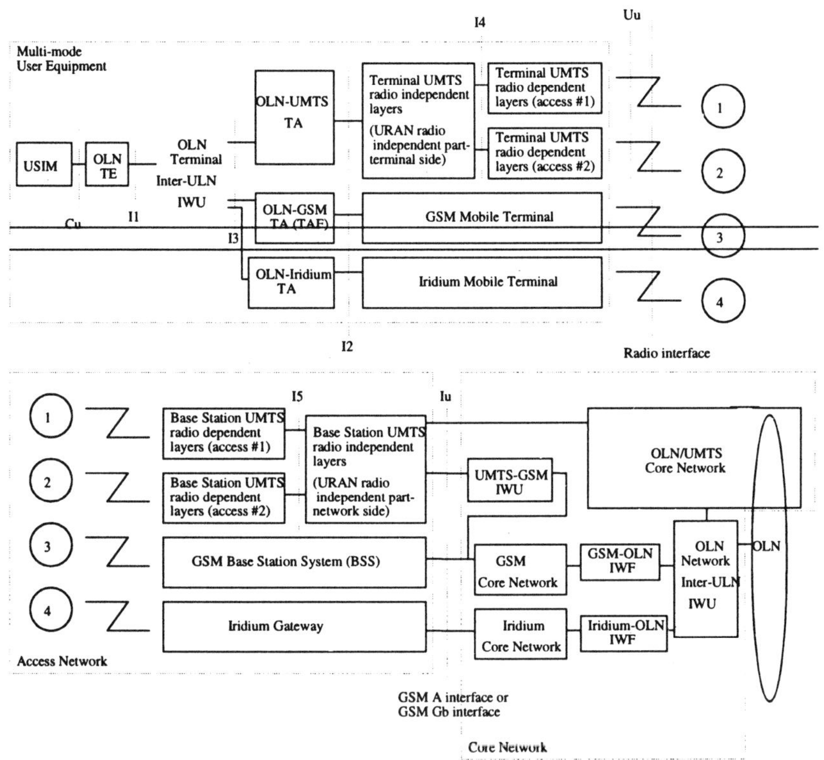

Figure 4.14 RAINBOW, ETSI, and INSURED concepts, combined to realize the second migration step towards S-UMTS.

Figure 4.15 depicts the four layer protocols of the full UMTS ULN, integrated with the considered OLN. The project INSURED recognized that a smooth migration from 2G to 3G was imperative. From the proposed approach discussed above, it could be seen that two systems, the GSM and a LEO satellite system, could be integrated not only at a network-features level, but also on a service-integration level by implementing necessary services foreseen in UMTS [25]. Although a fully integrated scenario seemed to be more suitable for UMTS, it must be noted that to reach all of its requirements is rather hard [26]. A fully integrated solution would have meant exploiting the fact that GSM and Iridium were compatible at a terrestrial network level [17]. Therefore, the demonstrator, representative of a UMTS environment, was proposed to show only the interworking solution that would allow testing of the feasibility of the interworking between two generic networks, which are not compatible at all layers. Actually, this has also been accepted as the most viable way towards UMTS. The complete demonstrator structure is depicted in Figure 4.16.

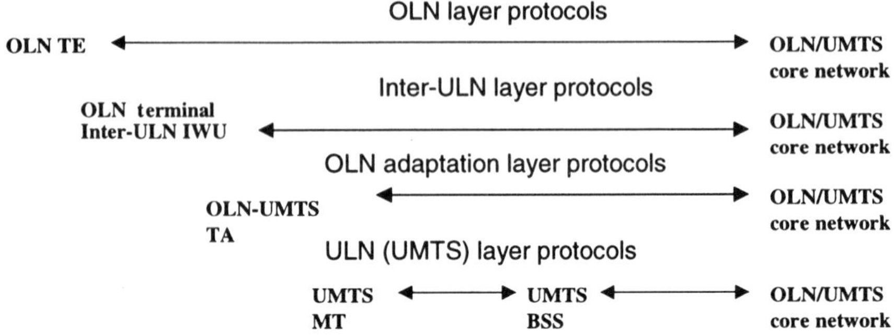

Figure 4.15 Four protocol layers for the full UMTS segment.

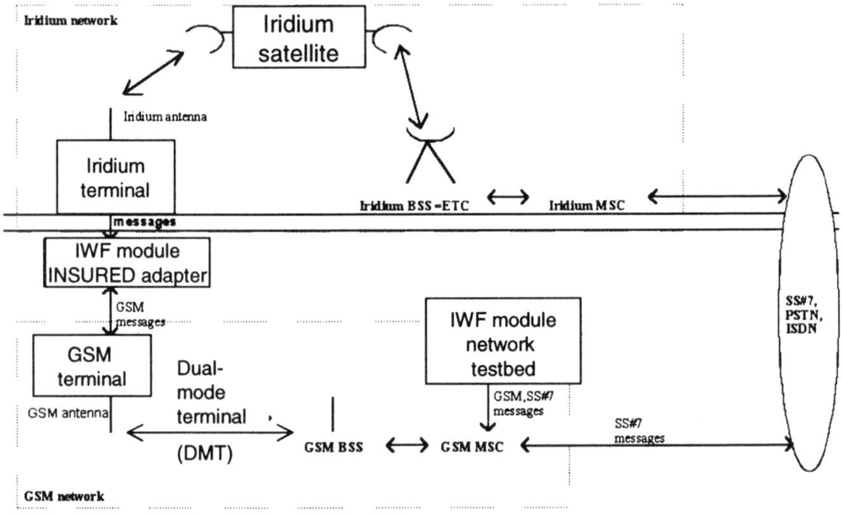

Figure 4.16 INSURED demonstrator architecture [17].

Figure 4.16 shows that the GSM network and the satellite system interact by means of two appropriate UMTS IWFs:

- *The INSURED adapter:* Interfaces a standard GSM terminal with a standard Iridium terminal to form a dual-mode terminal (DMT);

- *The INSURED network testbed:* Interacts on the network side with a standard GSM mobile-services switching center (MSC).

The two above-mentioned UMTS IWFs allow the ISHO from the GSM to the satellite network, and this was demonstrated by the project. INSURED did not foresee the implementation of the ISHO from the satellite to the GSM network because of certain practical/logistic restrictions, entailed by the interaction with the corresponding satellite system.

The project performed several ISHO trials. The procedure is depicted in Figure 4.17. Evaluation results of these field trials were used for the specification of future S-UMTS terminals. Some of the results included an ISHO-duration figure as high as 10s net.

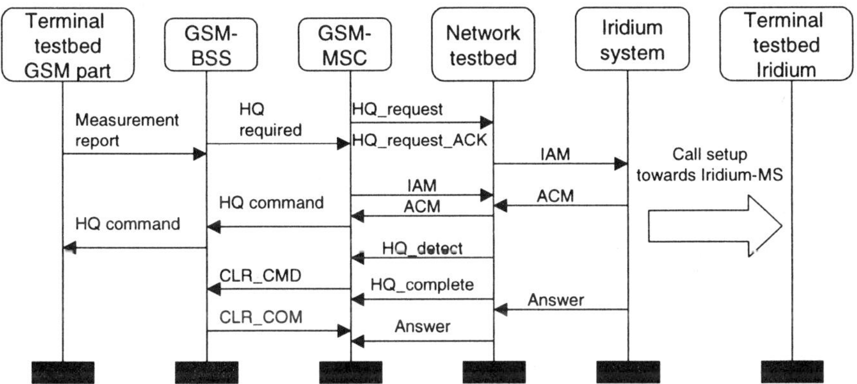

Figure 4.17 ISHO procedure.

It should be taken into account, however, that during the ISHO procedure, the call could survive on the old link (GSM link), which entailed that appropriate measurement processing on the GSM serving carrier had to be performed in order to predict whether an ISHO was really necessary. This should permit triggering the ISHO well in advance, such that the GSM link could still survive for several seconds. An enhanced version of the INSURED adapter should be in charge of deciding, on the basis of the processing of measurements performed on the GSM serving carrier, when to trigger the ISHO. Generally, the IWF approach showed the basic advantage of minimizing the impact on the procedures of the networks to be integrated.

4.2.3 Intertrial Testbed for Mobile Applications of Satellite Communications—The Project TOMAS

The ACTS mobile satellite projects targeted and deployed a wide range of services, representative of S-UMTS, concentrating on low-speed services provided by handheld terminals and their integration into UMTS. Because of the different focus, each project implemented only a subset [8]. The project TOMAS provided a testbed for high-speed services. The focus of the project was the implementation of such services, the realization of highly portable and even mobile terminal equipment, and their application in field trials using GEO satellites.

The project TOMAS followed two different approaches to implement 3G air interfaces. First, the transparent GEO space segment and fixed group segment were offered to cooperating projects to test modem and channel codec components in an end-to-end testbed. For example, there were planned trials with the SINUS project. Second, TOMAS evaluated the feasibility and performance of air interfaces proposed as a result of ACTS and non-ACTS R&D work for GEO applications. Within the constraints of the implementation into the fully operational at the time satellite systems (Inmarsat and EMS), candidate schemes were implemented using programmable high-speed terminal equipment.

The main objective of the project TOMAS was to establish an S-UMTS testbed for multimedia services with data rates up to 2 Mbps and to provide, evaluate, and enhance this intertrial testbed for joint trials [27]. For the evaluation of TOMAS services and applications, a cost-benefit analysis was performed.

TOMAS implemented a mobile satellite terminal (MST) that was designed to support UMTS bit rates from 1.2 Kbps up to 2.048 Mbps via GEO satellites [28]. It provided all low-, medium-, and high-speed bearer services specified for operation in the TOMAS S-UMTS testbed. The testbed also provided teleservices that were categorized as follows:

- Video/audio services using the H.324 encoding standard at bearer rates up to 384 Kbps;
- Multimedia services using the MPEG-4 encoding standard at bearer rates up to 384 Kbps;
- High-speed services with bearer rates up to 2.048 Mbps.

The restriction of the video/audio and multimedia teleservices to a maximum bearer rate of 384 Kbps was imposed by the processing capabilities of the source coding hardware available for the user terminals. With higher speed terminals, the maximum bearer rate could be increased up to 2.048 Mbps. The service specifications are given in [29].

The GEO space segments used in the testbed were Inmarsat III satellites and the EMS payload onboard Italsat F2. Both enabled the setup of high-speed carriers up to bearer rates of 2 Mbps; however, design constraints were imposed on the mobile satellite terminal (MST) in terms of antenna size and transmit power because of power limitations of the global and spot beams of these satellites. To establish the link to the satellite, LOS operation was required, which in terms of user mobility meant service access only in nonobstructed areas. For that purpose, TOMAS developed an architecture that foresaw physically separated user terminals and modem/RF units. Figure 4.18 gives an overview of the TOMAS S-UMTS testbed.

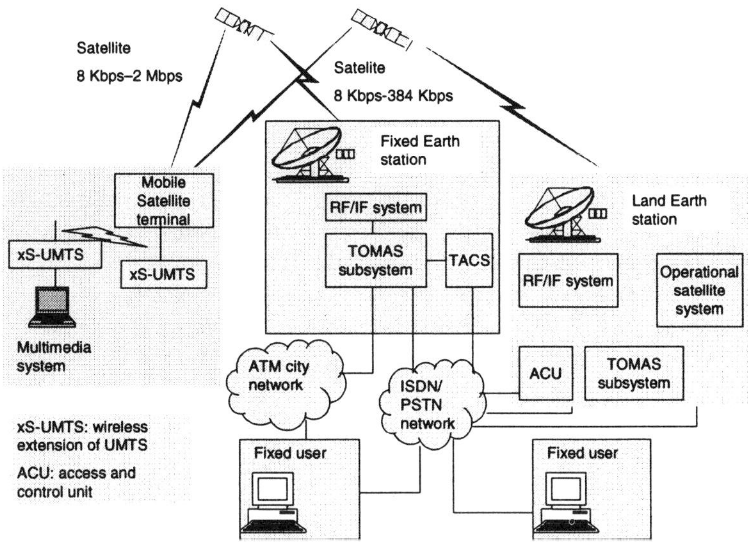

Figure 4.18 Overview of TOMAS S-UMTS testbed [30].

The overall TOMAS testbed setup was subdivided into five main units [31]:

- Mobile satellite communications equipment (MSCE);
- GEO segment;
- Fixed Earth station;
- Terrestrial network;
- Fixed user equipment.

4.2.3.1 Basic Configuration of the TOMAS Platform

The MSCE contained the mobile high-quality multimedia terminal, an optional radio link, and a mobile satellite terminal. In the primary configuration, the multimedia terminal used audio/video-coding schemes according to ITU H.324M standard. A detailed explanation of the implemented coding schemes and terminal is provided in [32].

The MSTs were based on modified Inmarsat satellite terminals with an attachable planar antenna [31]. The terminal size depended on the required data rate, while the terminals could be used for both space segments, due to the similar frequency range. In the course of the project, several upgrading trials were performed.

A particular potential for improvement was seen in the area of combined source and channel coding. The first TOMAS prototype terminal was equipped with an Inmarsat-B front-end, which could support all specified services. The hardware architecture of the TOMAS MST is presented in Figure 4.19 [28]. The RF front-ends were taken from available Inmarsat terminals. The RF carrier functions (transceiver tuning, HPA on/off and transmit level) were controlled by the main control unit (MCU) of the RF section, which was attached to the baseband processing unit (BPU) of the modem. The BPU executed the TOMAS MAC and UMTS adaptation functions resident in the MST. Further improvement was considered possible by the implementation of Inmarsat M front-ends with smaller antennas to increase the mobility, but with limited bearer capabilities. The modem was built on a double Pentium CPU board with a baseband processor, digital modulator, high-speed A/D converter, and analog IF interface add-on boards. The different interfaces to connect the MST to various types of user terminals included ISDN multi-BRI, USB, RS-232 and the universal serial interface (USI).

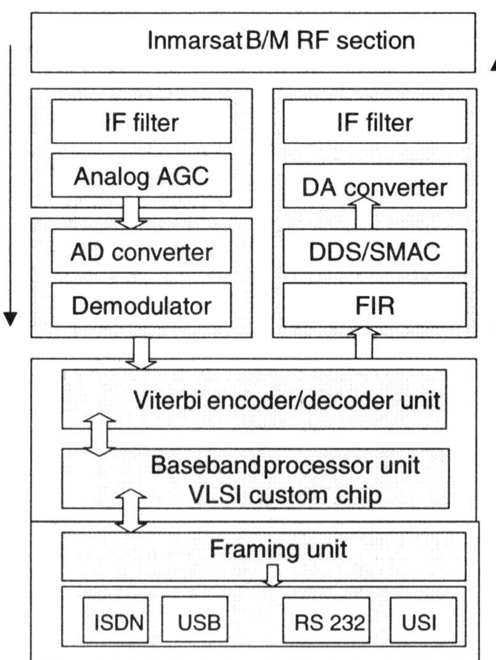

Figure 4.19 MST hardware architecture.

The RF front-end used for the TOMAS MST comprised three modules, an antenna, and a RF feeder assembly. The three modules included the following:

- The RF receiver section containing the duplexer, low-noise amplifier (LNA), and filter assembly;
- The RF transmitter section containing the multiplexer to separate the multiple signals connected on a single coax to the main control unit (MCU);
- The antenna control board containing the microprocessor to run the antenna system.

Various types of antennas were available for use with the TOMAS MST, providing gains up to 22 dBi at the L-band. It is important to mention that the gain determines the bearer capabilities of the terminal. Higher gain antennas supporting higher speeds are of a bigger size and as such reduce the mobility of the terminal. The user must make a trade-off in terms of supported QoS and mobility. The TOMAS trials showed, however, that the requirement for mobility was generally decreasing with the quality of the service used. For example, high-quality video services were normally used with the user/terminal being stationary, which is confirmed also by other projects [33].

The standard antenna type with the MST was a compact, low-weight, flat-panel antenna. The antenna elements were arranged as phased arrays. In the TOMAS testbed, right-hand circular polarization (RHCP) was used as required by the Inmarsat and EMS satellites. Typical antenna sizes for gains of 10 and 17 dB are 240 x 120 and 450 x 590 mm with 20-mm panels, respectively. The antennas used with the TOMAS MST were 770 x 770 mm; alternatively, a conventional parabolic type was available for trials. Both antennas provided a gain of approximately 22 dB at 1,630 MHz, which enabled all bearer services specified for testbed operation.

The variable rate modem connected to the RF front-end was realized as a PC-based system. The incoming IF signal was sampled with normalized rates of 9.6 Kbps or 5.333 Kbps at a carrier of 12 and symbol rate of 1. The modulator was implemented using DDS technology.

The MST provided different interfaces to connect various types of existing, non-UMTS terminal equipment as well as future user terminals with UMTS signaling capability using the UAL signaling protocol. The ISDN multi BRI interface provided up to 6-B channels with DSS1 signaling via the D-channel. The interface processor and framing unit mapped the B and D channels on the logical and physical channels of the air interface. For low-rate synchronous and asynchronous services, an optional RS-232 interface was provided. Particularly for the provision of synchronous bearer services, a USI for low, medium, and high bearer rates was provided by the project. In order to implement S-UMTS signaling between the user terminal and the MST, the USI comprised circuits for call control and bearer monitoring, thus adding S-UMTS signaling capabilities to existing serial interfaces.

Service and call control functions of the MST were assigned to the network and LAC/MAC layers of the terminal, which were implemented in the BPU. User interaction was provided through a graphical interface, implemented in the application layer of the user terminals, but also available on the display of the MST.

The MST architecture allowed the introduction of UMTS adaptation functions in the network layer, which preserved the identity among the higher-layer functions (e.g., bearer services, call control, and bearer monitoring). The lower-link and MAC layers were implemented according to the requirements for the services to be provided by TOMAS and the constraints of access to the operational GEO space segments, particularly the requirements for access coordination with the existing commercial services.

The bearer control and monitoring functions specified for the TOMAS testbed were derived from the UAL API proposed by the project OnTheMove [34]. Only functions relevant for satellite UMTS call control were retained, however, and their parameters were adapted or complemented according to the needs of satellite access (i.e., the parameters defined in the request specific bearer information function). The MST adaptation functions were defined by the mapping of the set of bearer control and monitoring functions and their parameters to the set of MAC functions specific to the TOMAS testbed [29].

The user terminals were physically separated from the MST. Because of that, access to an execution of the bearer control functions by the user terminal was defined by the UAL signaling protocol. It used ASCII strings and an asynchronous interface to exchange the signaling data between the user terminal (UT) and the MST; empty strings were defined for synchronization purposes of the signaling receiver.

The defined set of bearer control and monitoring functions was executed through the TOMAS graphical user interface (T-GUI). The T-GUI was a touch-screen version running on the service- and access-control unit of the MST; it existed as a Windows application program running on the UT. The control commands and parameters were exchanged with the MST in the format defined by the UAL signaling protocol.

The T-GUI provided two levels of access to the testbed, namely the following.

- Basic control and monitoring functions (i.e., service selection, dialing, and qualitative QoS);
- Access to all functions and parameters defined in the UAL signaling protocol, including output of detailed QoS information (BER and signal level).

An extended version of the T-GUI provided access to all MST functions for the system trials.

4.2.3.2 TOMAS S-UMTS Services

The services provided by the TOMAS system were defined following the needs of the cooperating user groups, including the construction industry, telemedicine organizations, and the media[35. The current trends in data and multimedia applications development were also considered. Table 4.6 gives an overview of the TOMAS services and their definition [30]. The services were divided into bearer services and teleservices. The TOMAS bearer services provided a bit-transparent connection between the MST and the fixed network termination [35. The transmission speed was selectable by the user before setting up a call and the supported speeds were in the range of 8 Kbps and 2 Mbps. The maximum speed available to a single user depended on the capability of his or her MST, the type of satellite accessed, and the current traffic load of the system.

Table 4.6

TOMAS S-UMTS Services

	Low/Medium/High Speed	*Video/Audio*	*Multimedia*	*IP Inter Connectivity*
Layer 1	Unrestricted data			Ethernet
Layer 2	User/application defined			802.2 MAC
Layer 3	User/application defined			IP
Layer 4	User/application defined	H.223	H.223/ T.120	TCP/UDP (user defined)
Layer 5	User/application defined	H.324M	H.324M/ MPEG-4	User defined
Data rates (Kbps)	8, 16,..., 64/ 8, 16,..., 384/ 8, 16,..., 2048	8, 16,..., 384	8, 16,..., 384	8, 16,..., 2048
Call control	T-GUI, T-MAC	ACI	ACI	ACI, dial-on-demand
UMTS adaptation	User channel routing and bearer control through UAL			
QoS	BER, delay, and higher order performance statistics subject to trials			

The TOMAS teleservices provided standardized transmission protocols and user interaction facilities for video, audio, and data applications. The TOMAS system supported teleservices for multimedia transmission based on H.324 and MPEG-4 standards and data services providing IP-connectivity between user terminals or LANs. The same type of connection between the MST and fixed networks, as for bearer services, was also established for the teleservices. Instead of a digital interface, the user was provided with video/audio inputs and outputs, such as camera interfaces, monitor outputs, headset ports, and video displays.

The H.324 and MPEG-4 transmission schemes have become industry standards for low-cost video/audio and data communications equipment for use in fixed and mobile networks. Both standards have been developed to achieve the maximum QoS at transmission rates below 2 Mbps and to maintain acceptable quality, also in error-prone environments like UMTS radio networks.

The layer 1 entries described the type of video, audio, and data interfaces provided by the terminal. Video overlay meant the display of their own and received video pictures in overlay windows on the screen of the service and access control unit of the MST, or on a separate PC unit connected to the MST.

Layers 2, 3, and 4 of the H.324 and MPEG-4 services comprised the multiplexing and transmission control protocols for the encoded video, audio, and data information. In the basic implementation, this included software based on the H.223 framing standard. TCP/IP and UDP/IP protocols (user-selectable) were supported in layers 2 and 3 for the MPEG-4 and H.324 services, respectively. The IP-connectivity service provided an Ethernet port to connect user terminals or LAN segments to the MST.

On the basis of the defined services, a cost-benefit analysis was carried out. The cost-benefit analysis is an acknowledged instrument of assessment for the different participants in the telecommunication market. The TOMAS cost-benefit analysis described two aspects of the specified and developed services. On one side, the so-called vertical chain network operator, service provider, and end user were followed with the objective to support the implementation into commercial S-UMTS networks. On the other side, the horizontal level, which showed the benefits of the services especially for the different user groups identified in the project, was investigated. The evaluation criteria for the cost-benefit analysis of S-UMTS applications and services are presented in Table 4.7. Figure 4.20 shows in detail the relations amongst the telecommunication participants.

Table 4.7

Evaluation Criteria for Cost-Benefit Analysis of S-UMTS Applications and Services [30]

Criteria	*End User*	*Service Provider*	*Network Operator*
Time/cost	Rapid setup; on-demand services; service and content-based billing	Rapid deployment; service and content-based billing	
Quality	QoS; user interface; data security	Quality of network; reliability; fraud management	High MTBF, low MTTR, billing and accounting; fraud management
Flexibility	Customized services; compatibility; mobility	Correspondence to clients requests; bandwidth on demand; choice of networks	Correspondence to clients' requests; flexible allocation of resources
External effects	Global connectivity; improved efficiency (time/quality); quality of life	Global coverage; competitive edge; market potential	Competitive edge; market potential, strategic alliances

The telecommunication market is characterized by a heterogeneous structure. Network operators provide fixed and mobile networks with different bandwidths, services, and cost structures to be used by the service providers, which run application-oriented services for the end user. Service providers take advantage of the networks of different operators. The requirements of the service provider are indicated by a service- and content-oriented contract. A similar relation exists with the end user, which also benefits, depending on the application, from different service providers. The TOMAS S-UMTS services were developed for these structures and offer benefits to the network operator, the service provider, and the end user. The cost-benefit model of the TOMAS project was intended to support users to find a solution for telecommunication systems depending on required services and applications. On the basis of the application for the mobile satellite communication, the analysis started with the examination of S-UMTS benefit criteria like mobility, setup time, or bandwidth.

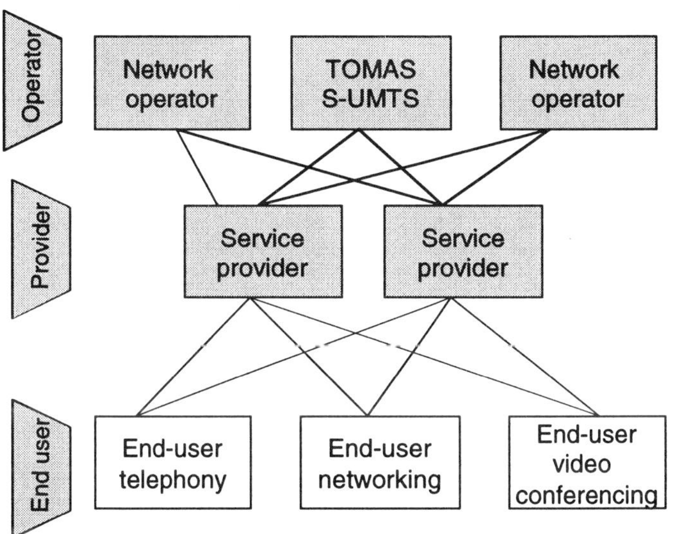

Figure 4.20 Telecommunication market structure [31].

The use of the TOMAS S-UMTS services was also indicated for the case of disqualifying criteria for the use of services based on other technology. The figures for the cost-benefit analysis were gained from different trial scenarios in real business applications. Even for the case of still-to-be-deployed services (e.g., mobile broadband satellite services), statements about the expected benefit for the future use of these services could be verified. Existing fixed and mobile services, which depending on the application and geographical location of the user could be in competition with the TOMAS S-UMTS services, are ISDN, ATM, GSM, and other satellite systems. The fundamental benefit that was expected of the TOMAS S-UMTS services lies in the rapid setup and global mobility of the satellite terminals, as well as in the provided bandwidth-on-demand and guaranteed QoS.

The cost, which is a critical factor, could also be broken in terms of mobile multimedia equipment, site setup, basic charges, and service charges. The results showed that there is a market position for the TOMAS mobile satellite services in the industrial sectors. Figure 4.21 depicts the classification of TOMAS scenarios in function of market maturity.

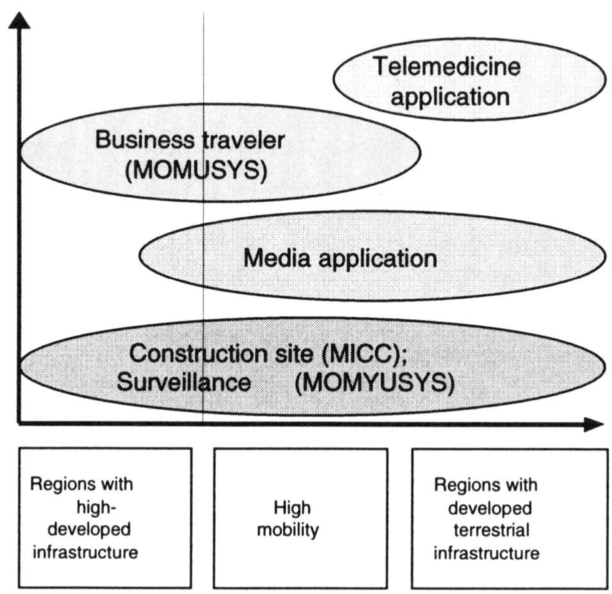

Figure 4.21 Classification of TOMAS scenarios depending on the market.

4.3 SATELLITE BROADBAND COMMUNICATIONS

Future broadband services must increase system capacity, reduce airtime service rate, and guarantee QoS [32]. Recent projections for the number of potential satellite broadband multimedia users up to the year 2010 are shown in Figure 4.22.

The following points should be mentioned:

- The total addressable market for broadband multimedia worldwide will grow steadily from 50 millions users up to 330 million in 2010;
- By then Europe represents about 30% of this total market, the United States another 30%, the rest is distributed over the other regions;
- The market that can potentially be captured by satellite systems amounts to 16%.

Considering the novelty of these systems, critical technological developments and demonstrations were considered necessary to start system and service operation in the best possible conditions. The expectations are that within the next years, a large number of new satellite systems for broadband multimedia communications will be developed for fixed, portable, and mobile terminals [35].

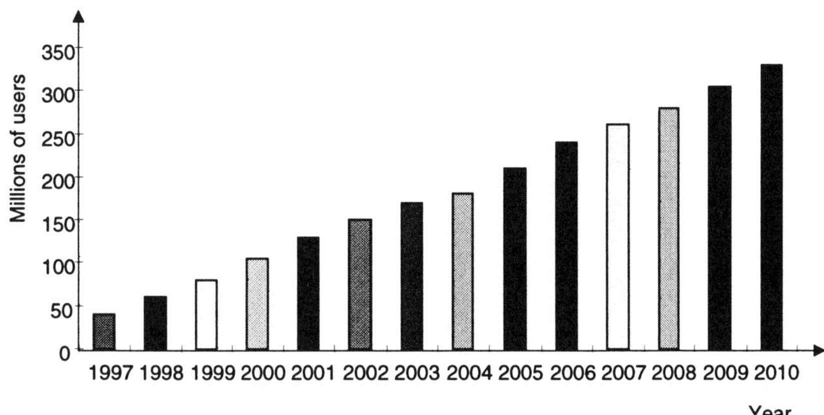

Figure 4.22 Potential broadband users worldwide (millions) [2].

To satisfy the expanded bandwidth requirements and to avoid the restricted capacity of lower frequency bands, these systems are supposed to operate at frequency bands above 18 GHz.

Within the ACTS program, channel measurement and multimedia service demonstration were conducted as part of an aeronautical measurement campaign for the satellite channel at Ka-band and an aeronautical multimedia service demonstration campaign for high data rate satellite services at K/Ka-band. The research was carried out within the framework of the project SECOMS/ABATE.

4.3.1 Satellite EHF Communications for Mobile Multimedia Services—The Project SECOMS/ABATE

The main objective of the project SECOMS/ABATE was the definition of the system elements and the development of related technologies for the future advanced satellite mobile multimedia service operating at Ka- and EHF- bands [36]. The latter will be capable of providing the widest class of broadband mobile services, in all environments, including the land mobile and aeronautical ones.

Portable and mobile terminals were foreseen as basic user terminal types to be used in the 20/30-GHz and 40/45-GHz frequency bands [37]. The system choices and developed technologies were validated by satellite trials.

The project was organized in three different phases. The first phase was devoted to basic system definition studies; the second phase referred to system feasibility studies; the third phase concerned terminal technological developments and was devoted to the definition and assembly of terminal prototypes and demonstrators.

One of the key issues relevant to the overall project activity was the selection of cost-effective solutions for unlimitedly accessible services with flexible adaptation to emerging services. The adoption of compact, user-friendly terminals to support broadband robust links was the major system design driver.

4.3.1.1 Mobile Multimedia Services and Requirements

Within the SECOMS/ABATE project, two main categories of mobile services were identified: the individual/personal and collective/group services. Depending on the application and user restrictions, a further distinction was performed between mobile and transportable terminals. In the course of the work, the project adopted a new methodology for market prediction as shown in Figure 4.23. The potential size of each user group and its market was evaluated taking into account several factors, namely the following:

- Profitability;
- Economy of the region;
- Market penetration and take-up rate of the services;
- Rate for the provided service.

Applying this methodology, the potential number of users (satellite terminals) expected to be served by the Ka-band segment was obtained (see Figure 4.24). It can be seen that the Ka-band user terminals were grouped within three different categories, whose main differences were in terms of uplink data rates, namely, SAT-A = 160 Kbps; SAT-B = 512 Kbps, and SAT-C = 2,048 Kbps. For the EHF band segment, only one satellite terminal at 64 Kbps was defined.

The obtained values were scaled-down considering the market penetration factor. Table 4.8 summarizes the final results in terms of the gross traffic capacity per satellite.

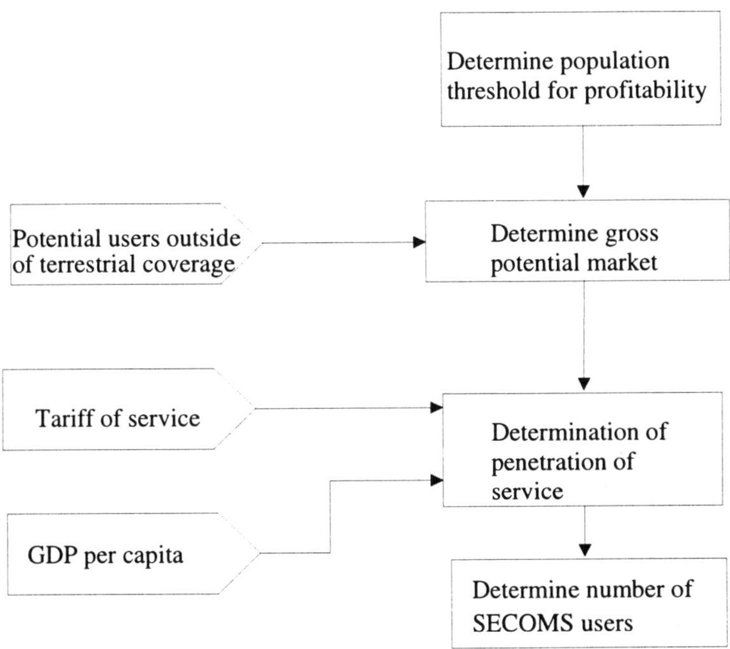

Figure 4.23 Methodology for the satellite market prediction.

Table 4.8

Net Information Traffic Requirements (Ka or EHF band)

	Ka-band		*EHF-band*	
	U/L (Mbps)	D/L (Mbps)	U/L (Mbps)	D/L (Mbps)
SaTs	411	1068	110	110
GTW/S	733	332	90	90
PS				
InSS	524	524	33	33
ISLs	131 and 33	131 and 33	33 and 33	33 and 33
Total	1832	2088	299	299

Referring to the Ka-band element, the evident unbalanced traffic capacities in the two traffic directions were because of the asymmetric nature of the required (point-to-multipoint) services.

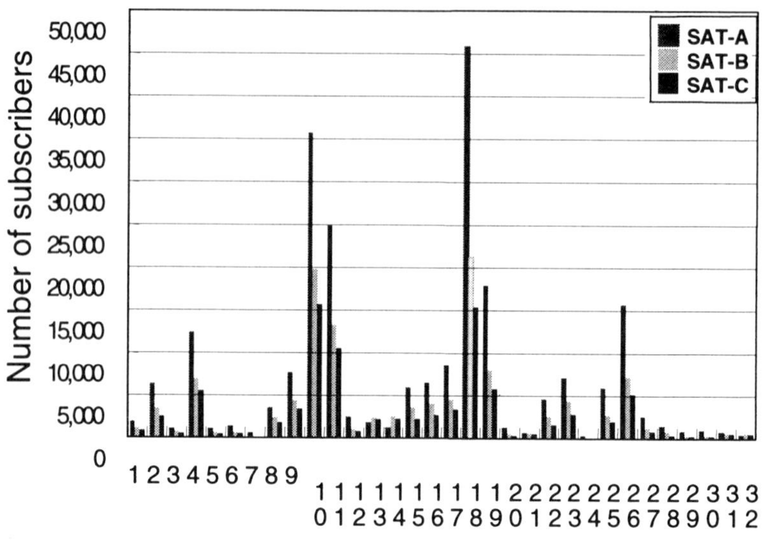

Figure 4.24 Terminal number for the reference-rate scenario in the Ka-band segment.

The reference architecture for the proposed system was the GEO-satellite network architecture; the enlargement of the project scope included various non-GEO satellite configurations. The results were used to compare the GEO and non-GEO architectures in the context of the SECOMS/ABATE project requirements, and they were summarized as follows:

- The GEO orbits proved better suited to provide multimedia services over a regional coverage area, while non-GEO solutions were preferred for a global coverage area;
- The GEO orbit was preferable for the provision of the SECOMS-target service;
- Payload-resources optimization was feasible with a GEO system;
- Non-GEO orbits required moving satellite tracking and a double beam antenna to guarantee soft handover without service interruptions.

Additionally, a brief investigation into other satellite systems operating in Ka/EHF bands was performed. The performance and some relevant characteristics were compared for planned GEO and non-GEO satellite systems (SECOMS/ABATE, TELEDESIC, and M-STAR) at Ka/EHF band. Based on the above-mentioned arguments the GEO configuration was selected as the most suitable for the SECOMS/ABATE project.

The system network configuration included the following elements:

- Ka payloads;
- EHF payloads
- Mobile/portable satellite terminals;
- Master control stations (MCSs);
- Fixed Earth stations (FESs): gateways, service provider stations, and interlink stations.

Figure 4.25 depicts the derived network architecture.

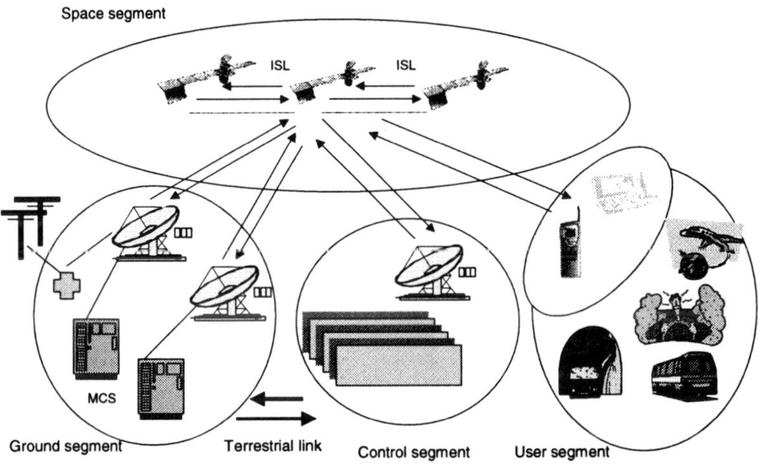

Figure 4.25 Network architecture and network elements.

Two subnetworks, which were interconnected through intersatellite links, were singled out. One included the elements working in the Ka-band, while the second one included the elements in the EHF-band. These two subconfigurations are presented in Figure 4.26 and Figure 4.27, respectively. All types of envisaged Ka-band satellite user terminals were grouped within three main different categories:

- SAT-A providing connections among low-medium data-rate users, asking up to 160 Kbps in the uplink and 2,048 Kbps in the downlink;

- SAT-B providing connections for medium-high data-rate users, asking up to 512 Kbps in the uplink and 2,048 Kbps in the downlink;
- SAT-C providing connections between high data-rate users, asking up to 2,048 Kbps both in the uplink and downlink.

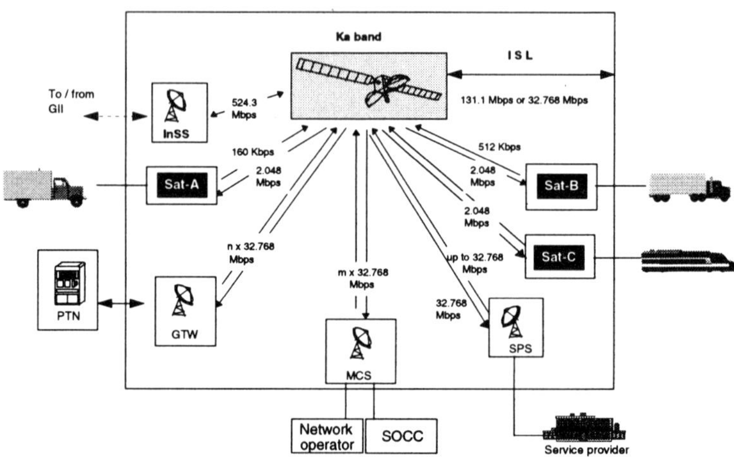

Figure 4.26 Ka-band network architecture.

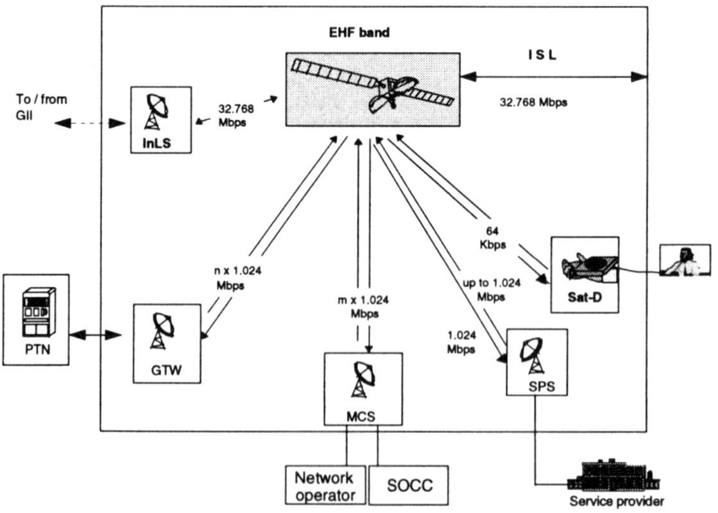

Figure 4.27 EHF-band network architecture.

Full interworking was provided among all SAT-types for any service. The demonstration trials were organized in three layers with the following objectives:

- Demonstration of bidirectional data transmission via Italsat satellite using two developed Ka-band prototypes;
- Demonstration of the performance of the retained access method (MF-TDMA, CDMA) and implemented electronics;
- Extrapolation of the results of the above-mentioned trials to the EHF band of interest through a characterization campaign.

The main system parameters for the two configurations are summarized in Table 4.9. The values are the result of the best trade-off analyses performed in order to optimize the space and ground segment resources.

Table 4.9

Main System Parameters

Parameter	*Ka-band*	*EHF-band*	*Remarks*
Satellite position	12°E	12°E	Study case only
Nominal satellite traffic capacity	2.1 Gbps	300 Mbps	—
Number of spot beams	32	64	—
Type of satellite terminals	3 (SaT-A, SaT-B, and SaT-C)	1 (SaT-D)	—
Type of FES	3 (GTW, SPS, InSS)	3 (GTW, SPS, InSS)	Interfacing the terrestrial net
SaT-A uplink information rate	16–160 Kbps	—	—
SaT-B uplink information rate	16–512 Kbps	—	—
SaT-C uplink information rate	16–2,048 Kbps	—	—
SaT-A, B, C downlink information rate	2,048 Mbps	—	—
SaT-D uplink inf. rate	—	4–64 Kbps	—
SaT-D downlink information rate	—	64 Kbps	—
FES information rate	32.768 Mbps	1.024 Mbps	Uplink and downlink
Frequency:			—
Uplink	27.5–30 GHz	43.5–47 GHz	
Downlink	17.7–20.2 GHz	39.5–40.5 GHz	
Coverage	Extended Europe	Extended Europe	—
Spot beamwidth	0.7° (at -3 dB)	0.49° (at -3 dB)	—
Polarization	Circular (RX opposite to TX)	Circular (RX opposite to TX)	—
Satellite G/T (at triple cross point)	15.7 dB/K	18.7 dB/K	—
Satellite EIRP (at tcp)	56 dBW 44.6 dBW	55.3 dBW 43.4 dBW	For mobile users, For FES (at tcp)
Link quality	BER < 10^{-10}	BER < 10^{-6}	

The project considered a limited number of GEO satellites. Compact terminals of various sizes and capacities were presented. The applications used for experiments are summarized in Table 4.10. Investigation of the control procedures of the network in order to support a great variety of user applications was carried out with special attention paid to the interconnection to the terrestrial networks, such as N-ISDN, B-ISDN, and the Internet [38]. The obtained results, including performance evaluation through simulation, identified the most critical design parameters affecting the performance of the system and suggested optimum solutions. Starting from the system procedures, both the logical communication channels existing amongst the SECOMS system entities and protocol stacks were identified. On this basis system modeling was performed. The results of the demonstrations showed end-to-end service with good quality and the capability of good antenna accuracy. The service field trials confirmed that the service on open area, rural area, and suburban area could be successfully accomplished, and that the system could play an important role in the development of a global information infrastructure, especially in the field of advanced Internet access.

Table 4.10

Applications Used for Experimental Environments [37]

Environment	*Trial*	*Service Type*	*Experiments*
Troposphere, #1: different meteo-conditions and elevation angles	Propagation measurement at 20/40/50 GHz via Italsat satellite station and beacon	—	Attenuation data measurement for the troposphere attenuation prediction model
Land mobile rural, #2: suburban, urban, and stationary; different elevation angles	Narrow-wideband measurements at 40/50 GHz, using a pseudo-satellite	One-way mobile link	Satellite channel characterization using an airplane-mounted transmitter and a mobile van
Land mobile rural, #3: suburban, urban, and stationary; different elevation angles	Narrowband measurements at 20/30 GHz, using a pseudo-satellite	One-way mobile link	Satellite channel characterization using an helicopter-mounted transmitter and a mobile van
Mobile link, #4: Laboratory-simulated channel	Laboratory end-to-end compatibility tests	Bidirectional link simulated channel	User terminal with gateway link via simulated satellite channel
Land mobile urban, #5: suburban, rural, and stationary	Land mobile satellite links, for satellite service demonstration, at 20/30 GHz	Interactive multimedia via satellite	Land mobile (van) user terminal with gateway link, via Italsat satellite

4.3.1.2 Aeronatical Channel Measurements and Multimedia Service Demonstration at the K/Ka band in the ABATE Project.

The aeronautical channel measurement campaign performed during the project was done to investigate the performance of the link [35]. Several field trials were carried out involving aeronautical and land-mobile channel measurements, and service demonstration campaigns at the K/Ka and EHF-band [37], [39]. The objective was to study the behavior of the mobile satellite channel, because it has a crucial impact on the design and development of satellite-based communications networks.

The use of K/Ka-bands for aeronautical mobile satellite communication has, apart from the broadband service provision, also the advantages of the spectrum availability and of the low dimensions and weight of the avionics [36]. In this context a subset of S-UMTS services was identified and two interfaces N-ISDN and B-ISDN were introduced. Figure 4.28 demonstrates the selected scenarios and how the aircraft LAN was configured as a mobile ATM node, interconnected to the satellite network through the radio interface [35].

The concept of office in the sky and multimedia-onboard service represented a clear perspective for the extension of all emerging services to personnel and passengers on board airplanes.

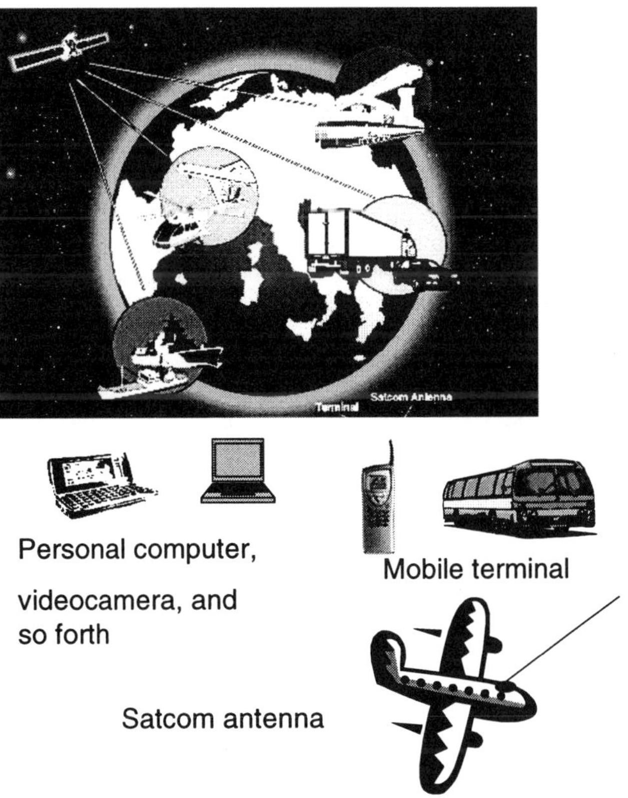

Figure 4.28 Applications scenarios for the SECOMS system.

Together, the two terminals form the technological kernel of the project ABATE provided that the demonstrator will use them through the existing space segment capacity. Furthermore, ABATE successfully demonstrated through flight trials some aeronautical applications oriented to public utility services.

The ABATE system network architecture was directly derived from the one defined in the frame of the project SECOMS. It consisted of only one segment (Ka-band) and foresaw the use of a constellation of colocated satellites interconnected through dedicated high-rate intersatellite links (ISL). The definition of the system architecture, best optimizing the performances of both the space and Earth segments, was obtained considering the following aspects:

- Use of multiple colocated GEO satellites, connected through ISL, to best adapt the ABATE system to the market demand and evolution;
- Use of a regenerative payload implementing onboard fast digital processing (OFDP), allowing direct interconnection among fixed and mobile users within the service area;
- Use of aeronautical satellite terminals with limited EIRP and G/T and characterized by a cost-effective design;
- Use of fixed Earth stations allowing the interworking of the ABATE network with the external public and private terrestrial/satellite networks;
- Mobility management requirements.

Point-to-point and point-to-multipoint interconnections among mobile users and mobile users-fixed stations located within the coverage area are requested by the supported multimedia services, which are also characterized by different real-time constraints and variable transmission rate demand. Regenerative payload architecture performing onboard digital processing (demodulation, switching, and modulation) was, therefore, addressed.

Finally, the solution based on a centralized ground station for system and mobility management, was selected as the most effective for the ABATE system. As a consequence the following network key elements were identified:

- Ka-band payload;
- Three types of aeronautical satellite terminals (SAT-A, SAT-B, and SAT-C) characterized by different uplink data rates and dedicated to aircraft having respectively a small, medium, and large number of passengers;
- Three different types of FES (gateways, service provider stations, and interlink stations).

Results from the detailed link analyses demonstrated the possibility to conveniently utilise the Ka-band frequency, especially when the planes were en route. The main characteristics of the ABATE target system elements are presented in Tables 4.11 and 4.12.

The setup for the channel measurements basically consisted of an antenna rack and a rack holding the control, monitoring, and downconverter units [35]. The receiver consisted of two-stage demodulators transferring the 20-GHz signal to a 2.15-GHz and a 70-MHz intermediate frequency band. Another in-phase and quadrature demodulator mixed the received signal into the baseband.

Table 4.11

ABATE Main System Parameters

System Parameters	*Values*
Frequency	20/30 GHz
Satellite traffic capacity	2 Gbps
Coverage	Multibeam
Number of spot beams	78
Type of satellite terminals	3 (SaT-A, SaT-B, and SaT-C)
Type of FES	2 (GTW, SPS)
Satellite G/T	15.7 dB/K
Satellite EIRP: For mobile user	43.4 dBW (SaT-A), 49.6 dBW (SaT-B/C)
For fixed stations	42.9 dBW (GTW), 44.6 dBW (SPS)
Link quality	BER < 10^{-10}

Table 4.12

Aeronautical Terminal Characteristics

Terminal	*SAT-A*	*SAT-B*	*SAT-C*
Uplink inf. rate	160 Kbps	512 Kbps	2048 Kbps
Receive frequency	20.1 GHz	20.1 GHz	20.1 GHz
G/T	5.5 dB/K	8.5 dB/K	8.5 dB/K
Transmit frequency	30 GHz	30 GHz	30.0 GHz
EIRP	38 dBW	44 dBW	50 dBW

The trials were performed during different flight scenarios and for different weather conditions. The results were then carefully analyzed. Two particular states could be identified, a signal received without shadowing and a signal that had been attenuated.

The goal of the multimedia demonstration campaign was to validate the aeronautical terminal prototype and the corresponding Earth-station equipment developed during the project. High-capacity bidirectional links were used to verify the behavior of aeronautical broadband services during operations. An in-flight demonstration of a wide range of multimedia services and telemedicine applications was accomplished. All flight trials for multimedia demonstrations were performed with a two-engine turboprop airplane (see Figure 4.29).

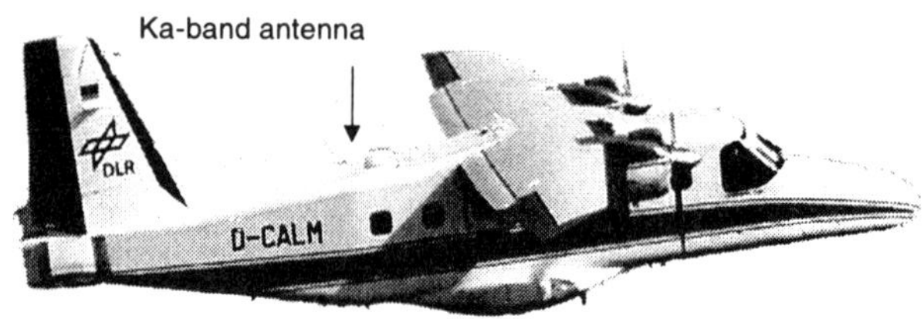

Figure 4.29 Flight-platform testbed.

The design and development of mobile aeronautical terminals for different application requirements, as well as the production, implementation, and testing of an aeronautical multimedia terminal prototype, were some of the main initiatives of the project.

Figure 4.30 shows the overall setup during a telemedicine demonstration to a medical expert team. The aircraft terminal baseband section consisted of a multiplexer that provided a variety of data interfaces and a user application PC. Two identical modems were used in the aircraft and the GES to feed the IF interface, and a multiplexer was necessary to support the required data stream. Optionally, two inverse multiplexers also made the digital data available outside the GES, but with the drawback of a reduced data rate.

A mechanically steered slotted array antenna was used for the aeronautical terminal. Mounted on top of the aircraft between the wings and the tail structure, the antenna consisted of separate arrays for receiving and transmission. A traveling wave tube (TWT) amplifier with a calibrated output power of 126 W was used to power the transmit-section of the antenna. The connection between the antenna and the high-power amplifier was made with a flexible waveguide. Cables were used for other RF- and IF-connection coax. The uplink and downlink frequencies of 30 GHz and 20 GHz, respectively, were converted in a two-stage up/down converter to the 70-MHz satellite modem output/input. The antenna, RF/IF converters, and modem are able to handle data rates up to 2 Mbps.

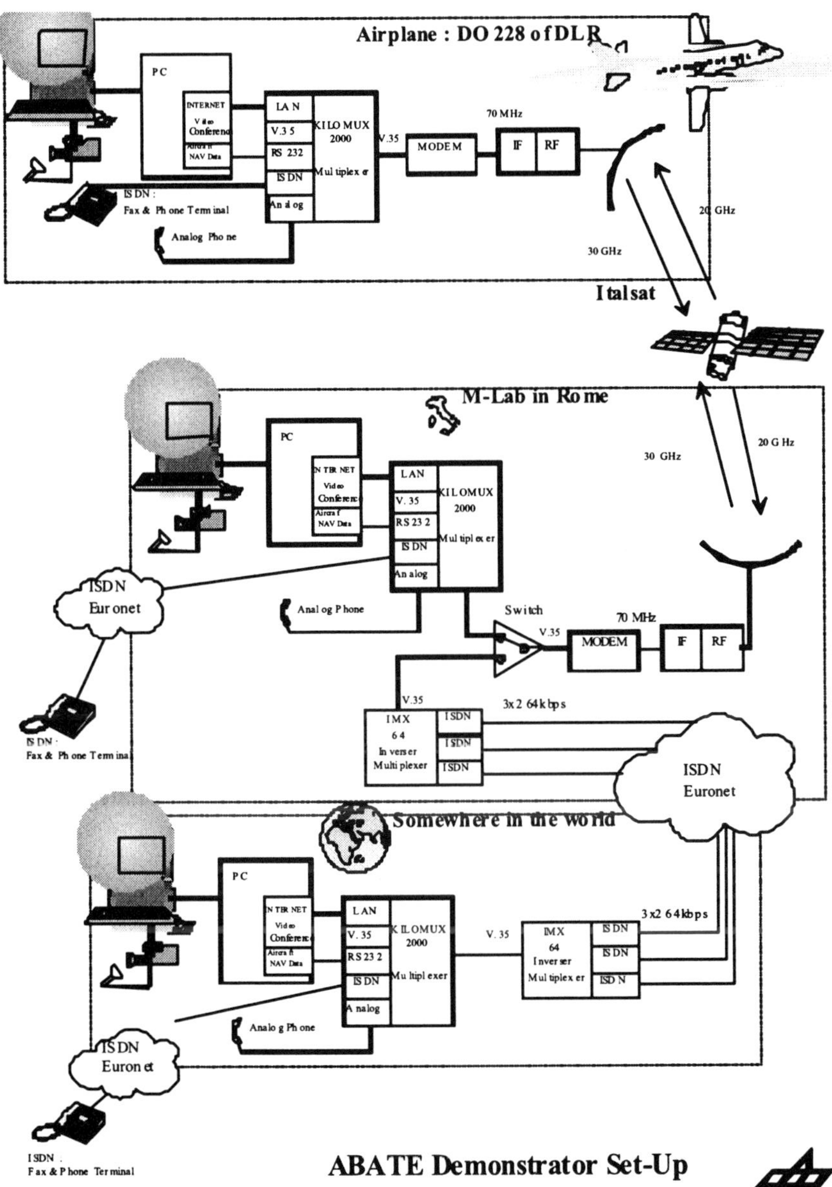

Figure 4.30 ABATE demonstrator setup.

The modem was running in BPSK mode and used a Viterbi FEC code. The baseband configuration adopted for laboratory tests is presented in Figure 4.31.

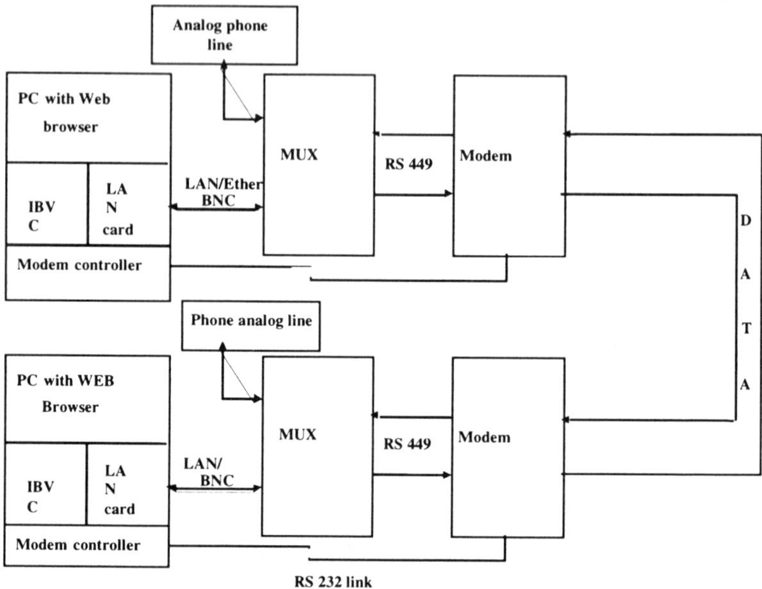

Figure 4.31 Baseband configuration of the end-to-end laboratory testing.

Doppler compensation was made in the modem. The modem was able to counteract Doppler shifts of up to 30 kHz. This was sufficient during all flight and on-ground maneuvers. Also, a Doppler buffer in the modem was implemented to compensate bit delays due to the aircraft's velocity.

A Doppler buffer size of 8,192 bits was used in the aeronautical modem. The frequency and bit-synchronization delay of the modem was less than 1 second and not optimized, regarding the frequency offset. A multiplexer was used to make a wide range of data interfaces available. During all demonstrations, the GPS and avionics attitude data were transmitted for display on the ground.

The goal of the multimedia demonstration campaign was to validate the developed aeronautical terminal prototype and corresponding Earth-station equipment. High-capacity bidirectional links were implemented to verify the behavior of aeronautical broadband services in operative conditions.

4.3.1.3 Measurement Results and Conclusions

The propagation and channel modeling activities of the SECOMS project involved big experimental and theoretical aspects. From the experimental point of view, various measurement campaigns were conducted with Italsat satellites and novel theoretical models were developed, compared with the data, and, in some cases, adapted to reflect their statistical properties. In particular, this applied to the prediction of attenuation, space-diversity advantage, fade duration, fade rate of change, and other parameters of use for system design. The measurement campaign had the following goals:

- To collect narrowband channel database measurements in a wide range of environments, elevation angles, and azimuth angles;
- To investigate the wideband characteristics of the land mobile satellite channel;
- To investigate antenna steering algorithms in the presence of shadowing and fading, using high-gain antennas.

Figure 4.32 depicts a narrowband measurement of the steered high-gain antenna in an urban environment. Severe shadowing and blockage characterize the power sequences. In shadowed conditions, the channel attenuation is at least 20 dB.

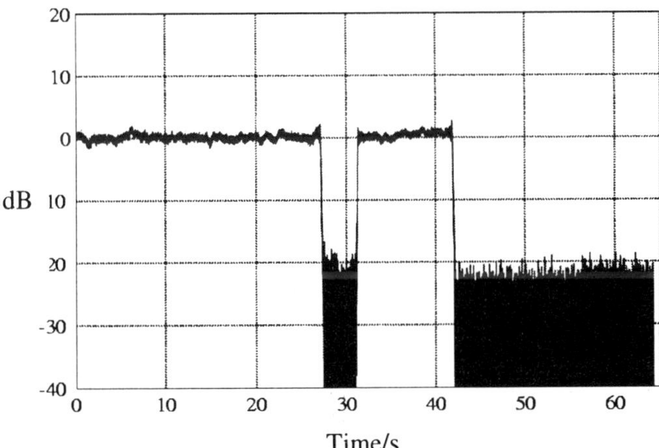

Figure 4.32 Power series in urban and rural environment.

Figure 4.33 gives the results for a wideband measurement in the urban environment. The echo attenuation is in the range 20–30 dB.

The project SECOMS/ABATE developed a target satellite concept, a 20-GHz land mobile array antenna and RF front-end prototypes. The project successfully executed airplane flight trials for 40/50-GHz narrowband- and wideband channel characterization tests. As a result, consolidation of the target satellite system design, finalization of the SECOMS demonstration phase for a multimedia land mobile link via Italsat, and demonstration of the ABATE trial for the aeronautical broadband satellite link were achieved. These results, together with technology developments, can be used as a basis for a recommendation of system architecture, terminals, and standards, relevant to the two-way mobile multimedia service.

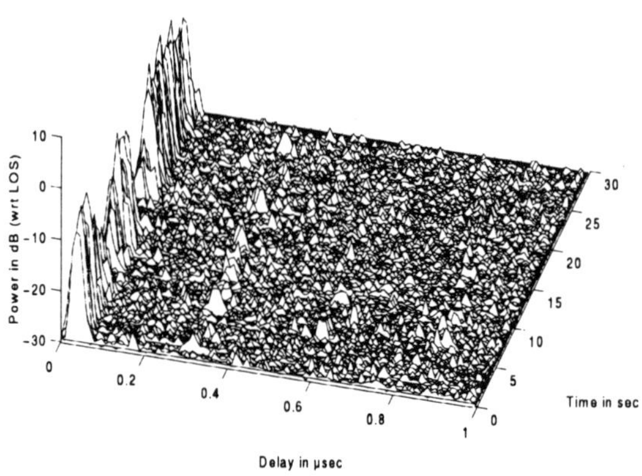

Figure 4.33 Wideband measurement for the urban environment.

4.3.2 Convergence of Internet-ATM-Satellite—The Project COIAS

Infrastructure technologies like ATM, cable network, satellite, and ADSL now provide very high-speed almost error-free communications to the office and to the home. Wireless access is also there for connecting portable terminals and nomadic users. This has provided the opportunity to build the information society, known as the global Internet infrastructure (GII), on top of efficient heterogeneous infrastructure networks, either wired or wireless, and widespread network and transport protocols.

The availability of high-speed infrastructure links is not sufficient for the deployment of the GII. The GII also has to guarantee a satisfactory level of QoS to make the new multimedia applications attractive, in terms of response time, guarantee of the service according to the provider's contract, a high level of interactivity, secure transfers, and transparency for the user accessing from different geographical areas using different types of terminals.

The number of applications that would use this GII infrastructure is unlimited. E-mail and file transfers will certainly benefit from high-speed links and efficient response time. New and interesting applications of the GII are interactive and multiparty games, telelearning, and tele-education. As a matter of fact, the available QoS for audio and video transfers clearly limits the field of potential applications. Enhanced Web servers associating high-speed access and multimedia capabilities will boost electronic commerce that is very sensitive to the presentation of the product to be sold. Because of the low connection cost, industrial applications would also take advantage of the GII and the associated technologies. The number of applications, like information systems for health care, military and factory command and control systems, air traffic management, distributed simulations, and electronic news distribution, has very stringent QoS, security, and real-time constraints. These applications would migrate over Internet protocols.

In terms of market, it is anticipated that the market for networked multimedia over a GII is likely to grow exponentially over the next few years and reach a size of trillions of ECU. This market includes the equipment necessary to bring the information society to the user (fixed or mobile terminals, home equipment like set-top boxes), infrastructure equipment, and, of course, services. Such market growth will only be possible if the level of QoS expected by the users is provided by the GII.

The new technology developments are aiming at providing enhanced services at the infrastructure network level. The project COIAS used ATM and satellite aiming at the implementation of a GII. One important feature of the ATM technology is the potential ability to provide point-to-point and point-to-multipoint connections with a specified QoS. From the operator point of view, ATM brings its scalability and flexibility in terms of traffic management in order to deliver to the user the exact expected level of performance.

The satellite technology, in a way complementary to ATM, can find its place in the Internet world for technical, economical, and strategic reasons. Satellite is preferred whenever the same information needs to be broadcast to many receivers disseminated over a wide area, or a high-speed but temporary link must be established. Other reasons can be that a fast network deployment over a wide area is required or a gradual increase of bandwidth requirement overtime is expected.

With the successful worldwide introduction of the DVB standard, satellite communications offer very high-speed transmissions for the receiver terminals at minimal cost. Other emerging satellite technologies such as onboard demodulation and multiplexing also reduce the size and cost of equipment at the transmit site.

The main goal of the COIAS project was to demonstrate how a GII could be engineered to get full benefit of ATM and satellite networks and at the same time satisfy the requirements of an increasing number of various new-generation applications. COIAS therefore contributed to the Internet-ATM-satellite convergence.

The objectives of a GII architecture are rather ambitious but not unimaginable. A large number of technologies that can make such architecture happen already exist. The project COIAS aimed at developing an essential technological element to help the convergence process. The approach is demonstrated in Figure 4.34. The project focused on developing and mapping the new generation of IPs onto data subnetwork technologies, such as ATM and satellite, while taking into account two main features for the information society, mobility and security. At the application level, the project COIAS studied the impact of using this new-generation IP.

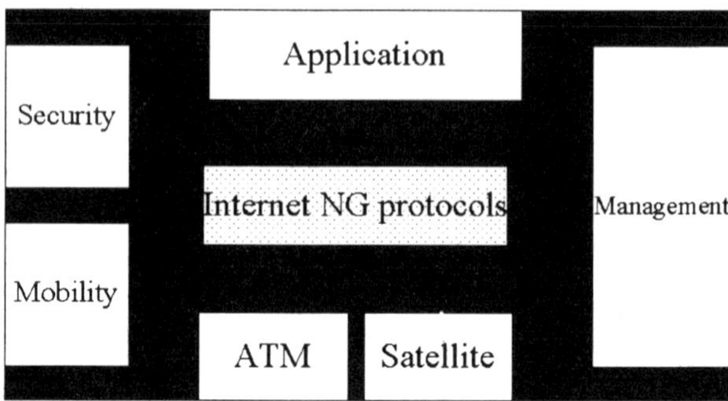

Figure 4.34 Framework of the project COIAS.

IPv6 is the core element of the new Internet technology, and IPv4-based networks will have to migrate to IPv6-based networks [40]. The COIAS project implemented IPv6 including security and mobility features. The COIAS network infrastructure was an IPv6-based network, and the idea was to provide important feedback before the completion of the IPv6 standardization process and help complete the IPv6 deployment.

There are several issues related to the management of the QoS over an aggregation of ATM and satellite networks. These issues deal with the following aspects:

- Mapping of the Internet IntServ model to the ATM QoS model;
- Making RSVP, the signaling protocol, run over ATM and satellite;
- Handling ATM VCs to optimize the network resources for given QoS;
- Handling many-to-many connectionless features of IPv6 and RSVP;
- Mapping the routing algorithms efficiently for the switching mechanisms;
- Optimizing the network resources to meet the required QoS;
- Measuring to prevent misuse/unauthorized use of network resources.

The goal of the project COIAS was to design solutions to solve these issues and to implement them in actual practice. The COIAS platform included up-to-date version of routers, IntServ mechanisms, RSVP functions, and IP satellite transmissions.

At the transport level, the problem to be solved was related to RM data transfer in the scope of the COIAS project, focusing on mobility and security features. The involvement of different technologies (ATM, satellite, and the

Internet) raised the need for a network management center to offer integrated management functionality. The network management component of the COIAS platform was mainly focused on the configuration, fault, and performance management areas.

The COIAS project demonstrated audioconference and videoconference interactive tools, Web browsers, distributed games, and distributed simulation applications. GII-representative applications, like mixing data, audio, video, and security aspects, were evaluated regarding the migration process between the current and the new generation of IPs and tested as an operational application over an Internet NG network using the media server.

The project COIAS participated actively in the standardization process towards the definition of an end-to-end seamless network. This led to the efficient dissemination of the R&D results and provided immediate feedback on the technical choices of COIAS.

COIAS intended to address important technical issues, such as Internet and ATM, QoS mapping, RSVP signaling over ATM and satellite links, IP flows aggregation, network resource optimization, load balancing between ATM and satellite networks, security for mobile access, and reliable multicast transport protocols. The project design phase concentrated on the Internet and ATM domain. The development phase provided IPv6 protocol stacks with added functionality (mobility, security, and QoS management), enabling existing applications to run and benefit from those new communication protocols.

The platform integration phase led to the availability of a European platform allowing the evaluation of a GII prototype network. The trial phase enabled performance analysis of the implemented solutions and assessment of the accuracy of the technical choices. This phase provided as well a complete analysis of the results and assisted in identifying the required evolution path of the GII design.

The strength of the COIAS project was the close cooperation between the technology providers and end users. The project addressed the following major issues of the GII:

- Convergence between the Internet and ATM, especially the availability of guaranteed QoS for the users;
- The ability to use switching techniques and short-cut paths between routers;
- The capacity to provide efficient multicast or broadcast communications in both reliable or unreliable modes;
- The use of satellite channels as part of the GII, and the way to manage alternative path routing between satellite links and ATM terrestrial infrastructure;
- Access for mobile or wireless host stations;
- The capacity to provide secure access to the users;

- The implementation of a certificate infrastructure to provide interdomain security features based upon different network layers;
- The availability of scaleable media servers.

Project work was divided into three major groups that concentrated on the following aspects:

- Management;
- GII;
- Validation platform and trials (VPT).

4.3.2.1 Global Internet Infrastructure (GII)

GII considered the aspects of service requirements, network architecture design, and standardization. The first phase of the COIAS project was devoted to defining the requirements of innovative applications requiring GII that guarantees an adequate level of service in terms of QoS, mobility, security, and the capability to transfer heterogeneous types of information over various kinds of transmission infrastructure according to application dependent criteria.

The network architecture design addressed the possibility for running IPv6 over ATM, mobile IPv6 and related problems in terms of security policy and key management, TCP and reliable multicast, network management, and the relation between the media server and the GII application.

Work was carried out closely with the following IETF working groups for standardization:

- IP next generation (IPng);
- Internetworking over NBMA (ION);
- Integrated services (IntServ);
- Integrated services over ATM (ISATM);
- TCP over satellite (TCPsat);
- Unidirectional link routing (UDRL);
- Resource reservation setup protocol (RSVP);

- Multiprotocol label switching (MPLS);
- IP routing for wireless/mobile hosts (MobileIPv6);
- Internet PKI (Pkix);
- Simple PKI (Spki);
- Simple key management for IP (Skip);
- DNS security enhancements (DNSsec).

4.3.2.2 Validation Platform and Trials (VPT)

VPT took care of the actual implementation and testing of the system. The most important task was to validate and demonstrate the R&D work done to meet the service requirements of the various technologies included in the scope of the project. A comprehensive platform distributed over Europe was put into place to support the trials in terms of the following:

- Service identification;
- Architecture definition;
- Implementation of the function identified in network architecture design.

The validation platform is shown in Figure 4.35. This platform was modified in the validation phase to meet the given requirements. All connections used an ATM terrestrial WAN interconnection and an existing satellite infrastructure based on the SP01 satellite network. It had three uplink stations in Nice, London, and Paris.

The platform development was devoted to seven main functions identified earlier for the GII requirements. The work was divided into the following activities:

- Protocol stack development;
- Security development software;
- Media server development;
- Multimedia application development and network management.

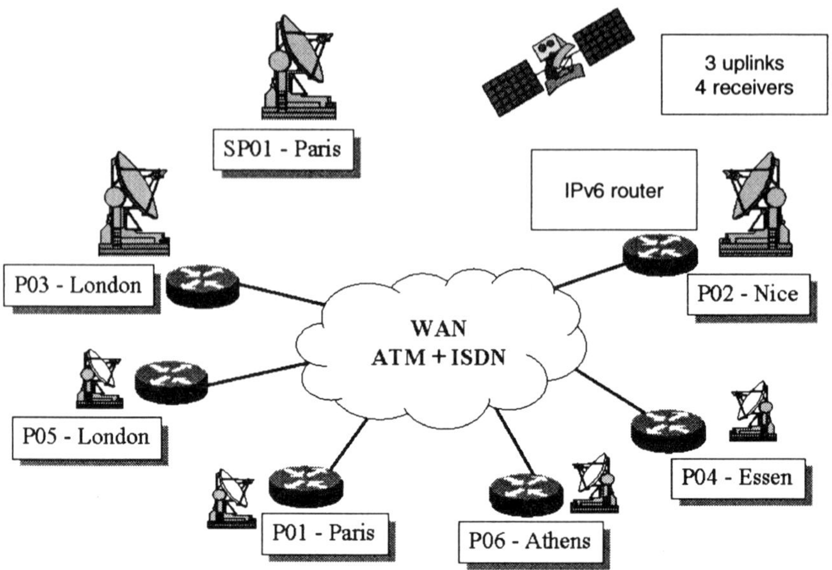

Figure 4.35 COIAS platform architecture.

4.3.2.3 Technical Aspects

Two main technical objectives were addressed by the project COIAS:

- *Development of novel technologies:* The focus was on the technologies required for delivering adequate services to distributed multimedia applications using the new generation of IPs above high-speed ATM and satellite links.
- *Demonstration of advanced multimedia applications:* The demonstration of such applications over a European platform connecting high-tech wired and wireless LAN platforms through available terrestrial ATM infrastructures and commercially available satellite links.

The project dealt with protocol definition for interactive multimedia, transparency, and interworking across heterogeneous platforms or QoS management.

The goal of the project was to harmonize the R&D efforts regarding satellite communications in the scope of the ACTS program or through other initiatives and to encourage the dissemination of the project results. The project COIAS contributed significantly to the study of applicability of the new IPs to satellite communications and the validation of a GII.

A special activity of the project COIAS was related only to interproject relationships. It identified connections with projects focused on other tasks and found ways of cooperation. For example, the COIAS platform was made available for studying the migration of multimedia applications over the new generation of IPs.

4.3.2.4 Industrial Aspects

The project COIAS was built around clear industrial objectives that guaranteed a direct use of R&D results. The idea was to develop Internet services over communication networks taking into account satellite communications, heterogeneous networks, mobility, and security and to validating these services through significant short-term industrial and multimedia applications.

Thus, a software stack for the new generation of IPs was developed that provided an opportunity to add some new functions to handle communications over satellite links; another study was on how to manage mobility and QoS over heterogeneous networks.

The project had an association with a major American player in the field of LAN interconnection. This company provided the IPv6 router software as support to the study of advanced Internet satellite services.

For the application scenario, QoS and management requirements were studied, and an application server to enable testing, together with test and integration expertise, was started. The results made possible the improvement of next-generation products and services.

4.3.3 Summary of Trials

The COIAS platform for comprehensive trials relied on ATM and satellite links located in France, Belgium, the United Kingdom, and Germany. IPv6 routers were used over the WAN links to interconnect local platforms located at the sites. These local platforms included LAN, ATM LAN, and WLAN. A three-month trial was defined to test the architecture first, with generic applications and then with the aeronautical application that was considered as representative of future GII applications mixing audio, video, and data in a mobile and secure environment.

The major issue of the project was determining how new IPs could best make use of available high-speed WAN links, such as ATM and satellite, to offer next-generation services to the user. The project achieved the desired objectives and, in particular, the following:

- Development of IPv6 software including security, mobility, and QoS management;

- Development of key management software for security management;
- Adaptations of existing applications to new IP stacks and comprehensive trial results for feedback on the GII network architecture design.

The COIAS platform was one of the first IPv6-based network architectures in Europe. The COIAS project achieved significant results in and outside Europe within this area with a major impact on VoIP, wireless IP, and QoS research.

4.4 CONCLUSIONS

The performed investigations proved that satellite communications could effectively complement terrestrial networks, wherever the latter either are not competitive (low-traffic density), not applicable (maritime and aeronautical services), or less/not developed.

For maritime and aeronautical communication services, where terminals are not strictly required to be portable, the technology of the GEO-satellite systems is probably the most suitable for present and future enhanced systems. GEO systems can also be effectively utilized for the establishment and development of S-PCS.

Other orbital configurations (LEO and MEO constellations) prove to be effective for the provision of PCS to handheld terminals. In contrast to GEO satellites, LEO spacecraft only fly across the service area at a lower-altitude level and therefore permit more effective communication performances with smaller/less complex user terminals, because of substantially lower link attenuation (15–20 dB, depending upon the orbit).

Already, the first fixed satellite CDMA-based phones for the Globalstar LEO mobile satellite system network had been introduced [41]. These fixed satellite phones were designed for users who lived in areas of the world where conventional wireline or terrestrial-based wireless communications solutions were not available, and they provide seamless and versatile communication.

Another step towards closing of the "data-communications gap" is the introduction of a new, IP-based global packet data capability, with a transmission rate of 64 Kbps. This new capability will enable companies around the world to extend their LAN and WAN capabilities into a global area network (GAN) in areas of limited or no terrestrial or cellular coverage.

REFERENCES

[1] Schwarz da Silva, J., et al., "Mobile and Personal Communications: ACTS and Beyond," in *Wireless Communications*, Dordrecht: Kluwer Academic Publishers, 1997, pp. 379–414.

[2] Satellite Working Group, "Report on Research and Development," *European Commission*, January 1998.

[3] UMTS Forum, "The Future Mobile Market," *Report Number 8,* March 1999, pp. 6–7.

[4] Efthymiou, N., et al., "Service-Dependent Resource Allocation in Hybrid Mobile-Satellite Networks," *Fourth ACTS Mobile Communications Summit,* Sorrento, Italy, June 1999, CD-ROM.

[5] Holzbock, M., E. Lutz, and G. Losquadro, "Aeronautical Channel Measurements and Multimedia Service Demonstration at K/Ka Band," *Fourth ACTS Mobile Communications Summit,* Sorrento, Italy, June 1999, CD-ROM.

[6] Losquadro, G., V. Schena, and M. Ferrari "Satellite Ka/EHF Communication for Multimedia-Mobile Services: Final Results and Relevant Exploitation of Broadband," *Fourth ACTS Mobile Communications Summit,* Sorrento, Italy, June 1999, CD-ROM.

[7] Schubert, A., "URAN Adaptation of the FDMA MAC Layer in the TOMAS/SUMO S-UMTS Testbed," *Fourth ACTS Mobile Communications Summit,* Sorrento, Italy, June 1999, CD-ROM.

[8] Guntsch, A., et al., "EU's R&D Activities on the 3G Mobile Satellite Systems (S-UMTS)," *Doc. No AC201/UHB/IHN/DN/006/a3.*

[9] Buracchini, E., "S-UMTS Activities Inside SINUS,"*ETSI* TC *SMG5,* Sophia Antipolis, France.

[10] Cullen, C., et al., "SINUS Satellite Requirements Towards FRAMES Multiple Access: FMA1 and FMA2," *Report SINUS/IPR/WP3400/AES/028,* Brussels, Belgium, February 1997.

[11] Efthymiou, N., et al., "SINUS Terrestrial Segment Simulator," *Fourth ACTS Mobile Communications Summit,* Sorrento, Italy, June 1999, CD-ROM.

[12] Faineant, V., et al., "Mobile Terminals and FES Specifications," *AC212 AES DNS DS P209.B1,* Brussels, Belgium, June 1998.

[13] Efthymiou, N., et al., "Intersegment Handover Implementation in the SINUS Project and Its Integration Within the RAINBOW Project," *Second ACTS Mobile Communications Summit,* Aalborg, Denmark, October 1997, Vol.1, pp. 121–126.

[14] AC212, "Terrestrial Segment Simulator Specification," *AC212 AES DNS DS P211 B1,* Brussels, Belgium, January 1998.

[15] AC015, "Baseline Document on High-Level Control Specifications for the RAINBOW Demonstrator," *AC015/BELL/CIT/DS/L/011/b1,* December 1996.

[16] AC015, "The RAINBOW Signaling Network Layer," *AC015/RAP1/UL/071/4.1,* August 1997.

[17] Brand, H., et al., "Intersegment Handover Results in the INSURED Project," *Fourth ACTS Mobile Communications Summit,* Sorrento, Italy, June 1999, CD-ROM.

[18] Delli Priscoli, F., et al., "An Enhanced S-UMTS Reference Model Based on a Practical Migration Path to S-UMTS," *Third ACTS Mobile Communications Summit,* Rhodes, Greece, May 1998, CD-ROM.

[19] Delli Priscoli, F. and A. Giralda, "INSURED General Concepts," *ACTS Mobile Communications Summit,* Aalborg, Denmark, October 1997, Vol. 1, pp. 175–180.

[20] ETSI UMTS 23.01, "General UMTS Architecture," *Draft Version 0.0.5,* June 1997.

[21] De Vriendt, J., et al., "RAINBOW: A Generic UMTS Network Architecture Demonstrator," *ACTS Mobile Communications Summit,* Aalborg, Denmark, October 1997, Vol. 1, pp. 285–290.

[22] Fleming, G., et al., "Architecture and Design of the RAINBOW Mobile Terminals, Base Stations and Mobility Servers," *ACTS Mobile Communications Summit,* Aalborg, Denmark, October 1997, Vol. 1, pp. 291–296.

[23] Saidei, A., et al., "RAINBOW Demonstrator Transport Chain," *ACTS Mobile Communications Summit,* Aalborg, Denmark, October 1997, Vol. 1, pp. 297–302.

[24] ETSI UMTS 23.20, "Evolution Of GSM Platforms Towards UMTS," *Draft Version 0.0.1,* May 1997.

[25] Obradovic, V., and J. Habetha, "The S-UMTS Functionality in the INSURED S-UMTS Demonstrator," *Third ACTS Mobile Communications Summit,* Rhodes, Greece, May 1998, CD-ROM.

[26] Dres, D., et al., "INSURED Project," *Report, European Commission,* Brussels, Belgium, 1998.

[27] Dreyer, K-P, and H-U. Marchione, "Evaluation Criteria and Cost-Benefit Analysis for the TOMAS S-UMTS Applications and Services," *Third ACTS Mobile Communications Summit,* Rhodes, Greece, May 1998, CD-ROM.

[28] Schubert, A., et al., "The TOMAS Mobile Satellite Terminal Implementation," *Third ACTS Mobile Communications Summit,* Rhodes, Greece, May 1998, CD-ROM.

[29] TOMAS Deliverable D 3.2, "Specification of S-UMTS Intertrial Platform," *Doc. No. AC201/UHB/IHN/DS/L/032/b1,* January 1998.

[30] AC201, "Intertrial Testbed of Mobile Applications for Satellite Communications," *Annual Technical Audit,* Brussels, Belgium, January 1998.

[31] Timm, A., "The TOMAS Intertrial Testbed and Supported Services," *ACTS Mobile Communications Summit,* Aalborg, Denmark, October 1997, Vol. 2, pp. 941–946.

[32] Mench, A., "Multimedia Coding for Mobile Terminals in the Framework of the TOMAS Project," *ACTS Mobile Communications Summit,* Aalborg, Denmark, October 1997, Vol. 1, pp. 61–66.

[33] EIES Deliverable D15, "Wireless Multimedia Needs and Scenarios,"*AC075/FTE/DS/R/15,* November 1996.

[34] Deliverable D4, "Specification of the Mobile API for MASE V.10," AC034/IBM/WP2/DS/P/013/b1, November 1996.

[35] Holzbock, M., E. Lutz, and G. Losquadro, "Aeronautical Channel Measurements and Multimedia Service Demonstration at K/Ka Band," *Fourth ACTS Mobile Communications Summit,* Sorrento, Italy, June 1999, CD-ROM.

[36] AC004, " SECOMS/ABATE Final Report," AC004/ALS/SNP/DS/P/062/b1, Brussels, Belgium, February 1999.

[37] Holzbock, M., et al., "Aeronautical Channel Characterization Measurements at K band," *Fourth Ka-band Utilization Conference*, Venice, Italy, November 1998, pp. 263–269.

[38] ASAP Project Web site, www.cpk.auc.dk/asap.

[39] Jahn, A. and M. Holzbock, "EHF-Band Channel Characterization for Mobile Multimedia Satellite Services," *Proc. of IEEE Vehicular Technology Conference,* Boston, Massachusetts, 1998, pp. 209–211.

[40] IPv6 Forum Web site, www.IPv6.org.

[41] http://www.qualcomm.com/.

Chapter 5

Networks

5.1 INTRODUCTION

UMTS was designed to have a terrestrial and a satellite component with a suitable degree of commonality between them, including the radio interfaces with the R&D effort concentrating on the following [1]:

- The level of UMTS support to ATM transmission technology of IBCNs;
- The compatibility of UMTS and fixed-network architecture;
- The location of intelligent functionality—IN and universal personal telecommunications (UPT);
- The level of integration of the satellite component of UMTS;
- The multiservice convergence philosophy of the UMTS radio interface;
- Support of IP.

A group of ACTS projects addressed the short- and middle-term scenarios for the evolution of fixed and mobile interworking and integration, and played an important role in the definition of UMTS. In particular, RAINBOW concentrated on the network concept and has been the major contributor to ITU-T. Its main contribution was the GRAN approach that separated radio-dependent and radio-independent (RD/RI) parts of the network, which facilitated the design of the required interfaces and their interworking with 2G systems (e.g., GSM). RAINBOW introduced the concept of a single generic functional interface between the UMTS radio access part and the core network. The same approach simplified the concept of integration of the satellite component of UMTS (S-UMTS), which was addressed in the SINUS project (see Chapter 4).

The network platform addresses all aspects affecting the network architecture of future mobile networks (UMTS, MBS, and WLAN—see Chapter 3) and their satellite components. It covers network architecture and planning, service definition and provisioning, impact of radio interface features on network aspects, protocol architecture, and interworking aspects.

5.2 TERRESTRIAL NETWORKS

The UMTS technical specifications concentrate on the following key features and requirements: access network, core transport network, security, operation and maintenance, services, and terminals. The evolutionary path of the network platform towards UMTS/IMT-2000 reflected on the standardization process in ETSI and other standardization bodies. Their partnership and standardization efforts continued within 3GPP [2].

UMTS network principles include a general UMTS architecture, the evolution of the GSM platform, the Iu-interface, the UMTS access stratum, and a framework of network functions for support of multimedia services in UMTS. Architectural concepts will continue to develop regarding high data rate and asymmetric data transmission and support of IP. Core issues handle mobility management, transport technologies and their relationship to higher protocol layers, and the definition of the Iu-interface and the split of the functionality between core and radio access networks.

The terrestrial network approach for 3G systems moved towards the globalization of network solutions and the provision of generic concepts for interfaces and interworking. UMTS is based on the UTRA and GSM core network. This is illustrated in Figure 5.1. This concept was later incorporated into the 3GPP framework. ETSI adopted the RAINBOW approach, which separated access (control and user data flow and control flow) and nonaccess strata, as depicted in Figure 5.2.

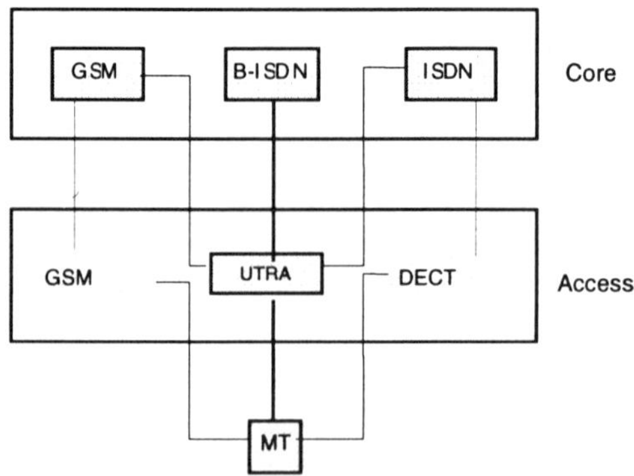

Figure 5.1 The UTRA concept [3].

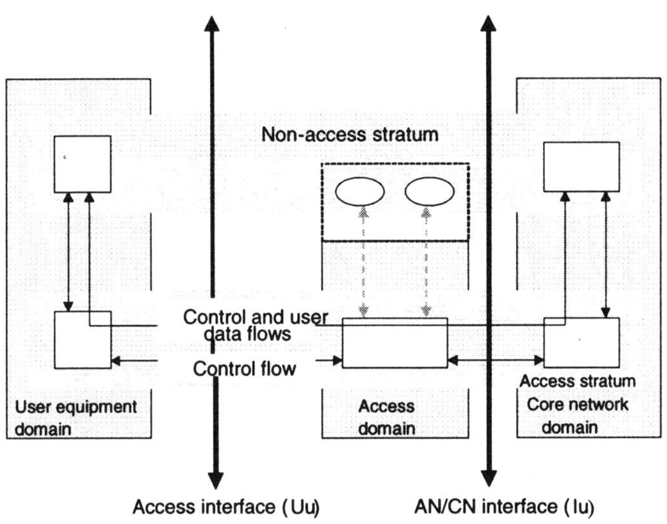

Figure 5.2 ETSI protocol framework.

The globalization approach was introduced in the network concept step-by-step. GSM, as a 2G system, was a starting point of this development. Network concepts towards 3G systems had to take it in account and to offer backward compatibility. The evolutionary path from the GSM network towards UMTS started with the GPRS overlay on the GSM backbone network, as depicted in Figure 5.3. Initially, ETSI SMG decided to evolve the GSM/GPRS core network in order to support both UMTS and GSM radio access by the evolved GSM-UMTS network. UMTS backbone evolved from this concept [4]. The UMTS network principles were formally approved by ETSI and the work now has been transferred to 3GPP [5].

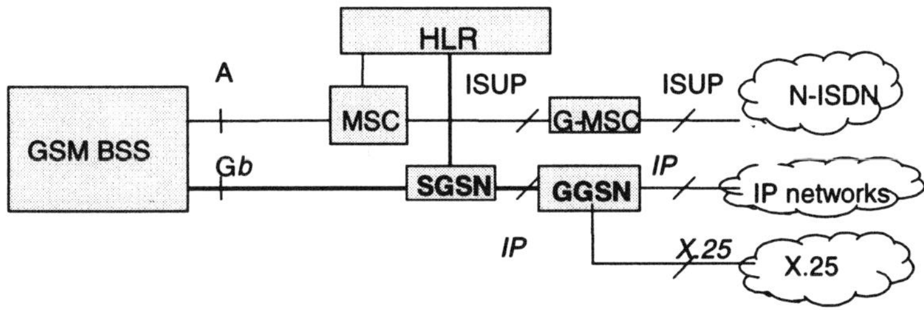

Figure 5.3 GPRS overlay on the current GSM network [4].

Important contributions to the standardization bodies and their decisions were provided from projects in the ACTS framework. The project RAINBOW proposed the approach to the interaction of the new UTRAN and core network (CN) with the existing GSM network.

This was done through IWF units, as shown in Figure 5.4. This was the basis of the GRAN approach that was later incorporated in UTRAN. It made UMTS appear simply as another overlay on the GSM network [4]. The RAINBOW project started solely as ATM-based but evolved to an IP-based UMTS CN (IPv4, mobile-IP, IPv6). Furthermore, the overlay nature of this network made it possible for the operators (especially the new ones) to skip altogether the deployment of GPRS, jumping directly into UMTS.

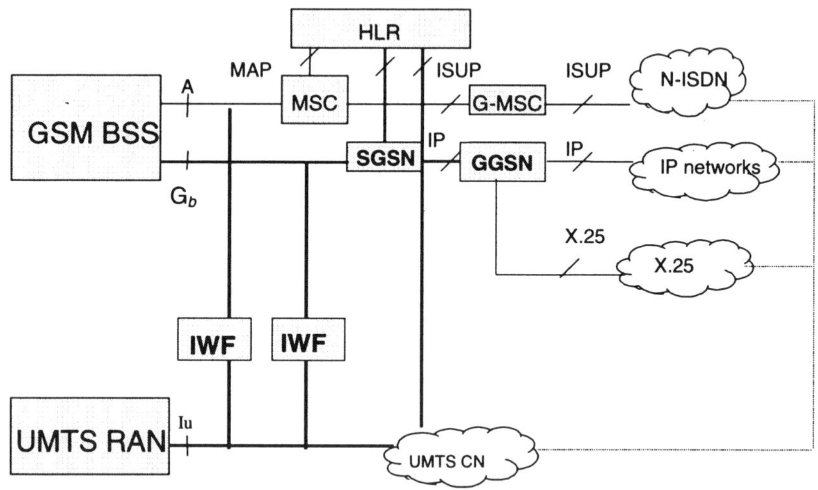

Figure 5.4 UMTS as an overlay on the evolved GSM network.

The network concept evolved towards the requirements of the future UMTS networks. The increasing demand for bandwidth, the changes in the regulatory environment and the introduction of new services have had a profound impact on the architecture and functionality of the future networks. That provided new opportunities and challenges for introducing, for example, ATM-based mobility support, photonic technology, and interworking of different network concepts. The variety of operators, networks, and access techniques and the obvious requirements for interoperability raised several crucial questions about issues such as building up generic architecture and access approach, protocol concept, core network evolution, network planning and management, and security. Several ACTS projects dealt with the terrestrial network issues and played a role in the definition of network topology and appropriate access techniques [6]–[7].

As it was already mentioned, the project RAINBOW studied a generic UMTS access infrastructure and proposed basic concepts for building a network platform. It was complemented by the RAISIN project, focusing on IP issues.

The introduction of 3G mobile systems requires effective software tools able to assist and support operators to design and plan UMTS networks. The project STORMS developed an open software platform, where additional modules could easily be plugged in. The planning strategy of STORMS started with traffic analysis and preliminary dimensioning supported by a geographical information system (GIS). After a preliminary evaluation of the radio-coverage pattern associated with each cell, the optimization procedure was performed in an automatic, generic-algorithm-based coverage optimization module. It optimized the number of radio base stations necessary to ensure a certain QoS requirement. The refined dimensioning module confirmed or corrected the previous decisions. The planning process was based on propagation, traffic, and mobility modules, which could cover all the cases (concerning environment, services, etc.) that were relevant to UMTS. The tools developed within the STORMS project played an important role in the implementation of UMTS and the performance evaluation of the proposed solutions.

Future mobile telecommunication systems will offer an increasing number of new and enhanced services to an ever-growing number of users, involving a variety of operators and service providers. It is essential for the continuing success of this process that security requirements are addressed in an appropriate and timely way. The ASPeCT project was initiated to ensure the feasibility and acceptability of new and advanced security features by implementing and running trials and demonstrations. It addressed the migration issue from existing mobile systems to UMTS and the security mechanisms (secret-key– and public-key–based) required for the future systems. The work of project ASPeCT concentrated on several features: authentication and security functionality required of future UIM; provisioning of end-to-end security services through TTPs; security and integrity of billing in UMTS; and analysis and trials of techniques for fraud detection and prevention. Some of the trials took place on a platform provided by the project EXODUS.

The latter project was focused on the evolution of the core network capabilities, which supported multimedia services (videotelephony, data access, and near real-time monitoring) with personal and terminal mobility. Service provision was envisaged in both wireless and wired environments with home-service availability. EXODUS studied an IN-based distributed mobility management paradigm and achieved improvements in the INAP and UNI signaling protocols, the service control point (SCPs), and the ATM switch with enhanced mobility support.

An interesting combination between fiber and wireless technologies for the access network was suggested by the ACTS project FRANS. It was focused on enabling optical support for radio technologies to facilitate a rapid rollout of IBC services before fiber is deployed to the home. The project developed and established field trials for an optically supported millimeter-wave link as a final drop to the subscriber to provide fast and flexible deployment of broadband interactive services. FRANS analyzed two specific technical arrangements of a hybrid, fiber wireless-access system, working at different bit rates and based on different multiple access techniques.

The heterogeneity of the network platforms and terminals is of vital importance for the provision of multimedia services. The ACTS program acknowledged this significance by adopting the work of the projects SUMO and VANTAGE.

The project SUMO was related to the projects SINUS and TOMAS as presented in Chapter 4; it aimed at identifying and demonstrating generic approaches to service provision and network control over S-UMTS. It developed a trial platform to support service provision through two different satellite access schemes. SUMO addressed interoperability between complementary satellite systems and terrestrial core networks.

The integration process benefited further from the work done in the scope of the project VANTAGE. This project addressed the complementary role of the two types of networks, terrestrial and satellite. In particular, the project demonstrated service flexibility of ATM and access flexibility of satellites. It implemented a novel architecture using a conventional transparent satellite with an Earth station (VSAT) functioning as a distributed ATM switch and moved the network boundary to service provision for remote and isolated users. The project VANTAGE was unique as it demonstrated the simultaneous work of an infrastructure provider and a trial conductor.

5.2.1 Radio Access Independent Broadband on Wireless—The Project RAINBOW

The project RAINBOW defined and implemented a generic UMTS access infrastructure involving the UMTS core network and the radio access parts [8]. RAINBOW developed a concept to cope with different radio-access techniques (long-term view) and guarantee a smooth migration from 2G to 3G systems (short-term view). This concept decoupled the radio access from the core network and enabled the connection of different radio-access modules (2G or 3G) to the same network infrastructure via a generic interface. It also split the radio-access network into a radio-dependent and a radio-independent part. This provided a modular approach (GRAN), which maximized the commonality within the access network. The main idea was to develop a future-proof approach that will satisfy the requirements of the different application and propagation environment in which UMTS would be used.

The RAINBOW project was proposed as a continuation of several studies performed in the RACE II program concerning UMTS functional and network architecture, namely the MONET project. In the radio interface domain it relied on the outcome of CODIT and ATDMA RACE II projects.

RAINBOW investigated conceptual and architectural issues by developing a demonstrator to prove the feasibility and evaluate the complexity of the solutions. The RAINBOW demonstrator, built up on an emulated platform, could model the radio, traffic, and user mobility context. It demonstrated various architectures and topology options, relevant transport and control functions, and migration requirements. Joint demonstrations with other ACTS projects, such as FRAMES (see Chapter 3), FIRST (see Volume Two), and MULTIPORT (see Volume Two), were performed.

The RAINBOW project had cooperation activities with [9]:

- *COBUCO and EXODUS:* Protocol harmonization for DECT migration;
- *SINUS/SUMO:* Protocol adaptation to the satellite segment via reuse of the demonstrator software;
- *OnTheMove:* Integration with a client-server architecture;
- *INSURED:* Harmonization of UMTS reference model.

The RAINBOW concepts, and namely GRAN, were presented and promoted in ETSI SMG and ITU-R TG8/1 with the purpose of obtaining a unified view of the evolution of the standards towards UMTS.

The RAINBOW project investigated architectural and integration issues through the implementation of the transport and mobility control functions expected for UMTS. This implementation was based on a conceptual design developed in RAINBOW that allowed the generic UMTS access infrastructure to cope with different innovative radio-access techniques (long-term view) and to guarantee a soft migration from 2G to 3G systems (short-term view).

The main objectives of the RAINBOW project were the following.

- To demonstrate the feasibility and evaluate the complexity of the UMTS access infrastructure able to cope with different UMTS innovative radio access techniques and to present a solution for the migration from 2G mobile systems (GSM and DECT) to UMTS [10];
- To contribute to the UMTS standardization process in ITU;
- To study the possible solutions for the integration of the UMTS radio access system in the B-ISDN and IN context for both transport and control procedures;

- To explore the impact of multimedia services and variable bit rate techniques on the transport and control procedures of the UMTS radio access system.

5.2.1.1 Technical Approach to the Concept of RAINBOW

The RAINBOW approach was based on the definition of a generic interface between the access and the core network and the split between radio-dependent (RD) and radio-independent (RI) functions in the radio access system (RAS) [11]. Decoupling radio access from the core network allowed independent design and enhancement of these two parts. It provided the possibility to use more than one radio interface at the same time, and to optimize these interfaces to different environments. The generic interface concept, as depicted in Figure 5.5, had the ability to combine different access modules with the same core network, and to combine different core networks with the same radio-access module. This gave operators the freedom to select their own evolution path towards UMTS, and still ensure a convergent solution, which maintained a high degree of compatibility and interoperability. It also provided the smooth evolution from GSM and from a fixed network towards a convergent, yet flexible UMTS solution. The RAINBOW protocol-structure concept is depicted in Figure 5.6.

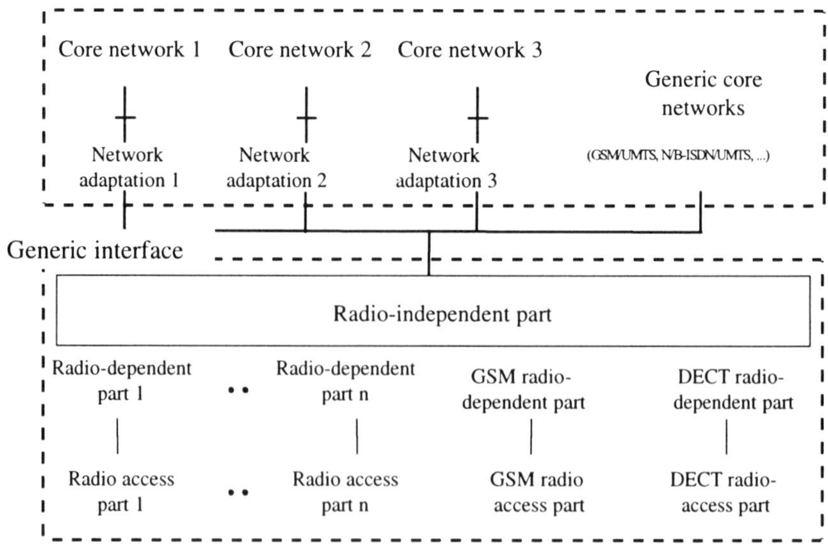

Figure 5.5 RAINBOW system concept.

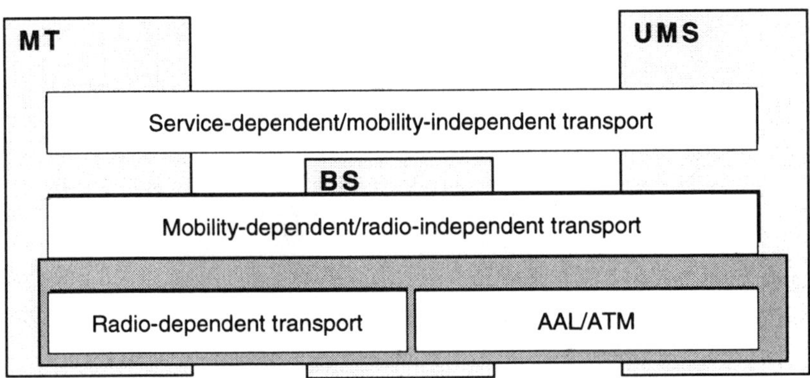

Figure 5.6 RAINBOW protocol structure.

In the development of the RAINBOW architecture and protocol concept a division between control and transport functions was used in the design of the considered nodes. The transport plane was responsible for the transfer of user data and control messages between the nodes. Many functions performed for the transport of control and signaling were different and so the transport plane was subdivided into user data transport and signaling transport functionalities [12].

The transport functions of the RAINBOW architecture were divided into three groups:

- Basic transport (radio-dependent and fixed-network-dependent);
- Mobility-dependent/radio-independent transport;
- Service-dependent/mobility-independent transport.

The RAINBOW transport plane architecture is presented in Figure 5.7.

For each radio-access technique a different radio adaptation layer (RAL) is required. The mobility adaptation layer (MAL) dealt with handover, macro-diversity, and synchronization. The RTE-enhanced logical layer (RELL) improved the services of a radio interface emulator to provide the functionality similar to that provided by the 3G radio interfaces under investigation (i.e., CODIT and ATDMA). The mobility-interworking unit (M-IWU) was responsible for the interworking between the RAL and the fixed network. It also provided the necessary synchronization mechanisms.

A third category of transport functions was built on the mobility-independent functions of the underlying layers. The service adaptation layer (SAL) and the signaling network layer (SNL), existed in this category. SAL was responsible for speech coding, transcoding, and user plane interworking. SNL was present in all nodes and was used for the transparent routing of signaling messages.

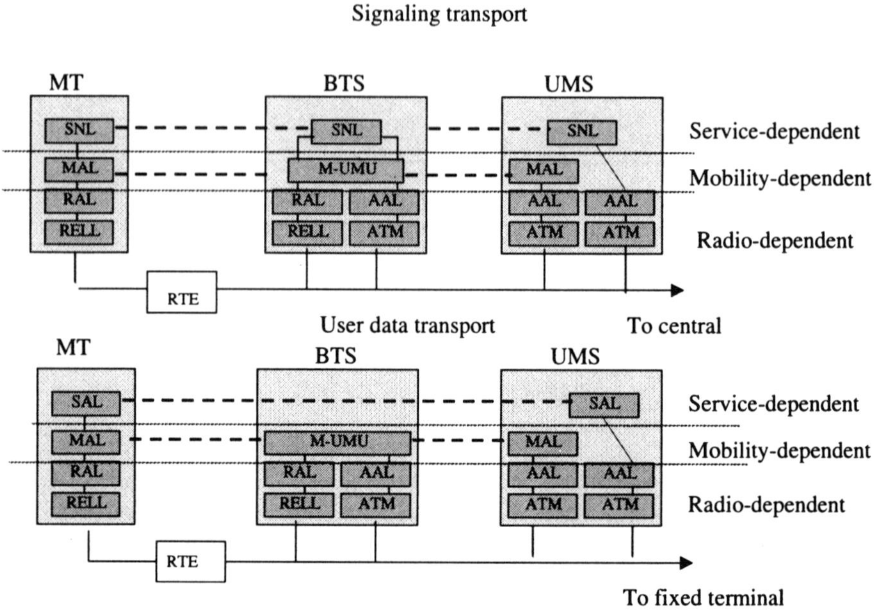

Signaling transport

MT BTS UMS

SNL SNL SNL Service-dependent

MAL M-UMU MAL Mobility-dependent

RAL RAL AAL AAL AAL

RELL RELL ATM ATM ATM Radio-dependent

RTE

User data transport To central

MT BTS UMS

SAL SAL Service-dependent

MAL M-UMU MAL Mobility-dependent

RAL RAL AAL AAL AAL

RELL RELL ATM ATM ATM Radio-dependent

RTE

To fixed terminal

Figure 5.7 The RAINBOW transport plane architecture.

The following ideas developed in RAINBOW complete the projects approach:

- RD/RI split;
- IN functional model;
- ATM transport;
- B-ISDN call control.

The division of RD and RI functionality provided a modular approach and was by ITU-R. It identified the functionality, which was part of the common infrastructure for several radio-access techniques, and specific functionality, which needed to be added/replaced when introducing a new radio technology. The RD/RI classification reflected radio efficiency and network efficiency requirements in the access and network parts, respectively.

The IN functional model for mobility management was based on work carried out in standardization bodies such as ETSI and ITU. The mobility management included location management and handover control. The project performed conceptual studies for service and mobility provision and investigated the most suitable protocol for the requirements of UMTS.

The transport technology, initially chosen in the RAINBOW project, was ATM because at the time ATM seemed to be the transport technology solution for the future IBC. ATM was also used as the reference point between the access network and the core network (Iu reference point) and for the packet-switched part. AAL5 was selected as the most appropriate adaptation layer. Within the project, however, other ATM adaptation layers were considered depending on the type of data—AAL2, AAL1, as well as other transport technologies (frame relay, IP) [13].

The call-control, signaling basis used in the RAINBOW project was the B-ISDN call control Q.2931, as defined by ITU-T. Some modifications were made to take into account mobility, which was not supported by Q.2931. Additional modifications dealt with service aspects and channel coding. RAINBOW assumed the split between the call and bearer control, so some additional bearer control signaling was required. A lighter version of the protocol was used for the demonstrator purposes. Also, some analysis with purposefully designed bearer control protocol that managed the ATM resources and with possible B-ISDN/IP evolutions paths was performed.

5.2.1.2 Demonstrator

The main objectives of the RAINBOW project were achieved through a testbed-demonstrator implementation of the UMTS access infrastructure with relevant transport and control functions [11]–[14]. The interworking/integration with IBC was achieved by using an existing ATM switch at a national host to provide B-ISDN control capabilities and ATM transport in the core/access network.

The testbed was working in real time in order to perform the explicit laboratory trials with voice and data traffic. It was built up on an emulated platform to reflect real scenarios modeling the radio and channel traffic and the mobility context [15]. Migration solutions were implemented by relevant architectural solutions and specific laboratory demonstrations adopting 2G systems (GSM and DECT) on the UMTS access infrastructure. Innovative radio access techniques, developed by other ACTS projects, were managed by the generic mobility control functions.

Architecture

The RAINBOW demonstrator functional architecture is shown in Figure 5.8.

The network entities provided in the demonstrator architecture were the base transceiver station (BTS), cell site switch (CSS), UMTS mobility server (UMS) access, mobility and service control point (MSCP) access, local exchange (LE), UMTS mobility server (UMS) core, mobility, and service control point (MSCP) core [16]. The CSS and LE were the access and the core network components of standard fixed network switching infrastructure. The CCS was added to allow the different nodes of the demonstrator to be configured for different demonstrations and transport services.

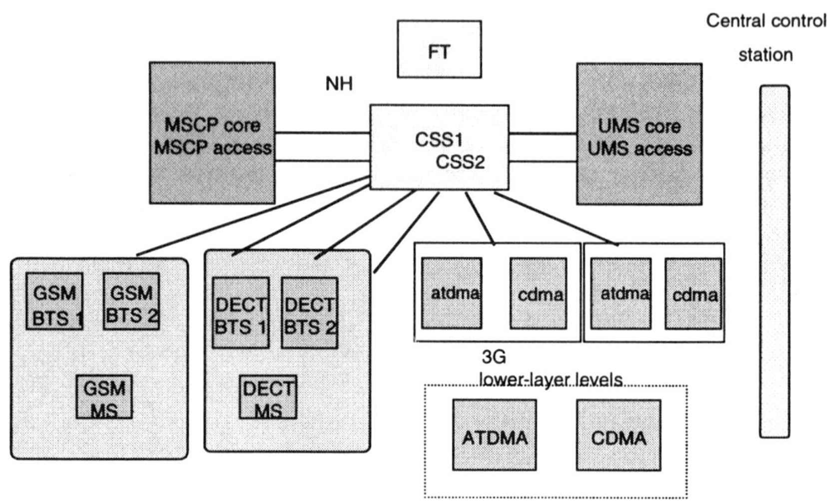

Figure 5.8 The RAINBOW demonstrator functional architecture.

The MSCP supported the service-control function (e.g., mobility management, and handover initiation) in the access and core network. The UMS was introduced in both the access and core network for taking care of special functions, which were required by mobile networks, but could not be easily provided with a standard switching network (ATM). The UMS was considered as a solution to support 3G mobility in B-ISDN.

To achieve the goal of radio independence, the nodes of the RAINBOW demonstrator were designed with a high level of radio independence. All the fixed network nodes (e.g., MSCP, UMS, CSS, and LE) were completely radio-independent. These fixed network nodes were only affected by the services to be supported, the applications that were used to support them (e.g., codec type), and architectural choices made in the network configuration (e.g., higher-layer macrodiversity provision). This functional architecture concept was used to develop the physical demonstrator concept, which is depicted in Figure 5.9. For the 3G radio interface, a radio emulator was used to interconnect the MT to the BTSs. Real radio interfaces (GSM, DECT) were used in the case of the 2G systems.

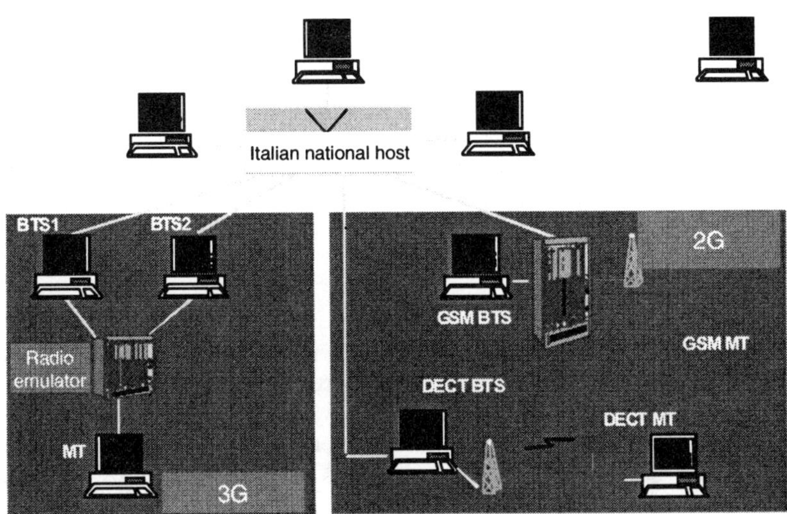

Figure 5.9 RAINBOW demonstrator physical architecture.

The RAINBOW demonstrator was used to demonstrate the following features of next-generation mobile systems:

- Radio independence;
- Topology and functional allocation;
- Migration;
- Handover and macrodiversity.

It served to test various architecture and topology options and various functional allocation options by means of a flexible architecture that could be easily reconfigured.

The functionality of the demonstrator was modeled by dividing the functionality of each node into control and transport functional groupings, creating two largely independent layered-network protocol stacks [5]. The control plane and transport plane processes communicated using XDR-encoded messages.

The smooth migration from 2G to 3G systems was addressed in RAINBOW. In order to demonstrate the feasibility of such migration, RAINBOW included DECT-access networks (DAN) and GSM base station system (BSS) in the demonstrator. For DECT the integration strategy was adopted, while for GSM the interworked scenario was considered more feasible. A central part of the migration scenarios was the use of IN functionality for the provision of mobility.

Macrodiversity and its support in the fixed network were one of the more innovative and relevant features demonstrated in the project [17]. The macrodiversity was one of the most restricting features in the development of a radio-independent network. The RAINBOW approach was based on results of the RACE II CODIT project adapted to the RAINBOW architecture. The RAINBOW macrodiversity solution was not fully compliant with the UTRAN approach [18]. However, many of the macrodiversity aspects tackled in the development of the RAINBOW demonstrator were relevant to UTRAN.

Joint Demonstrations

Within the mobile and personal communications domain of the ACTS program, contacts between different projects that addressed complementary issues were established in order to define and perform joint UMTS demonstrations. The generic approach used in the RAINBOW demonstrator provided a strong platform for trials and demonstrations of concept and performance issues addressed in Volume 2. A special interest group (SIG2) dealing with hardware trials was created involving representatives of the projects FIRST, FRAMES, MultiPort, and RAINBOW. A common proposal for joint demonstrations was agreed upon by these projects, developing the scope and workplane of the envisaged trials [20]. The joint demonstration based on the RAINBOW platform will also be addressed in Section 5.2.2.

The RAINBOW project reflected some of the ACTS program strategies in the area of mobile communication for implementing the UMTS-system demonstrators. It developed the conceptual studies of the UMTS access network and proposed a reference configuration for the UMTS system with a single generic functional interface between the UMTS access part and the core network and the separation of the RD/RI parts. These two concepts were submitted and adopted by ETSI SMG and ITU-R to obtain a unified view on the evolution of standards towards the UMTS.

The other important achievements were the following.

- The design of a real-time emulator;
- The description of the trials and the identification of the innovative features that the UMTS will provide;
- The complete specification of the demonstrator architecture in terms of hardware/software platforms and interfaces between the various entities of the testbed.

The results provided by RAINBOW had a strong impact on network operators, manufacturers, and academic groups involved in implementation and development of UMTS concept.

The project provided the precise knowledge of possible UMTS network configurations and their relevant flexibility in terms of service/environment applications that could undoubtedly benefit the network operators. Migration paths from 2G to 3G systems were also defined and investigated.

Manufacturers could verify the basic characteristics of the developed UMTS entities. The radio-access segment of UMTS and its feasibility and complexity were also verified. RAINBOW addressed the network performance requirements and the relevant impact on the control structure of the implemented entities.

The modeling of innovative radio access techniques, traffic studies, and new telecommunication processes, and the definition and performance evaluation of UMTS architectures were undertaken by the project RAISIN to be discussed in Section 5.2.2.

5.2.2 Extension for Trial Integration—The RAISIN Demonstrator

RAISIN was the extension of the project RAINBOW concentrating on new issues to be considered for UMTS. An attempt was made to prove the concept of a UMTS generic network architecture by performing a joint demonstration with the project FRAMES. The project RAISIN implemented a connection-oriented approach within the existing RAINBOW demonstrator. The project RAISIN followed two core objectives. The first one involved proof of flexibility of the RAINBOW demonstrator and its concepts by adapting it to support a real radio interface, which was aligned to target the UTRA interface [21]. The second objective related to the performance of a detailed investigation of the support of the Internet in the UMTS system. This objective was particularly relevant to the explosive growth of the Internet and its currently perceived importance in UMTS.

As an extension to RAINBOW, the project RAISIN used its demonstrator to implement the desired objectives. The first objective of the project was verified by implementing a joint demonstration with the project FRAMES [22]. The second objective of investigating UMTS/the Internet was undertaken by a combination of demonstrations and concept studies. Interswitch mobility-management procedures were also considered within RAISIN with an emphasis on handover control protocols. These aspects of handover, which could not be considered in RAINBOW because of time limitations, were specifically investigated here. Because of the increase in the relevance of the Internet to UMTS, and telecommunications in general, a significant portion of the total RAISIN effort aimed at UMTS/Internet activities.

5.2.2.1 Joint Demonstration with FRAMES

The following objectives were targeted for joint demonstration between the projects RAINBOW and FRAMES:

- Capability of RAINBOW to handle different radio interfaces with the same radio access network;
- Higher bit rate transmission than GSM in outdoor environments;
- Flexibility between environments to be provided by UMTS.

The UMTS networks were able to cope with different (UMTS) innovative radio access techniques and to provide a solution for the migration from 2G systems (e.g., GSM, DECT) to UMTS [23]. The project RAINBOW demonstrated the feasibility and evaluated the complexity of a UMTS access network. Because of absence of available implementations, CODIT and ATDMA radio interfaces were emulated in the project. Therefore, it was necessary to implement the generic mobility control functions, designed and implemented in the project RAINBOW, with an actual radio interface designed and implemented in FRAMES and representing the TD/CDMA mode of UTRA [24].

The demonstration of high bit rate services was implemented in RAISIN. The flexibility between environments, which was an important consideration, was demonstrated by conducting the trials for the indoor and outdoor environments.

5.2.2.2 Joint Demonstrator System Architecture

The existing RAINBOW demonstrator, which contained a generic RAS infrastructure capable of accommodating innovative radio access techniques, was used for the joint demonstration. It was built on the radio-independent nature. The RAINBOW radio path emulator and lower protocol layers were replaced within the joint demonstrator by the radio interface demonstrator provided by FRAMES. The FRAMES demonstrator had a mobile station and a base station. This provided the radio lower layer functionality necessary to allow the provision of call management and mobility procedures. To facilitate the interconnection of the two demonstrators, a generic interface was defined between the RAINBOW and FRAMES demonstrators. Within the respective projects an adaptation layer was developed to interwork to/from the generic interface to demonstrator specific functionality. The adaptation layer required for the FRAMES demonstrator was referred to as the external adaptation layer (EAL), while the adaptation layer required within the RAINBOW demonstrator was referred to as the FRAMES adaptation layer (FAL). The relationship between the RAINBOW/FRAMES demonstrators is shown in Figure 5.10.

Both the MT and BTS consisted of higher layer components provided by RAINBOW and lower layer components provided by FRAMES. These components were interconnected through ATM. The higher layers of the BTS were connected to the fixed network part of the RAINBOW demonstrator. The RAINBOW end user could communicate with an end user terminal or service center. The joint trials consisted of speech and data services demonstration. A CODIT codec at 16 Kbps was used for speech service demonstration. To demonstrate data services a WWW session was used at different data rates (64 Kbps and 144 Kbps). A call had to be established between the RAINBOW/FRAMES MT and a fixed terminal for each of the above services.

Both the MT and BTS consisted of higher layer components provided by RAINBOW and lower layer components provided by FRAMES. These components were interconnected through ATM. The higher layers of the BTS were connected to the fixed network part of the RAINBOW demonstrator. The RAINBOW end user could communicate with an end user terminal or service center. The joint trials consisted of speech and data services demonstration. A CODIT codec at 16 Kbps was used for speech service demonstration. To demonstrate data services a WWW session was used at different data rates (64 Kbps and 144 Kbps). A call had to be established between the RAINBOW/FRAMES MT and a fixed terminal for each of the above services.

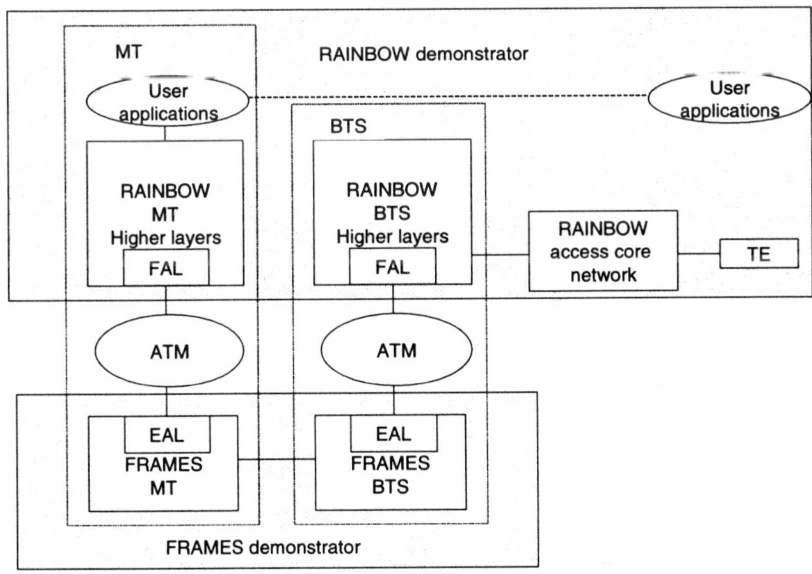

Figure 5.10 Interconnection of the RAINBOW/FRAMES demonstrators.

5.2.2.3 UMTS Internet

Internet Access in RAINBOW

Internet access in RAINBOW was implemented via an IWF, which was external to the UMTS system (see Figure 5.11). A UMTS data call was established between the MT and the home Internet service provider because the UMTS and IP networks were completely separated. RAINBOW implemented a connection-oriented approach with a remote access to the "home" Internet service provider (ISP) via the fixed network, where the UMTS operator was not an ISP. The IP application that was used within this demonstration was the WWW browsing.

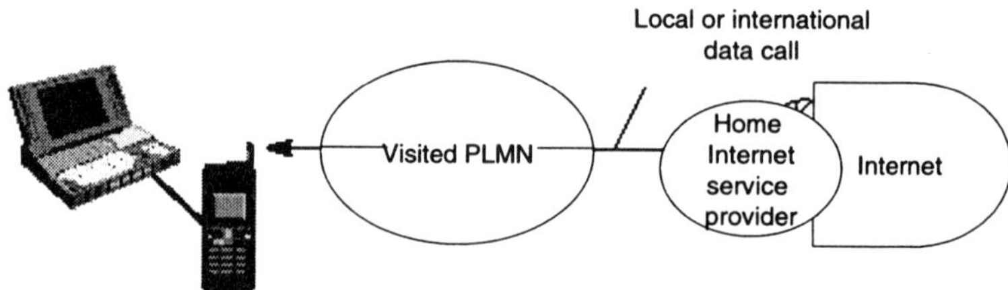

Figure 5.11 Internet Access in RAINBOW.

Internet Access in RAISIN

RAISIN implemented a more integrated approach to Internet access within the existing RAINBOW demonstrator (see Figure 5.12). It involved a connection-oriented approach with a local access to a local ISP via the UMTS network. The IP application that was used for this demonstration was WWW browsing.

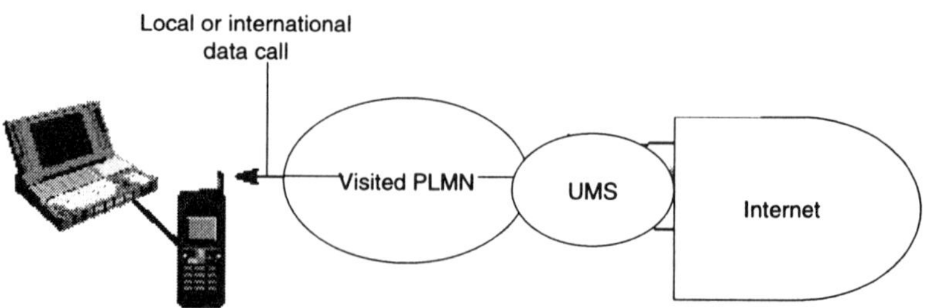

Figure 5.12 RAISIN-implemented Internet access.

On the radio interface, the data call was handled differently for different types of data in an optimized way. For the ATDMA approach a version of the packet reservation multiple access (PRMA++) was implemented and the radio channel resources were released between subsequent packets [25]. In the CDMA case the variable bit rate feature was exploited. This meant that when the bit rate was zero, then no data was transmitted.

UMTS Internet Studies in RAISIN

Some important issues that were not considered in implementing UMTS Internet were addressed in RAISIN project. These issues involved alternatives for support of Internet services, transport of UMTS services in a UMTS network, relation between the mobility procedures, and the impact of radio environment on the behavior of the TCP/IP protocol.

There are different alternatives for the support of Internet services. Provision of Internet service integration with provision of circuit-switched services over the same UMTS access network was one critical aspect of study. In the project RAISIN, studies were undertaken for finding the various possibilities to support IP services over UMTS. Two extreme possibilities in this direction that were considered were the cases where UMTS radio access with IP technology only provided the network functionality and where complete UMTS network was providing users with a mobile link to a dedicated Internet access point.

The transport of UMTS services in a UMTS network was another critical issue. IP applications could be viewed as any other data service and transported over the radio link and over ATM in the fixed part of the UMTS network without considering any IP functionality within the UMTS network. On the other hand, all services including voice and other real-time services were transported using IP over ATM and IP network functions within the UMTS network. RAISIN considered different kinds of solutions where both IP and ATM transport techniques were used. The solution was developed based on solutions from both extremes.

Mobility procedures for UMTS and the mobility procedures for the Internet world based on mobile IP (IPv4 or IPv6) were also considered in RAISIN. The provision of QoS has an important role to play in the IP world. An evaluation was made, therefore, on the impact of the radio environments on the behavior of the TCP/IP protocol and the necessary modifications to TCP/IP protocol for use in the mobile IP environment.

5.2.2.4 Advanced Handovers

Handover is needed between any environment, operator domain, or service provider. In the project RAINBOW, only the inter-CSS handover and the inter-BTS handover were studied. In RAISIN, however, the RAINBOW studies were extended and the handover of multiple bearers and the handover between two different LE based domains/networks were included.

Multibearer Handover

There is a need for providing multiple bearer connection across the radio interface to the user because a single radio bearer cannot support high bit-rate user services (HBRS) in UMTS. HBRS can be provided by multiple bearer connections through multiplexing and demultiplexing functions. Multiplexing and demultiplexing functions always exist in the mobile terminal and they are located in the network at the BTS or at the LE level.

Multiplexing/Demultiplexing at BTS Level

In this scenario, it is assumed that the BTS is assigned the multiplexing/demultiplexing role. This implies that the support of multibearer service is restricted to the setting up of the appropriate number of bearers only on the radio interface. In the fixed part of the access network the different data streams on the radio interface are merged to a single ATM connection. In this way, the support of multibearer service by the fixed part of the access network is not perceptible.

Multiplexing/Demultiplexing at LE Level

This assumes that the different radio bearers are mapped onto separate ATM connections in the fixed network part, up to the LE level. Different options for supporting multibearer services are shown in Figure 5.13. Each stream undergoes independent processing in the lower layers.

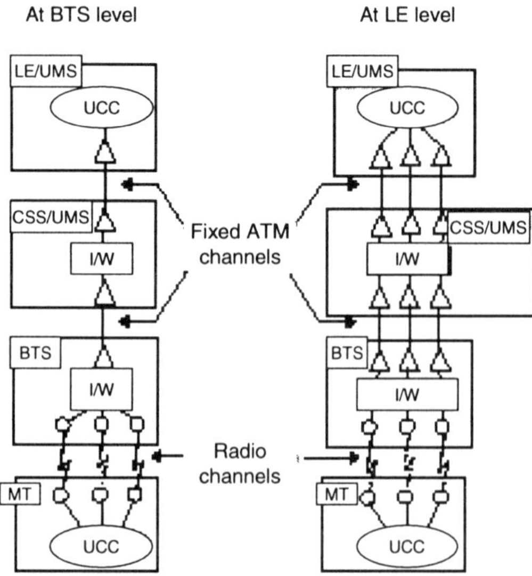

Figure 5.13 Options for supporting multibearer services.

InterLE Handover

Different options for multibearer have been investigated, analyzed, and compared with regard to their impact on handover flexibility, complexity, and bandwidth management. The following objectives of the interLE handover were investigated in the RAISIN project:

- To extend the handover control protocols, developed in RAINBOW, to support two RASs of different environments (e.g., from business to public and vice versa);
- To investigate the key signaling protocol requirements of the UMTS core network in order to support such a handover.

When the LE to which the MT was connected changed during a call/service, the bearer connections with the other parties in the call/service could be maintained with a different mechanism. For example, new bearer connections towards the fixed network can start from the new LE and the optimal path through the network selected. Alternatively, an "anchored," GSM-like approach can be adopted. Some signaling relationships between the MT and the network, which are relevant even after interLE handover, should be transferred instead of reinitiated [25]. RAISIN investigated the following generic issues in the context of handover:

- Definition of alternative B-ISDN/IN core network architectures that could be employed for the target UMTS;
- Evaluation of the capabilities offered by the existing network-to-network interface (NNI) signaling protocol standards (e.g., ATM and IN) for the support of the interLE handover;
- Examination of the feasibility of the "anchored" and the "cross-over point discovery" approaches with an aim to defining a suitable NNI handover mechanism;
- Adaptation of the handover protocols developed to date to the above mechanism.

Three key areas were addressed in the project RAISIN. First, a joint trial was performed between RAINBOW and FRAMES. The trial added significant value to the RAINBOW concepts. It proved that the RAINBOW demonstrator could be adapted to support an innovative radio access technique similar to the TDD option in the UTRA.

Secondly, the following critical issues for UMTS were considered and their solutions obtained for the RAINBOW demonstrator:

- Solutions for support of Internet services and evaluation of their impact on the UMTS architecture as seen in RAINBOW;

- Identification of the relationships between IP technology and UMTS functionality in these solutions;
- Integration of providing Internet services and their use in circuit-switched services over the same UMTS access;
- Identification of common and complementary aspects of GSM/UMTS mobility support and Internet mobility support (mobile IP).

Last but not least, two critical aspects of handover (i.e., interLE handover and multibearer handover) were investigated and RAINBOW protocols modified to encompass these types of handover.

5.2.3 Software Tools for Optimization of Resources in Mobile Systems—The Project STORMS

The aim to integrate mobile network planning in a global process towards 3G UMTS resulted in the launch of the project STORMS. The intention was to develop an integrated tool applicable to UMTS and comprehensive of all steps of the planning process.

STORMS defined, designed, and developed a complete suite of network-planning software prototypes addressing UMTS and starting with a definition of the categories of UMTS environments and the identification of the most appropriate propagation model [26]–[27]. A critical review of existing results from other studies, projects, and initiatives (e.g., RACE CODIT, MONET, and COST 231) was carried out and the planning methodology was focused on CDMA and ATDMA [27]. After ETSI accepted the UTRA concept, a generic air-interface dimensioning methodology for WCDMA and TD-CDMA was specified [28].

STORMS developed also a piece of hardware, which acted as an automatic subscriber unit. This unit was used for measurements on a live network, both for objective and subjective parameters. It can also be used to compare the measurements with the simulation results.

The project STORMS developed close collaboration with the other ACTS projects [29]. The most important cooperation was established with TSUNAMI II (adaptive antennas), SUCOMS (superconductivity), and RAINBOW (innovative access networks). STORMS participated in the BAM chain, where a common area of interest was the possibility of sharing an access network infrastructure among different service providers distributing fixed and mobile services. The results of STORMS contributed to the activities undertaken in the standardization groups (mainly ETSI SMG2 and ITU for the air interface development).

Although STORMS was mainly focused on 3G mobile systems, a certain degree of "backward" compatibility was, however, maintained as joint planning of UMTS and GSM networks is envisioned. STORMS aimed at defining, implementing, and validating a software tool capable of supporting the planning of 3G mobile networks, UMTS in particular.

The primary, easily measurable objective was the implementation of a prototype capable of handling all currently defined and known environments, features, and services that included voice, SMS, data, and video. For both systems and services, the reference scenarios were based on already deployed networks such as GSM, DECT, and HIPERLAN.

To enrich the tool, the project aimed at extending the UMTS network to face both an increase of subscribers and an addition of new services and innovative features (advanced radio schemes and new technology functionally). Such flexibility was approached through appropriate software tools within the software-planning phase. The STORMS tool was designed and implemented as an open software platform capable of hosting external modules and proprietary procedures tailored to specific needs of potential users.

In order to achieve these objectives, STORMS concentrated on the following objectives:

- Characterization of the UMTS environment and identification of UMTS service scenarios (creation of a comprehensive library/data base);
- Mapping of the UMTS environment and services onto user/network requirements;
- Mapping of UMTS environments onto propagation, traffic, and mobility models (development of algorithms and development of software code);
- Dimensioning of radio resources to meet preassigned QoS performance objectives (development of optimization algorithms);
- Fixed-link optimization for optimal UMTS nodes interconnection (development of optimization algorithms);
- Implementation of a UMTS simulation model to evaluate network performance and validate assumptions agreed in the design and planning phase.

5.2.3.1 Technical Approach—The UMTS Planning Methodology

STORMS started by defining a good planning strategy. The generic planning process of STORMS consisted, as shown in Figure 5.14, of certain steps: the initial dimensioning, the automatic radio coverage optimization, and the refined dimensioning. The STORMS platform had incorporated a powerful simulator to assess the variety of network performances.

Initial Dimensioning

The initial dimensioning phase performed initial, rough network dimensioning used to start up the planning loop. It was responsible for developing tools necessary to access the traffic distribution on a per-service, per-pixel, and per-layer basis [if hierarchical cell structures (HCSs) were adopted]. This stage performed a preliminary, approximated identification of reference traffic environment ranging from indoor coverage (picocells) to regional coverage supplied by satellite systems, integrated with a terrestrial cellular structure. The measurement results of the demographic and the geographic database were used in preprocessing to estimate the traffic distribution as a function of a particular environment (population density, street density, telephone service user's distribution, data services user's distribution). The traffic segregation criteria regarded either service requirements (minimum C/I or bit rate) or capacity requirements (traffic density).

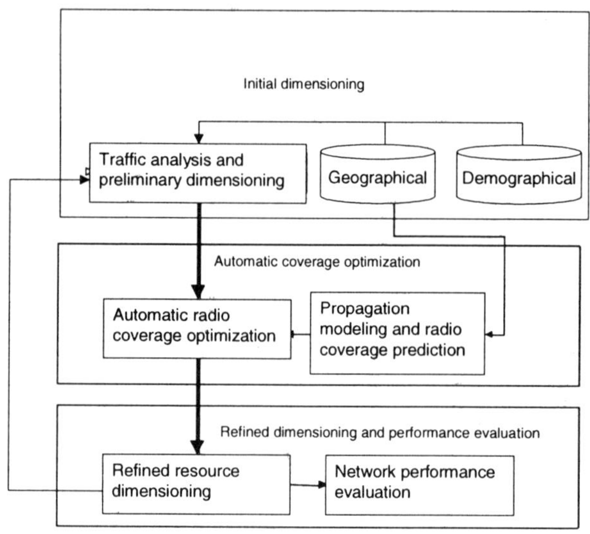

Figure 5.14 Global network planning process proposed by STORMS.

The expected results of this module was the provisioning of some estimates related to the maximum cell sizes that could be tolerated in the planning area as a function of the traffic intensity and services mix. At this initial stage no coverage prediction was performed yet.

Traffic analysis provided input to the preliminary dimensioning. The major objective of this phase was to identify the minimum density of the required BTS per subarea. The cell layout was not yet available and the preliminary dimensioning phase was a simplified process based only on traffic data. Initially, areas (per layer) where traffic demand could be assumed uniformly distributed were identified as subareas. The traffic offered by each subarea was compared to the capacity of typical BTS configurations, which were available in the tool libraries. This determined the minimum BTS density per subarea.

This strategy fitted both WCDMA and TD-CDMA. The capacity-dimensioning for WCDMA derived the maximum cell size fulfilling two constraints, in the uplink, interference/noise limitations, and in the downlink, the hard limit imposed by the channelization codes. TD-CDMA initial dimensioning was based on the available traffic maps. The maximum cell size was derived from total load calculated for every service and every pixel of the area under consideration. This determined the minimum BTS density per area.

Automatic Coverage Optimization

The automatic coverage optimization stage comprised two phases: prediction and optimization. In the first phase, using an appropriate object-oriented module, coverage prediction was performed. Within it, the BTS transmitted power was adjusted aiming at maximum coverage with respect to minimum BTS density (specified by the initial dimensioning). This addressed the propagation issues (geographical and morphological area characteristics), capacity-related constraints (the minimum BTS density), service characteristics (radio-level QoS requirements), and system features (e.g., adaptive antennas).

The coverage calculations were followed by radio-optimization software, which determined the best position of the base transceiver stations (BTSs). STORMS developed a fully object-oriented platform on which all types of propagation models could be developed. This software was called the topological geographical information system (TGIS). It contained many modules, which could be used to efficiently apply ray-tracing algorithms. Within STORMS, a whole set of relevant propagation environments (macro-, micro-, and picocellular) and propagation conditions (2D irregular terrain, over-roof propagation in urban macrocells, horizontal propagation in urban environment, 3D propagation in complex environment, building wall penetration) were tackled.

The prediction phase was followed by an optimization process where the requirements in terms of fixed network infrastructure corresponding to the radio channel dimensioning are assessed and optimally evaluated according to cost/effectiveness criteria. The radio coverage optimization module is considered essential for starting any planning activity and is based on generic algorithms. An initial redundant set of BTS specified by the operator is the starting point of radio coverage optimization.

The optimization-procedure objective was to determine the minimum subset of BTS, which guaranteed the target-coverage degree at minimum cost, while respecting the capacity constraints of the initial dimensioning phase. The optimization modules use the previously mentioned propagation libraries. The block-diagram of the optimization process is presented in Figure 5.15. This stage produced a radio-network configuration and the relevant multilayer layout. It should be noted that STORMS was designed to cope with different air interfaces, and from a planning and optimization point of view, the design of a multisystem network was absolutely equivalent to that of a multilayer network.

Refined Dimensioning and Performance Evaluation

When the cellular layout was available, a refined dimensioning took place [30]. STORMS developed a generic, radio-resource dimensioning method, which is depicted in Figure 5.16 and validated for the main UMTS air-interface proposals, namely UTRA. In the case of new types of air interfaces it provided maximum reuse of existing modules and easy customization.

The generic approach included tasks such as traffic load scenarios, radio resource demand per BTS, air interface capacity, radio-resource allocations phase, power control handling tool, calculation of carried and overflow traffic.

In particular, the WCDMA and the TD-CDMA dimensioning process knew some specific planning steps due to the air interface specifications [28]–[31].

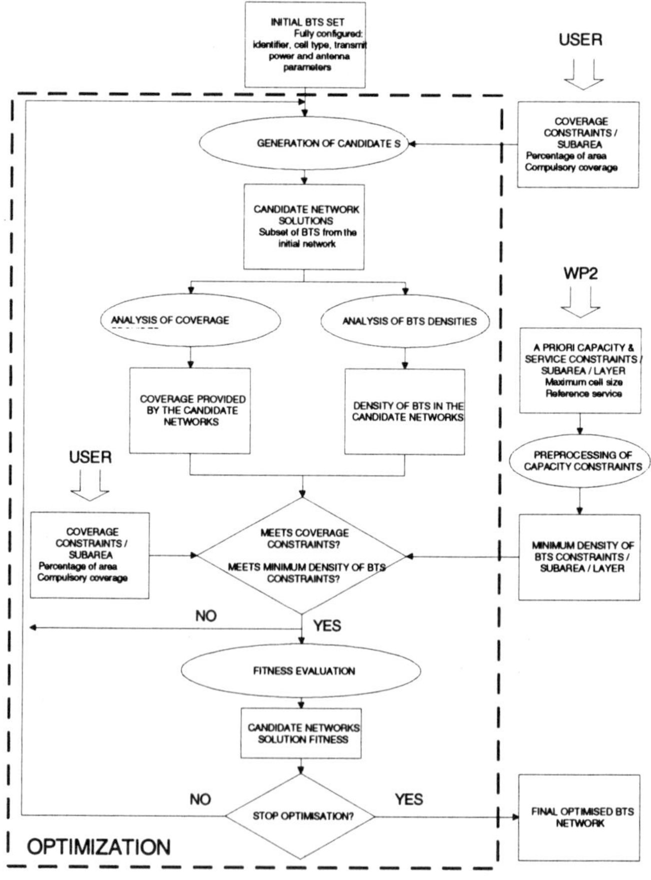

Figure 5.15 Block diagram of the algorithm for optimized extraction of a final BTS network.

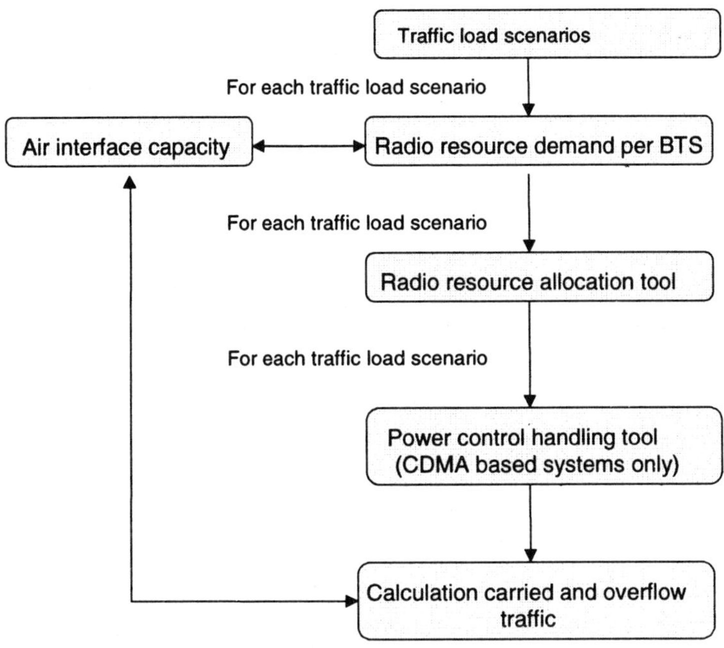

Figure 5.16 Generic refined dimensioning process for a single-cell layer.

TD-CDMA

The structure of the TD-CDMA interface was TDMA with up to eight CDMA codes in each timeslot. The main advantage of this scheme was the flexibility for asymmetrical traffic because of the dynamic assignment of uplink and downlink slots. In this case DCA schemes were allowed; the resources estimations regarded frequency carriers (as a function of service mix and traffic volume); air-capacity estimations dealt with multibearer capabilities and hard-capacity limit; radio-resource allocation coped with frequency planning; no power control was implemented. The TD-CDMA dimensioning process is depicted in Figure 5.17.

WCDMA

In case of the WCDMA air interface, macro- and microlayer dimensioning was performed in parallel, as the assumed hierarchical structure was not modified later in the process. The dimensioning phase actually performed a detailed capacity analysis, which started with the BTS network in the coverage optimization phase. The target of the analysis was to verify if the capacity demands were satisfied. The key feature of the analysis was the power control under realistic traffic and propagation conditions. Additional features as microdiversity could be modeled as well.

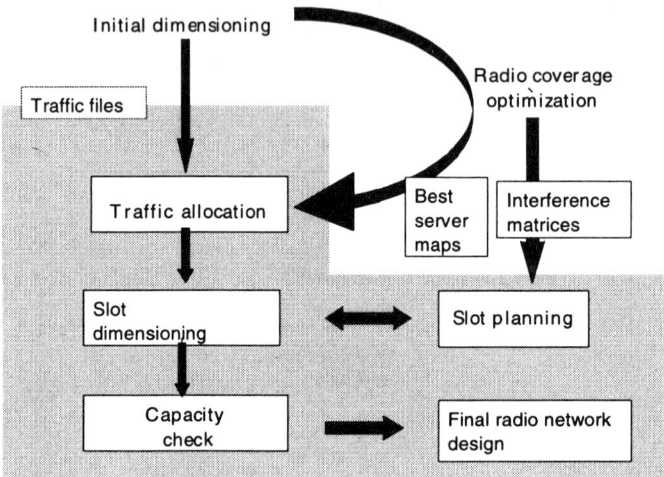

Figure 5.17 TD-CDMA dimensioning.

The WCDMA accurate dimensioning had to consider jointly the up- and downlink limitations. In the uplink, the capacity-limiting factor was the interference detected by the BTS. The interference evaluations included the power-control mechanism, the peculiar characteristics of the examined air interface, and microdiversity. The uplink dimensioning is depicted in Figure 5.18.

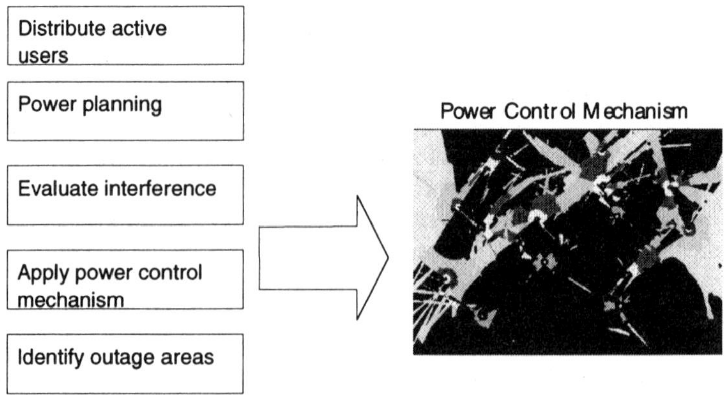

Figure 5.18 WCDMA dimensioning (uplink).

For the downlink analysis, STORMS focused on the hard limitations imposed by the number of channelization codes. Channelization codes were framed into a tree structure, where the orthogonality was ensured at each level. If a parent code was assigned, all the "children" codes could not be allocated in the same cell due to the orthogonality principle. High bit rate services needed short codes (with a low spreading factor), while low bit rate services needed long codes (with a high spreading factor). The upper and the lower bound in the service allocation were the following: 128 channels at 8 Kbps (telephony) could be assigned using SF = 128; 4 channels at 384 Kbps (interactive multimedia) could be assigned using SF = 4. In STORMS, the *equivalent telephony channels* concept was implemented to elaborate whether the codes assigned to a cell could handle the expected mix of traffic. WCDMA downlink dimensioning is depicted in Figure 5.19. The loaded network capacity is met only when both uplink and downlink analysis concludes successfully.

Network Performance Evaluation

Network performance evaluation was performed through the following procedures:

- Fixed-infrastructure optimization;
- Network-performance assessment.

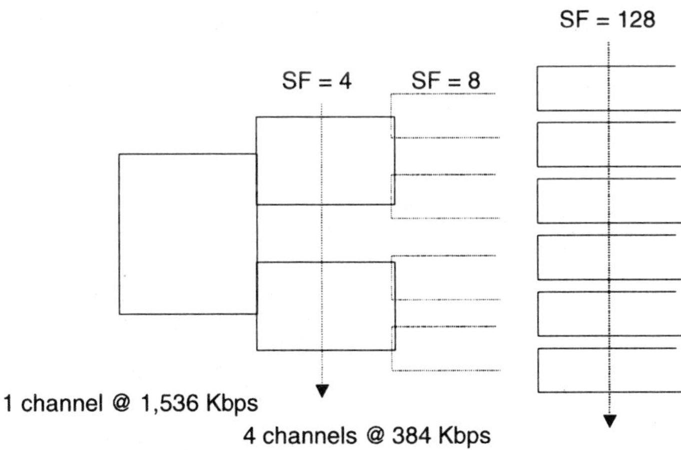

Figure 5.19 WCDMA dimensioning (downlink).

Fixed-Infrastructure Optimization

In STORMS the fixed-infrastructure optimization considered two aspects: access-network optimization and location/paging areas dimensioning.

Once the radio-part planning was completed, operators need to optimize the fixed network infrastructure. That involved the estimation of the optimum number and allocation of BTS controllers (CCS), LEXs, and the most cost-effective interconnecting network. STORMS developed appropriate software tools to handle these topics. The software module originally developed in STORMS was improved in the final part of the project by adding the satellite segment. As a result, in the final configuration, some terrestrial BTSs were removed since the traffic offered in the relevant areas could be appropriately handled by the satellite system. Figure 5.20 and 5.21 illustrate this configuration procedure.

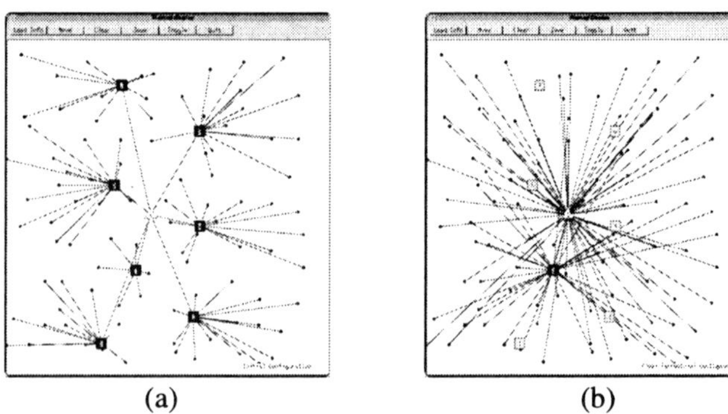

Figure 5.20 (a) Initial network configuration (not optimized) and (b) final network configuration (optimized).

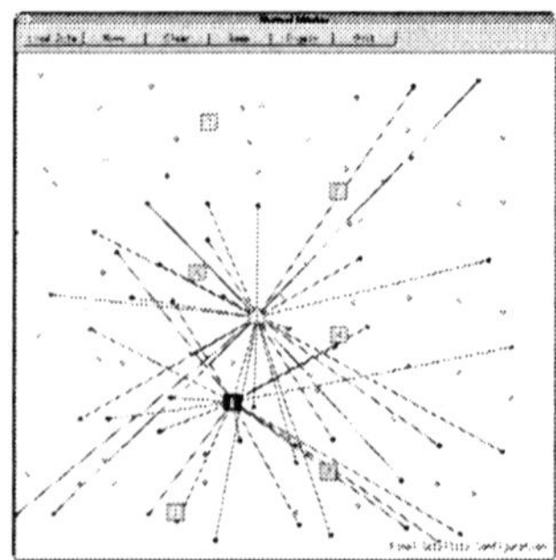

Figure 5.21 Final network configuration with satellite segment.

The fixed network optimization also comprised location and paging areas planning. This resulted in a lower network cost, a higher exploitation of deployed equipment (CCS and LE), and a minimization of the signaling overhead due to mobility management procedures (location updating and paging). Intelligent paging adopted in the UMTS concept could further minimize the signaling by separating the location updating (LU) and paging (PA) areas into two different logical layers.

Network Performance Assessment

Before network implementation, evaluation of the expected UMTS network performance is necessary. STORMS developed a simulation module fully integrated with the previous planning stage [32]. The input to the simulator is the proposed network configuration, and it provided the statistical behavior on a per-node basis and could recommend the improvement of the dimensioning stage. The simulator could also be used to evaluate the network sensitivity to a traffic pattern deriving from new services or a new traffic policy.

The STORMS simulator is a discrete-event driven structure that consists of four components: environment modeling libraries, event generator, event handler, and statistics, as shown in Figure 5.22.

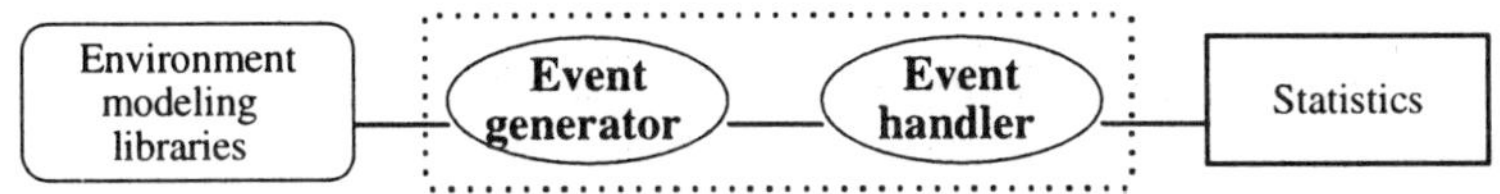

Figure 5.22 Four components in the UMTS simulator.

The main features of the modeling libraries are the user behavior (traffic and mobility) and propagation conditions. The event generator plays the role of the scheduler, in order to produce the desired test case. The event handler is the core component of the simulator. It consists of models that represent the UMTS network entities. Each network entity comprises the necessary functionality dictated by the UMTS operations and procedures and can include the intelligent mechanism, while covering a number of sub-classes that correspond to environments, in which the respective entity exhibits a different behavior. The fourth component of the simulator collects statistics from various network elements, in order to evaluate the overall performance of the given UMTS network configuration. The simulator was used to validate the previous planning phases and to experiment with the concepts that can only be studied through simulation (difficult for analytical modeling). The simulation tool was applied to various test cases to validate both the tool and points.

One of the main achievements in the project STORMS was the definition of a complete methodology to plan a UMTS network, which was mapped into specifications used to develop a software prototype. The tool available at the end of the project, a working software prototype, considered all elements of a 3G mobile network (traffic distribution, base-station positioning, radio coverage, resource dimensioning, fixed infrastructure optimization, and network simulation). After the ETSI discussion on UTRA, a software module was developed capable of dimensioning and optimizing a WCDMA and TD-CDMA network.

Another important achievement was the radio-coverage optimization tool. It could speed up and facilitate the delicate process of locating sites and computing the relevant coverage. It automatically planned the BTS maximum transmitted power and suggested the most appropriate locations.

One significant achievement was in the computer simulation area. The object-oriented approach allowed future users to easily make upgrades necessary to follow the UMTS specification evolution.

The main objective of STORMS was the identification of a novel planning and dimensioning procedure suitable for 3G mobile systems. STORMS considered different aspects in the UMTS network structure: propagation modeling, radio-coverage optimization, radio-resource dimensioning, and fixed-infrastructure optimization. In order to achieve a valuable product, capable of proving attractive in planning UMTS, STORMS implemented advanced information technology techniques such as parallel computing, heuristic approaches and client-server applications. STORMS developed a generic dimensioning platform to host different dimensioning modules addressed to the various air interfaces that could be for 3G mobile systems.

The project investigated the most appropriate procedures for planning and dimensioning the network where the UTRA air interfaces were used. It provided an automatic radio-coverage optimization tool that simplified the dimensioning task.

Within the automatic radio-coverage optimization procedures, STORMS developed the generic optimization algorithm for the selection of optimal BTS configurations. This algorithm followed the incremental and modular approach. The complete optimization product was divided into two distinct modules:

- APPEALS, which contains the core of the algorithm and the associated data structures;
- ParaGene, a genetic algorithm that is a specification of the former C++ classes, taking into account the specific optimization problem tackled by STORMS (i.e., the dimensioning of a mobile cellular network).

Both packages are generic and modular enough to be reused easily for solving similar problems.

STORMS developed the access network optimization tool. It was supported with a simple database system plus a built-in geographic information system (GIS). It allowed the operators to conduct studies and experiments for the identification of the most cost-effective grouping in the network infrastructure (BTS, BTS controllers, switching centers), migration to novel transport mechanisms (SDH, ATM), and introduction of new technologies, such as GPRS.

STORMS cooperated in the definition of the fill-service access network (FSAN) concept and the conclusion for PON implementation to interconnect UMTS/MBS remote antennas distributing the RF signal directly over fiber (radio over fiber).

Another important achievement of STORMS was the development of a network simulator. Again, the object-oriented approach was used for the design and implementation of the simulator. It was supported with the environment modeling libraries that could be enriched or even changed to simulate other traffic and mobility models. The simulator tool developed in STORMS was not applicable only to the 3G mobile systems but could simulate already existing systems as well.

The radio-planning methodology of STORMS was applied to some concrete and realistic cases, to investigate the improvements in the BTS concept. It was exploited in cooperation with TSUNAMI II to assess the benefits of the proposed adaptive antennas and with SUCOMS to assess the benefits from the introduction of high-temperature superconducting devices in future BTS technology [33].

5.2.4 Advanced Security for Personal Technologies—The Project ASPeCT

An increasing variety of mobile services is already being offered by a variety of public and private network operators in both indoor and outdoor environments to an ever growing number of users. To allow a cost-effective introduction of UMTS, migration/evolution scenarios have been defined within ETSI. The need for enhanced security features in UMTS has led to the definition of specific security objectives. Evolving security requirements for the UMTS were addressed through the ACTS project ASPeCT [34].

ASPeCT focused on security aspects related to UMTS as standardized by 3GPP and ETSI. Some security features, which primarily addressed the vulnerability of the air interface, were already provided by the existing mobile and cordless systems (GSM and DECT) [35]. However, the future UMTS/3GPP networks required more sophisticated and advanced security protection. This, in particular raised questions about the possible functionalities of future user service identity modules (USIMs); multi-application smart cards to access UMTS/3GPP networks, providing authentication procedures, fraud detection, end-to-end security requirements, security and integrity of billing, and so forth. The ASPeCT project, by providing valuable input to the standardization of UMTS security by ETSI, made significant contribution to enhancing security of the next generation of mobile communication systems. The project also performed a set of trials and demonstrations, carried out in conjunction with the project EXODUS.

The project defined the specific and sensitive issues of performance, ease of integration, and compatibility of new security systems for 3G mobile systems [36].

The general objective of the project was to study the feasibility and acceptability of new and advanced security features in existing and future personal communication networks, based on trials and demonstrations.

The more detailed objectives necessitated investigation, implementation, and testing solutions in the following areas:

- Migration of security issues from existing mobile systems to UMTS;
- Fraud detection and management in UMTS;
- TTPs for end-to-end security services in UMTS;
- Capabilities of future USIMs;
- Security and integrity in billing in UMTS.

The work undertaken in ASPeCT was divided into three main areas [36]:

- Work focusing on authentication and the USIM;
- Work on TTPs and their use to support a secure billing protocol for value-added services;
- Work on fraud detection in mobile systems.

ASPeCT developed an authentication framework for UMTS [37]. The main purpose of the authentication framework was to provide a flexible procedure for user-network authentication allowing a number of different mechanisms and algorithms to be incorporated, with smooth migration between them. The key points were the following.

- A concept of used capabilities classes;
- User-network (mutual) authentication using the negotiated mechanism;
- Automatic establishment of network-operator (NO) and service provider (SP) roaming agreement where required.

The negotiation of the mechanism to be used was performed between the different entities (SIM, NOs, SPs, and TTPs). The interworking between the NO and SP involved the certification authority (CA).

To facilitate roaming with many NOs and SPs, roaming agreements had to be set up dynamically as required, after the initial authentication. The *capability classes* identified whether the particular authentication mechanism was supported by the USIM of a user. The capability of USIMs was to assist the smooth migration from the 2G SIM concept.

The UMTS authentication framework was implemented in combination with *public-key*—(with USIM) or *secret-key*—(without USIM) based authentication mechanisms. A new protocol was developed for authentication between the user and the network. Its design exploited the advances in two fields:

- Cryptocontroller smart cards (which have a coprocessor that efficiently supported public-key cryptographic mechanisms);
- Elliptic-curve cryptosystems (which permitted the use of smaller cryptographic parameters).

Other important issues addressed by the project, were the improvements in smart-card technology. ASPeCT carried out some work in user authentication using voice biometrics and smart cards. Biometrics demonstrations showed that voice could be used to authenticate users to their smart cards, with the support of the terminal. Three possible approaches were investigated: free speech input, prompted text, and pass-phrases. The biometrics user authentication was compared to PIN-based access control mechanisms and showed that the voice recognition was the most promising approach, especially if combined with some security mechanisms.

Secure Payment for Value-Added Services (VASs)

A significant difference between current 2G and future 3G systems, such as UMTS, will be the variety and number of VASs that the user will be able to purchase over the network. These services will be provided by an ever-increasing number of competitive value-added service providers (VASPs), serviced by a large number of public and private network operators. This complex mosaic of network relationships raised the importance of reliable charging procedures, which will permit billing for value-added services to be made reliably and with minimal risk of fraud.

Extensive investigations have been undertaken into the use of TTPs by mobile communication networks. TTPs allow the user to establish confidential channels with other users, possibly in different countries and it should satisfy law-enforcement requirements at both national and international levels, by allowing the recovery of confidentiality keys under appropriate control. The ASPeCT TTP approach provided a mechanism to support end-to-end confidentiality of communication. An important feature of this mechanism was that some information used to generate the shared secret key, was escrowed to the TTP. An interception agent could obtain the information, which was then used to try to decrypt targeted communications. TTPs acted as certification authorities (CAs) for entities such as mobile users and VASPs for the value-added services, paid to the VASPs.

ASPeCT implemented a payment mechanism [37]. It was used to pay for the provision of value-added services that provided information to the users based on

WWW technology. The novelty of the ASPeCT approach was the integration of tick-payment and authentication protocol proposed for the mobile UMTS system, and the payment scenario for basic and value-added services in UMTS [38], [39]. The billing model is presented in Figure 5.23. The charge for using VAS was composed of a basic charge for the provision by the network operator of the communication link between the user and the VASP.

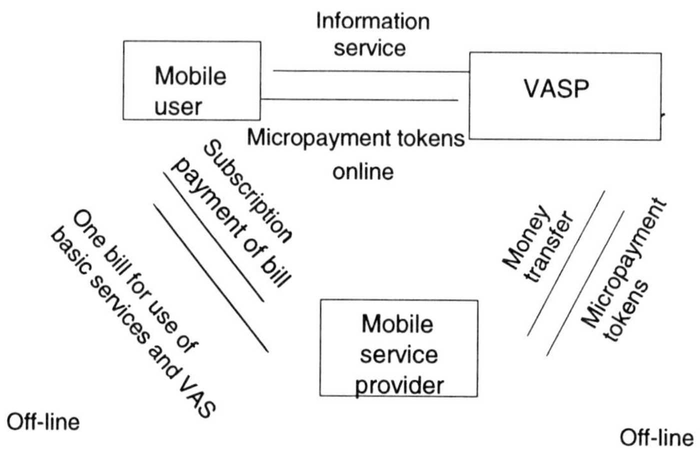

Figure 5.23 The ASPeCT billing model.

The payment method developed in ASPeCT was flexible, efficient, and secure. The method has potential application to billing for any telecommunications services.

Fraud Detection in Mobile Communication Systems

Prevention and detection of fraud is an important goal for network operators. Sophisticated fraud detection techniques can assist in early detection of such activity and reduction of network vulnerability.

In the ASPeCT project, several approaches were adopted to detect and identify patterns and instances of possible fraudulent behavior. The information on the behavior of a user was monitored over two windows: the short-term window was called the current user profile (CUP), and the long-term window was called the user profile history (UPH). Major results of the project included the following three fraud detection tools [40]:

- A rule-based tool;
- A neural-network based tool using supervised learning;

- An unsupervised learning tool utilizing neural networks.

Each of the proposed tools used profiling techniques and based its decision on a differential analysis. A rule-based tool was a white box approach that allowed detecting the frauds with a low rate of false alarms. It used the profiling strategy, and the rules for triggering an alarm were designed manually by an expert. A supervised neural network was implemented to avoid the necessity of an expert design of the rules. This concept dealt with determining the optimal neural network that would minimize the probability of false alarms. The unsupervised neural-network tool did not need any prior knowledge of fraud and was based on dialed-number analysis to detect changes in the user behavior and an international dialed-number analysis to trace specific changes in behavior of a user making international calls.

In the final stage of the project, these three tools were combined into an overall fraud-detection system. This integrated tool, together with its data handling module and its Web-based GUI front-end, was named BRUTUS. It was demonstrated on available data [41]. The GUI provided access to records of suspicious users and allowed the operator to look at calls of individual users, to judge whether they were fraudulent, and to take further actions. BRUTUS was a hybrid detection tool that enabled the profiling of both network subscribers and network traffic.

The project conducted demonstrations and trials on the UMTS security mechanism. The objective was to prove that the security mechanisms defined for UMTS were executable in real time and on a physical network. The trial platform architecture is shown in Figure 5.24.

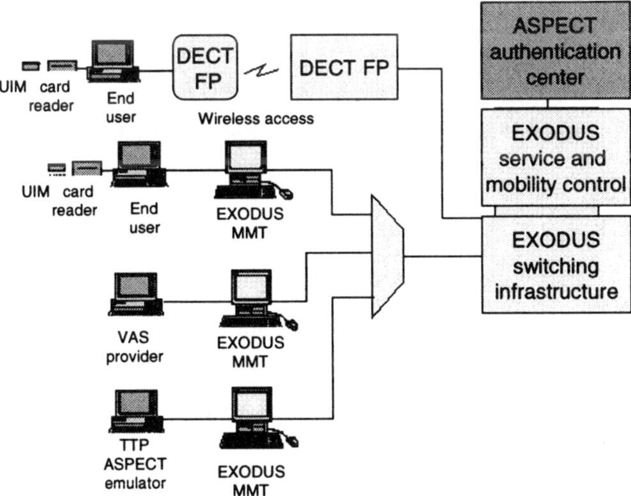

Figure 5.24 The trial-platform architecture.

Four trials took place. Two of them used the experimental UMTS platform provided by the EXODUS project and included the following:

- A trial of migration towards UMTS security, using an ASPeCT mutual-authentication protocol between multi-application smart cards (USIM) attached to EXODUS terminals, and an ASPeCT authentication center, attached to the EXODUS SCP;
- A trial of TTPs for managing keys, and secure billing services, using tick payments, applied to the provision of VAS over UMTS. EXODUS provided the data channel between user, value-added service provider (VASP), and TTP.

The stand-alone trials included the following:

- A trial of ASPeCT fraud detection tools—functioning in a real network environment—whose usage data are made anonymous;
- A trial of vocal password-based user authentication.

Prior to the trials the feasibility of the proposed advanced security features was confirmed through various demonstrations.

The work undertaken in ASPeCT was important for the incorporation of security concerns in the smooth-migration path to UMTS. The project addressed many important topics and provided theoretical and experimental results that allowed:

- Comparison of different techniques for fraud detection;
- Investigation of the legal and presentation issue relating to fraud detection;
- The demonstration of a user identity module (UIM) for UMTS;
- Advances of user-to-UIM authentication based on biometrics technique;
- Demonstration of end-to-end security services with real users;
- Demonstration of secure billing of mobile users for access to VAS.

The project proved the feasibility of end-to-end security that will clearly benefit the operators and providers of future mobile telecommunications networks. Results in fraud prevention will have an important impact on the economics of telecommunications. The development of secure billing solutions offered benefits to all involved parties. Manufacturers of terminals and smart cards can also use the results that were achieved in ASPeCT through experiments and demonstrations.

The standardization of the security architecture continued in the follow-up project, USECA [42].

5.2.5 Enhancements on ATM-Based B-ISDN—The Project EXODUS

The project EXODUS aimed at identifying a smooth transition path from disparate fixed and 2G mobile systems towards UMTS. It was focused on the core network capabilities that supported terminal and personal mobility for specific application sectors (e.g., health services and business travelers) [43]. The project was based on the theoretical studies performed in RACE II and was more oriented to implementation and field trials, in order to implement and deploy the UMTS platform.

EXODUS contributed to building up the interoperability between the fixed and mobile networks. Introducing IN capabilities in this concept, the project addressed the required common definition of bearer, call, service, and data functionality, and protocols for use over future mobile and fixed personal networks.

The project objectives included the following:

- A demonstration of terminal and personal mobility, spanning both public and private operating environments, using systems supporting voice, data, and image applications;
- Identification, implementation, and demonstration of IN functionality for meeting UMTS requirements;
- Identification, implementation, and demonstration of ATM-based B-ISDN to support mobility management.

To fulfill these objectives, the project evolved the experimental platforms with real users [44]. To support UMTS service capabilities, several new identifiers and features were introduced. They were handled by special-purpose signaling entities in the terminals and in the core network. These included the following:

- The terminal and user identities;
- The point of attachment identifier;
- The service identifier;
- Support of simultaneous multiple ATM connection establishments for multimedia applications;

- Intraswitch handover;
- Multiuser registration of one terminal.

The project started from the narrowband experimental platform, based on the IN capability that supported personal and terminal mobility for voice and specially enhanced data services in different environments (public, private, and residential). In this phase DECT was the considered access network.

Phase two addressed mainly the migration of the overall platform towards UMTS and the relevant impact on core network (CN), INAP, and UNI signaling protocols. The phase two platform considered broadband core network, with an advanced IN infrastructure that supported both personal and terminal mobility for multimedia services. The service-control point (SCP) was evolved to support new services, while the ATM switch included enhanced functions for mobility support. UNI signaling protocol was extended too, taking into account new services requirements and mobility aspects. The access network was enhanced DECT.

The previous project results from RACE II on UMTS, IN and B-ISDN, and available IN and B-ISDN standard guidelines. The key issues in EXODUS were the following:

- Focus on the CN;
- Strong orientation towards experiments and use trials;
- Involvement of a large number of real users in service trials;
- Close link with national host initiatives;
- Strong participation in the standardization process.

The interest was focused on the control plane of the core network and on INAP modifications required for mobility support. The B-ISDN signaling deployed distributed databases and distributed processing capabilities and relevant protocols. For the radio access network, existing technologies with functional enhancement were used.

To adjust for the multiplicity of terminals (fixed, mobile, wireless), the core network provided UNI points to a common protocol stack, as shown in Figure 5.25. These UNI points were ATM-based, supporting AAL5 traffic and providing a uniform API to user and signaling applications [45].

The initial experimental platform, defined in phase I, involved in a two-site field trial over 100 real users to verify the usability and perceived QoS (see Figure 5.26 [46].

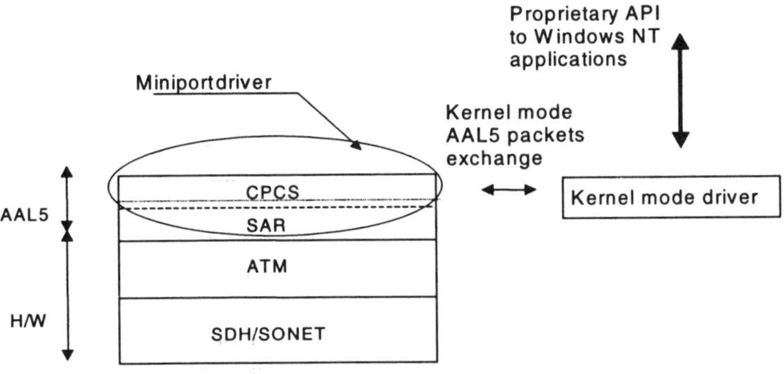

Figure 5.25 Protocol stacks for the UNI points of the EXODUS UMTS network

The tests included both voice and data services. A full operative network was based on n-ISDN with IN support for terminal and personal mobility. The platform consisted of the following:

- Two LEX equipped with DECT base station to provide suitable coverage in the two sites;
- One intelligent node to provide both mobility and service management;
- One data server to support data services (and to allow access to the Internet);
- More than 100 DECT terminals, four of which were equipped for enhanced data services.

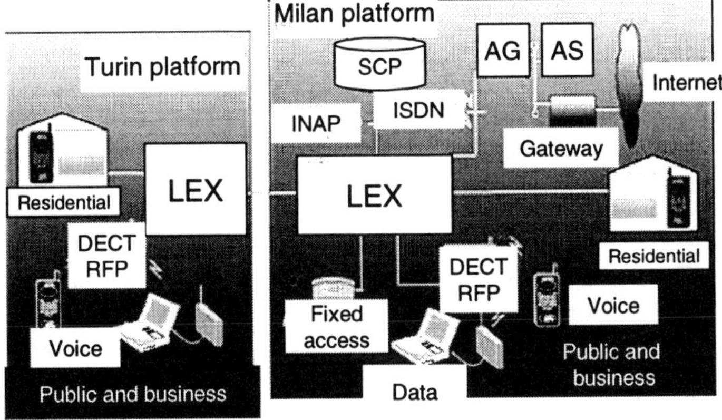

Figure 5.26 EXODUS phase-I platform.

The deployment and successful completion of phase I experiments provided confidence for the further trials and demonstrations continued in the next phase of the project.

The experimental platform in phase II was realized with three experimental islands (two sponsored by national hosts) that provide ATM functionality with base station infrastructure for wireless access and multimedia (see Figure 5.27). The experiments were based on the UMTS core network developed according to the EXODUS approach.

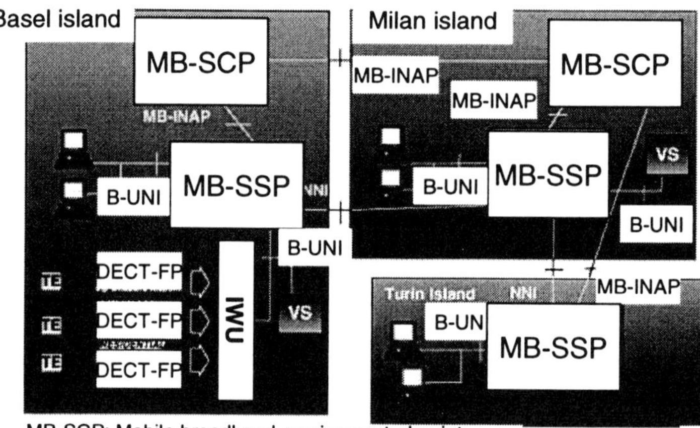

MB-SCP: Mobile broadband service control point
MB-SSP:Mobile broadband service switching point
VS: Video server

Figure 5.27 EXODUS phase-II platform.

This platform consisted of the following components:

- Three mobile broadband service switching points (MB-SSPs);
- Two mobile broadband service control points (MB-SCPs);
- At least two DECT-to-UMTS IWUs;
- At least two multimedia DECT terminals and base stations;
- At least eight fixed broadband terminals equipped to support personal mobility in both trials;
- A multimedia information retrieval server;
- Broadband router to interwork the EXODUS networks to the University hospital in Basel (only in the health care trial).

The phase-II experiments included a health care trial and a mobile multimedia trial. The health care user trial deployed multimedia and services for tele-observation in the hospital. The mobile multimedia trial deployed applications such as multimedia data retrieval, direct playing, and collaborative work (video telephony and concurrent data exchange), and demonstrated the advantage of mobility to real users.

EXODUS established cooperation with the ASPeCT project, during both the health care and mobile multimedia trials. EXODUS provided the network, and ASPeCT provided the equipment for authentication. EXODUS also hosted a demonstration of ASPeCT-trusted third-party (TTP) applications.

Some results obtained in EXODUS were the basis for new projects. The EXODUS core network was the basic platform for the trials that continued in the ACTS project ACCORD, which built the variety of access networks, including satellite access.

A set of results was obtained in terms of systems, subsystems, components, and methodology. EXODUS performed the mapping of certain mobility management information flows to the GSM-MoU model [47].

The main achievements in phase I that faced the complexity of the integration activities included the detailed technical specifications, implementation of all network elements, and detailed planning of the user trial. It showed that DECT could be used to provide mobile voice and data coverage in all environments. From the operator's point of view, the advantage was very low impact on the existing network infrastructure and complete reuse of the existing infrastructure, including base station. It allowed the fast deployment of services over the whole coverage area.

Phase II completed the phase I service description that was implemented in user trials (user and mobility services) [2]–[44]. Within this phase, the EXODUS functional architecture concept was developed and description of each layer of signaling stack at IN interface was produced. The project achieved important results regarding protocols; it defined the enhanced protocol called mobile broadband-INAP (MB-INAP) that supported mobility over broadband services, modified the B-ISDN UNI protocol to support mobility management procedures, and broadband NNI protocol to support MB-SSP to MB-SSP and MB-SCP to MB-SCP signaling. EXODUS produced and designed specifications for mobile and fixed terminals and other network entities, produced requirement specifications for software applications, IWFs between DECT and B-ISDN, implemented and realized the MB-SCP prototype, and enhanced protocol modifications.

In phase II, the following user services were demonstrated and tested:

- Database access;
- Multimedia information retrieval;
- Broadband video telephony.

The project EXODUS fulfilled a number of goals, showing the feasibility of a UMTS oriented approach for mobile systems and demonstrating, on a significant platform, a set of multimedia services with different mobility patterns. It demonstrated the upgraded IN capabilities and enhanced B-ISDN signaling in a variety of trials. EXODUS produced a set of results in terms of systems and subsystems, components, and technology, with a positive impact on the performance of European businesses, and contributed to standardization bodies.

5.2.6 Demonstration of a Hybrid Radio-Optical Fiber Loop for Interactive Broadband Services—The Project FRANS

The future infrastructure of the IBCNs will support the wide range of new technologies, emerging services, and multimedia applications with strong demand for bandwidth. Future 3G mobile systems will require the support of an integrated optical network infrastructure. Within this process, the role that fiber radio systems can play in the access network section of the IBC network was outlined and contrasted to the other access technologies by the project FRANS [48].

The project FRANS demonstrated a fiber-supported, millimeter-wave radio final-drop solution to permit wireless broadband access to the fixed, fiber-based network [49]. Fiber-supported millimeter-wave systems, as addressed by FRANS, were emerging as an attractive option for the delivery of broadband data to the users. The project was concerned with field trials demonstrating an optically supported millimeter-wave radio link as a final connection between the optical network and the customer premises [50]. The project objectives were to provide the following.

- Fast and flexible deployment of broadband wireless access in the micrometer/millimeter-wave range;
- Signal transport to and from antenna unit via fiber;
- Optical signal appropriate for radio channel;
- Feeder transport of intermediate or radio frequency (IF or RF);
- Small, simple, easy-to-maintain antenna unit.

It merged the existing passive optical networks (PONs) with a photonic technique for generation, modulation, and transmission of information bearing millimeter-wave signals to a hybrid fiber-radio customer access system [51]. Two trials were performed using different bit rates and based on different multiple access techniques.

FRANS performed extensive studies of the system network architecture and service deployment scenarios to achieve optimum system parameters and design. During the project lifetime, two different approaches were developed and realized:

- A 622-Mbps downlink with 40-Mbps aggregate TDMA uplink (~2 Mbps per user), both with a millimeter-wave radio part working in the 30-GHz band;

- A 155-Mbps downlink with millimeter-wave radio interface at 30 GHz and CDMA uplink (1-5x2 Mbps per user) with a microwave radio part at 2.5 GHz.

The downlink was based on a principle developed in the earlier RACE project MODAL. Two optical spectral components, separated by the required millimeter-wave frequency, were generated and modulated in a central unit. They were distributed via a passive optical network (PON) to antenna units close to the users. In these units, the two components were mixed in a PIN photodiode and a modulated millimeter-wave signal was generated and radiated. The TDMA uplink used a different millimeter-wave frequency, remote down-conversion in the antenna unit, and microwave transport over the PON to the central unit. The CDMA uplink used a microwave carrier frequency, which was transmitted directly or over a fiber optic connected remote antenna, to the central unit.

Several key issues were addressed during the project:

- Selection of the modulation scheme for broadband/subscriber multiplexed signals;
- Design of highly independent fiber and radio links;
- Realization of multiple access uplinks or a hybrid fiber-radio network with an antenna unit of low complexity;
- Ruggedized realization of an advanced subsystem for field application.

The work in the project was split into several topics:

- Enabling technologies and submodules: optical generation and detection of millimeter waves, MMICs, MICs, antenna unit, and CDMA processing;
- Network development: millimeter-wave up- and downlink, radio/fiber microwave links, access adaptation;
- Customer premises equipment: millimeter-wave and microwave subsystem, modulation, demodulation, signal conditioning, and network termination;
- Field trials: integration with host, service deployment.

FRANS conducted two trials on the developed experimental platforms. The first trial demonstrated the feasibility of an optical/radio system, and verified the quality of service that could be achieved comparable to that of an all-optical system. The signal processing was performed in the central unit.

The second trial was performed as a joint trial with the ACTS project ATHOC. It allowed the comparison between hybrid fiber/radio (HFR) and hybrid fiber/coax (HFC) technologies, and demonstrated the interworking between them.

The trial used a 622-Mbps, radio-over-ATM PON platform. The block diagram can be seen in Figure 5.28. Services involved in the trial were MPEG-2, Internet, and B-ISDN.

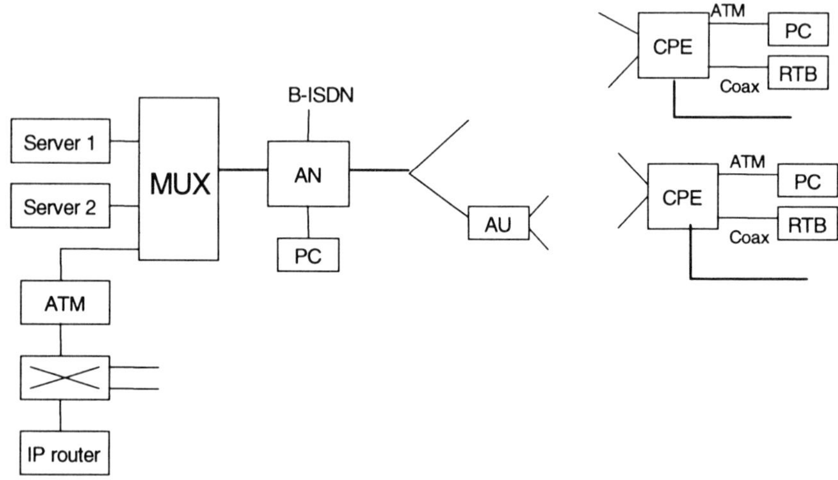

Figure 5.28 Block diagram of the radio-over-ATM PON field trial.

Two MPEG programs were multiplexed and delivered to the customer, who selected the program through the set-top box (STB). Internet access was possible through IP-router connection. Customer-terminal equipment was comprised of an STB and a personal computer. The fiber section had a maximum length of 6 km. The radio-cell diameter was 500 m maximum. Technical parameters involved in the trial are presented in Table 5.1.

Table 5.1

Characteristics in Service Trial 1

Characteristics	*Downlink*	*Uplink*
Bit rate	622 Mbps	40 Mbps
Modulation	16 QAM	FSK
Multiple	TDM	TDMA
Radio drop carrier frequency	27.875 GHz	29.26 GHz
Optical feeder carrier frequency	RF 27.875 GHz	IF 615 MHz

The second trial was a multi-34-Mbps/multi-2-Mbps joint trial with ATHOC. It complemented the HFC access of ATHOC with a HFR access of FRANS and allowed communication between HCF and HFR access areas, as depicted in Figure 5.29.

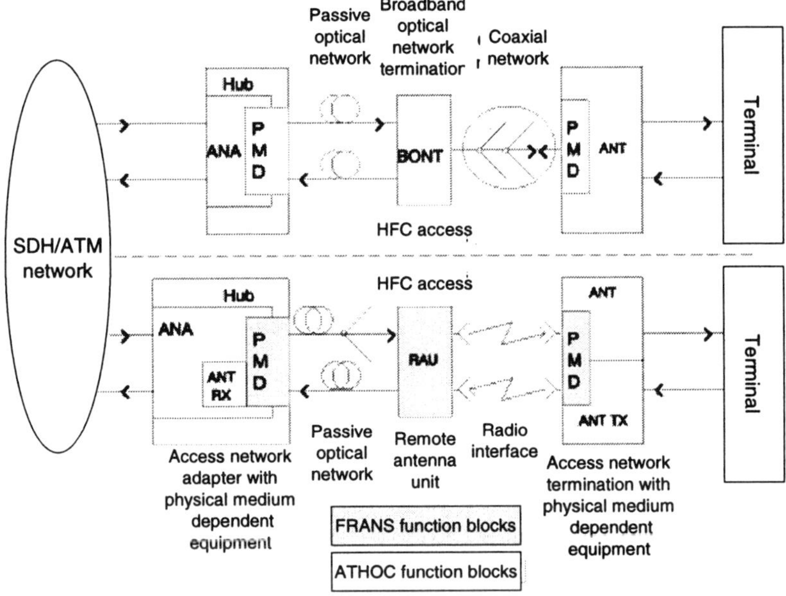

Figure 5.29 Block diagram of the joint FRANS/ATHOC field trial.

The following services were used in the trial: TCP/IP-based ones (hypertext or e-mail) and native ATM applications (VOD on customer/medical information and teleworking). The technical characteristics of trial 2 are presented in Table 5.2.

Table 5.2

Characteristics of Service Trial 2

Characteristics	*Downlink*	*Uplink*
Bit rate	Nx34 Mbps	Nx2 Mbps
Modulation	QPSK	QPSK
Multiplex	FDM	CDMA
Radio drop carrier frequency	42 GHz	2.2 GHz

The main achievements of the project were the system concept and specifications, as well as the specifications of the interfaces to the trial platform. It also realized the first components and design of subsystems.

Important achievements were the first realization of QPSK transmission (27.98 GHz, 25 Mbps) over 12-km standard single mode fiber and 1:8 splitting, and hybrid fiber-radio over 49.2-km installed single mode fiber loop, using 60-GHz carrier frequency and a 140-Mbps data rate.

The project FRANS demonstrated the integration between the optical network and the wireless techniques that could provide the broadband services to the customer. The achieved results contributed to functional specifications of IBCN and could facilitate the introduction of the IBC services. Fiber-supported millimeter-wave radio technology was a promising technique. It provided a rapid broadband drop into areas where considerable time and financial resource would have to be invested to achieve similar wired connection results. Moreover, it led to an upgrade path to full FTTH, which was the eventual goal of the access network development.

5.3 SATELLITE NETWORKS

The globalization principle in networking concepts also applied to the satellite segment. The significance of wireless and mobile technology and its ever-growing importance to the telecommunication world raises the evolved requirements of the mobile users to avail themselves of the full range of broadband multimedia services provided by the global information infrastructure, whether wired or wireless- connected. Some of the issues regarding the development of satellite networks have already been considered through the work of projects included in Chapter 4. It is obvious that, in the medium and long term, several satellite systems will operate worldwide. The standardization process of satellite concepts and systems are not as advanced as for the terrestrial networks. However, definition work for a possible S-UMTS system has been carried out by the European space agency (ESA) and by the SINUS project (see Chapter 4).

The basic concepts introduced by RAINBOW and SINUS for network interoperability provided the direction for further integration of terrestrial and satellite networks. The GRAN concept introduced by RAINBOW, and followed by SINUS, allowed seamless service provision and ISHO.

The heterogeneity of the network platforms and terminals, which will coexist in the future, points out the importance of network/terminal provision of multimedia services. It concentrated the interest on development of APIs capable of delivering service to the user that will support the underlying network and the terminal. This was the focus of the ACTS project SUMO.

The SUMO project, related to SINUS and TOMAS (see Chapter 4), aimed at identifying and demonstrating generic approaches to service provision and network control over S-UMTS. It developed a trial platform that allowed supporting a service provision through two different satellite access schemes.

SUMO addressed interoperability between complementary satellite systems as well as with terrestrial core networks, automatic selection of the access network depending on the service requested and the mobile environment (e.g., urban, suburban, rural), bandwidth-on-demand and common signaling for service provisioning and network control for different access schemes.

The integration process moved towards growing interest in implementing the flexibility of ATM concept in satellite environment. That gave rise to several projects addressing these issues. The ACTS project VANTAGE, through a series of trials, addressed the complementary role between the satellite and terrestrial networks. The VANTAGE project united the service flexibility of ATM and the access flexibility of satellites. It implemented a novel architecture using a conventional transparent satellite, with its Earth station (V-SAT) as a distributed ATM switch, and moved the network boundary to remote and isolated users.

The OAM functionalities in the combined terrestrial-satellite environment are essential for proper network operability. The network management concept was addressed in VANTAGE. This project proposed an SNMP and CMIP-based network management (VNM) approach.

This section will give an overview of these two projects that look into compatibility and interoperability of terrestrial and satellite networks. They played an important role in the network-globalization process.

5.3.1 Resolving the Issues in Interoperability Between Complementary Satellite Systems and Terrestrial Networks—The Project SUMO

For truly global penetration of the UMTS/IMT-2000, particularly in the absence of complete 2G terrestrial coverage (e.g., GSM), a satellite component is also necessary for providing fallback service. Satellites provide direct access to users independent of their geographical location, a key means to global coverage. The next generation of S-UMTS system based on LEO, MEO, and GEO constellations is being prepared for provision of multimedia capability and full internetworking. To provide mobile multimedia services seamlessly via satellites as well as terrestrial networks, full interoperability needs to be ensured. Satellite integration with the radio-terrestrial part required a maximum of commonality in the access network part and dual (or multi) mode terminals.

A possible solution was considered in the ACTS project SUMO. It was based on the achievements of the SINUS and TOMAS projects (see Chapter 4) and its main purpose was to demonstrate and evaluate the operation of multimedia services over a hybrid S-UMTS network. The SUMO testbed architecture was designed to be consistent with the ETSI UMTS architecture concept. The role of the satellite component in UMTS was specially addressed, considering different types of implemented services.

The project SUMO aimed at identifying and demonstrating a generic approach to UMTS service support and network control, focusing on the satellite segment [52]. To fulfill this goal, it defined the following objectives:

- Interoperability among the complementary systems (e.g., LEO-GEO) both real and simulated, and with the terrestrial core networks (e.g., B-ISDN);
- Advanced S-UMTS services, based on bandwidth-on-demand flexibility of communication channels for application services and definitions of the common signaling for service and network control among different access schemes and according to the GRAN concept;
- Automatic selection of alternative access network resources in an integrated UMTS system (e.g., different satellite or terrestrial networks) depending on services and mobile environments (e.g., urban, rural).

The project created an advanced S-UMTS testbed to demonstrate the above techniques. SUMO's testbed was an improved version of ACTS testbeds developed in SINUS and TOMAS.

The project SUMO took the results of SINUS and TOMAS a step further, and developed the advanced service provision and interworking on the testbed for use in the integrated multisatellite and terrestrial service trials [53]. Those two projects addressed the same issues through different approaches. TOMAS adopted a rather pragmatic approach by implementing a fully functional and portable multimedia terminal able to communicate over a GEO satellite.

The TOMAS system was based on a FDMA radio-access scheme. SINUS, on the other hand, developed a laboratory testbed with prototype terminals capable of demonstrating the advanced protocols required for integration of terrestrial UMTS together with various S-UMTS constellations. SINUS developed a WCDMA radio-access system, similar to what was later on adopted in ETSI for terrestrial UMTS.

Based on these ideas, SUMO verified the generic approach considered by URAN and enabled further refinements of the S-UMTS service management functionality. SUMO developed testbed architecture according to the concepts adopted by ETSI for UMTS systems. The physical UMTS architecture, with appropriately defined domains, is depicted in Figure 5.30. It was built on modular principle to allow different UMTS access and core networks to be used.

SUMO used alternative CDMA and FDMA satellite technologies, in the access segment, and considered different satellite constellations (GEO, MEO, and LEO). FDMA was used in global mobile GEO systems, and could achieve bit rates up to 2 Mbps. CDMA used for LEO and MEO constellations, was a more appropriate solution for flexible management of system traffic with user date rates up to 144 Kbps.

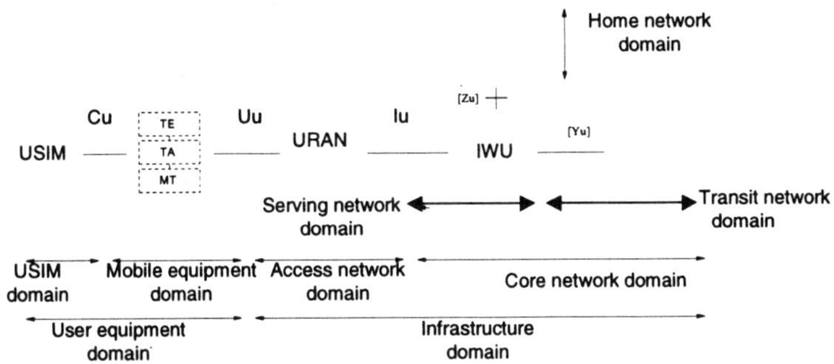

Figure 5.30 UMTS domains and reference points.

UMTS end-to-end services could be provided largely and independently of the type of satellite technology, due to the accepted GRAN concept, as a basis for its own adaptation functionalities and protocol specifications. This is depicted in Figure 5.31.

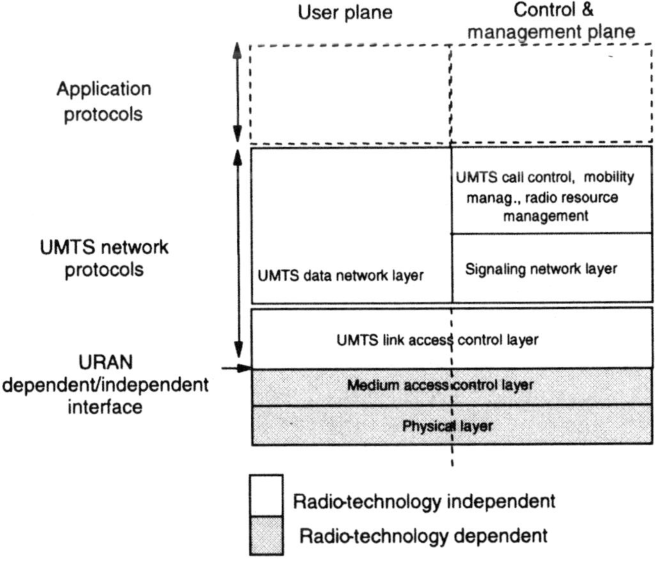

Figure 5.31 URAN protocol layers.

According to these principles, SUMO built the testbed architecture that allowed transparency of access technologies to the UMTS user. The testbed architecture is depicted in Figure 5.32. The USIM domain was implemented through the software functions in the MT/TA/TE. FES was multi-mode. The link-simulator was used to evaluate effects of inter-segment handover.

According to the accepted approach, and based on SINUS and TOMAS results, SUMO provided several hardware and software developments for the testbed. One of the main technical developments was the URAN adaptation for satellite FDMA.

FDMA modems needed adaptation via their MAC to provide the common interface to the LAC. Also some higher layer modifications could offer a more generic LAC interface. CDMA lower layer function was already built into URAN [55].

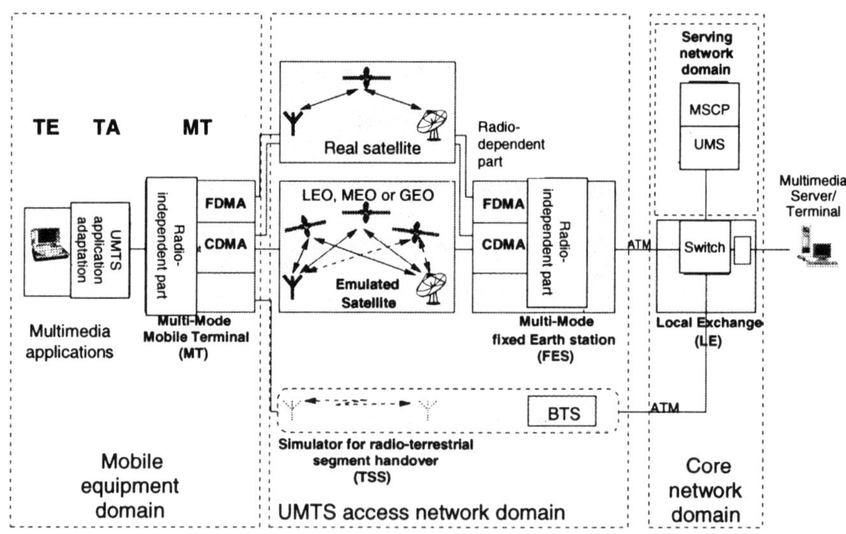

Figure 5.32 Testbed architecture mapping into UMTS concept.

The possibility to choose between different networks in order to meet application requirements in a seamless way provoked implementation of service and network adaptation functions in the SUMO application adapter (SAA). The alternatives in access segments of the UMTS (e.g., satellite FDMA, satellite CDMA, and terrestrial radio), and presence of 2G systems, may offer different bearer services, transport, and signaling characteristics. The SAA provided the user with simplified access to S-UMTS bearer services developed in SUMO, and simulated access and adaptation to T-UMTS and a range of potential mobile networks. SAA centralized call control and monitoring functions and provided a harmonized user interface [56].

The SAA consisted of two main layers, as depicted in Figure 5.33:

- Service-adaptation layer (SAL);
- UMTS-adaptation layer (UAL).

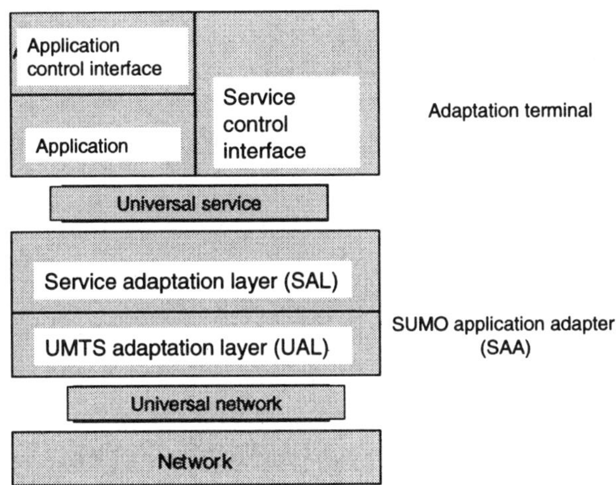

Figure 5.33 Realized adaptation structure

The SAL hosted all functional entities that are concerned with application adaptation and service control. The UAL included functional entities that are concerned with user traffic and control adaptation to different networks. The concept of service control interface (SCI) was based on a set of service control commands displayed graphically on the AT.

The main functionality of SAA consisted of adaptation functions, implemented in these two levels. With the service control interface (SCI) and the control functions of the SAA, the application could adapt to the appropriate network in order to fit to the user demands. More detailed information about SAA can be found in [53]. Another technical development of the project was the simulation of satellite interference.

The satellite channel emulator (for LEO, MEO, and GEO constellations) was enhanced with emulation of interference arising from other satellites, not only with the same constellations (especially CDMA case), but also with other systems. SUMO also developed advanced network services to provide increased capability for user applications. Advanced S-UMTS service provision could be achieved through the above-mentioned network and service control functions (SAA). These include [57] the following:

- Bandwidth-on-demand (BOD);

- Service-dependent satellite resource selection (SDSRS).

BOD intended to allow the user flexibility to request a bearer service to suit his needs in terms of throughput, cost, delay, and QoS. BOD was implemented at call setup and for a given access network.

The SDSRS service could be considered as an extension of BOD. It intended to allow the user greater choice between services from different satellite access networks available at the MT, and according to type of application. The advanced UMTS services were evaluated through several trials that implemented the above-mentioned approaches.

For the testing of the SUMO system and its UMTS capabilities, demonstrations and trials were planned and carried out [58]. SUMO aimed to demonstrate and evaluate advanced S-UMTS features, such as the following.

- *Service integration.* Compatibility of different physical channels (LEO, MEO, GEO, with appropriate access schemes—i.e., CDMA, FDMA) with a common range of bearer services and teleservices;
- *Service-dependent resource selection.* Between different satellite (and terrestrial) UMTS segments, with selection between different orbital configurations;
- *Bandwidth-per-channel resource allocation.* Matched to application request (BOD) and with data rates up to 144 Kbps, and potentially up to 2 Mbps;
- *URAN approach.* Interfacing and interoperability (including signaling) of mobile terminals with different access network segments distinguished by heterogeneous lower OSI layers (e.g., transmission, access);
- *QoS measurements.* For bearer services and teleservices for each of the above features.

For this purpose, the SUMO trials were divided into three different trial configurations.

The main objective of the first trial configuration was to evaluate the performance of the TOMAS multimedia application in operation over the SINUS S-UMTS terminals in an emulated satellite environment. It implemented the development of application adaptation software (SAA). In addition, the satellite channel emulator (from SINUS) was completed with a new functional unit that introduced the effects of satellite interference (inter- and intrasatellite interference). The trial configuration is depicted in Figure 5.34.

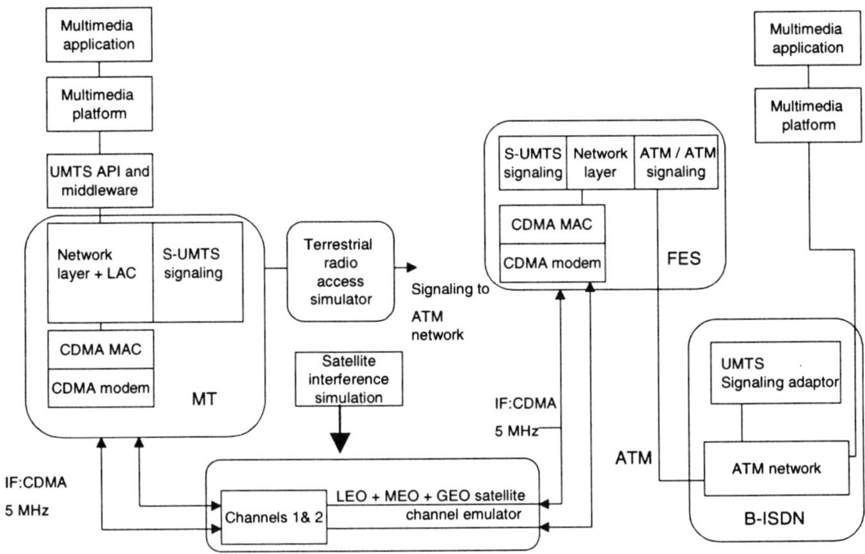

Figure 5.34 Trials configuration (1).

The objective of the second trial was to test the URAN interface. The SINUS signaling protocol was designed in accordance to the URAN principle SUMO performed a trial to verify that SINUS implementation was truly URAN-compliant. The satellite channel emulator and the CDMA modems used in the first trial were replaced with FDMA modems and the live GEO-satellite link, used in TOMAS (see Figure 5.35). Thus, the second trial configuration used FDMA+GEO satellite physical layers, with the LAC- and higher layers from CDMA, and implemented the integration of UMTS signaling (including call control, radio resource management, ISHO), developed for CDMA/URAN with the FDMA access scheme of the GEO satellite. In addition to validating the URAN concept for different satellite access schemes, this trial performed testing of some QoS parameters for UMTS bearer services. SUMO used several multimedia transmission standards (MPEG-4, H-324M) and applications (conferencing, information retrieval) and compared the results with those for SINUS and TOMAS, in order to demonstrate how standard multimedia applications performed in typical satellite multimedia environment.

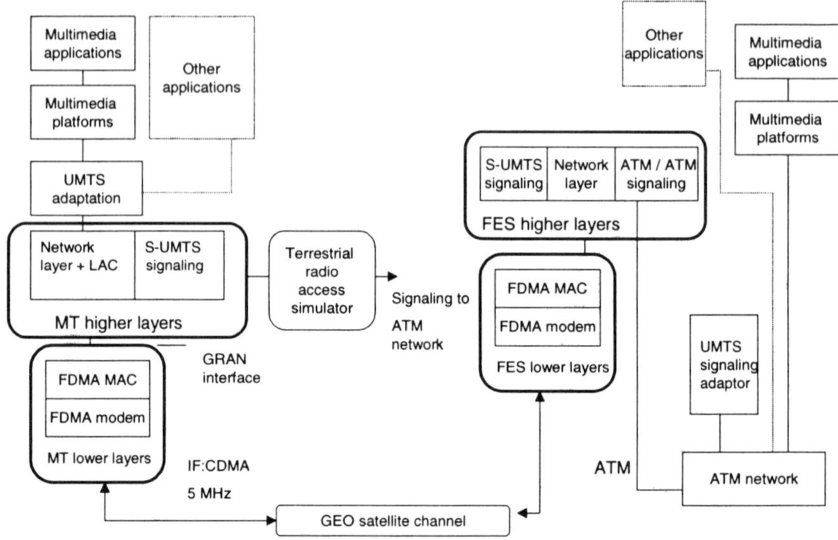

Figure 5.35 Trial configuration (2).

The final setup of the testbed included the MT and FES terminals from SINUS, now equipped with CDMA modems. The GEO satellite was added in parallel to the first-trial configuration.

The satellite channel emulator was enhanced to take in account the effects of interference from different traffic sources. This trial demonstrated the following.

- Interconstellation (GEO-LEO, GEO-MEO) satellite-resource allocation,
- Validation of the satellite simulator, by comparing the results from the emulator and the live link (with identical parameter settings).

The final trial configuration is depicted in Figure 5.36.

Some more important achievements could be summarized as follows:

- Interworking and integration of satellite-oriented testbeds;
- Comprehensive testing of a range of mobile multimedia services through LEO, MEO, and GEO configurations and under realistic communication conditions (simulated and live environments);

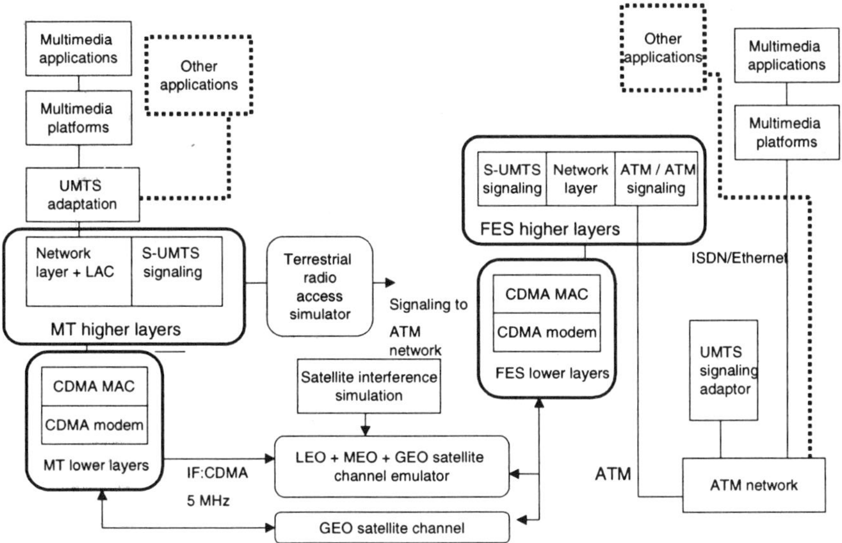

Figure 5.36 Trial configuration (3).

- QoS measurements using above applications under a range of trial network considerations (e.g., BER, delay, service reliability and availability, and handover transparency);
- Demonstration and evaluation of advanced S-UMTS features: service integration, service-dependent resource selection, and BOD.

The SUMO testbed is a powerful platform for further demonstrations and testing of satellite-based communications.

5.3.2 Service Platform for Other ACTS Trials and Applications—The Project VANTAGE

The integration issue between the satellite and terrestrial networks was also addressed in the ACTS project VANTAGE [59]. The ATM-based B-ISDN was at the time considered the most promising technology to provide services to future applications and it promoted integration processes of various network solutions. VANTAGE was concerned with the concept of a GEO satellite, VSAT small-Earth stations and the use of ATM transport technology. Its target was to provide trials to users of IT applications, so that they could experiment their distant interconnections, specific applications, and networks.

The ultimate goal of VANTAGE was to seamlessly integrate a satellite-based system in a terrestrial network. The project aimed to show that the elements of the satellite-based system could be managed using standard terrestrial procedures and protocols and that the satellite subsystem required no special concessions.

The strategic objective was to prove, by demonstration, that satellite could complement the terrestrial network on an equal-partner basis, and they are not the "poor relation," as it was often assumed [60]. This strategic objective was translated into the engineering objectives to unite the service flexibility of ATM and the access flexibility of satellites in order to provide pan-European interconnection service. The project performed a series of trials with a variety of Earth station technologies and access techniques and aimed to do the following.

- Extend the core network to remote and isolated users;
- Explore the possibilities offered by ATM of providing a unitary network over a heterogeneous mix of transmission technologies;
- Demonstrate the seamless extension of core networks via satellite to isolated communities with a uniform level of service;
- Offer the use of VANTAGE infrastructure to other ACTS and non-ACTS projects for developing, testing, or demonstrating new applications and services;
- Explore the requirements and techniques for satellite bandwidth allocation on demand.

These objectives were met by running a series of three major trials, in which remote users were connected by satellite. Satellite connectivity was ensured via use of various Cadenza configurations, which provided switch-in-the-sky capability over a transparent satellite.

The VANTAGE project was unique amongst the ACTS projects, in the sense that it was simultaneously an infrastructure provider and trials conductor. VANTAGE had internal and external users. The internal were the consortium partners, whose interests were related to performance and management of satellite networks infrastructure in general. The external ones used the VANTAGE trial platform to perform their own trials. VANTAGE implemented a novel architecture using conventional transparent satellite, with its Earth stations, as a distributed ATM switch (see Figure 5.37). This provided ATM technology in satellite connections, made available also to remote and isolated users.

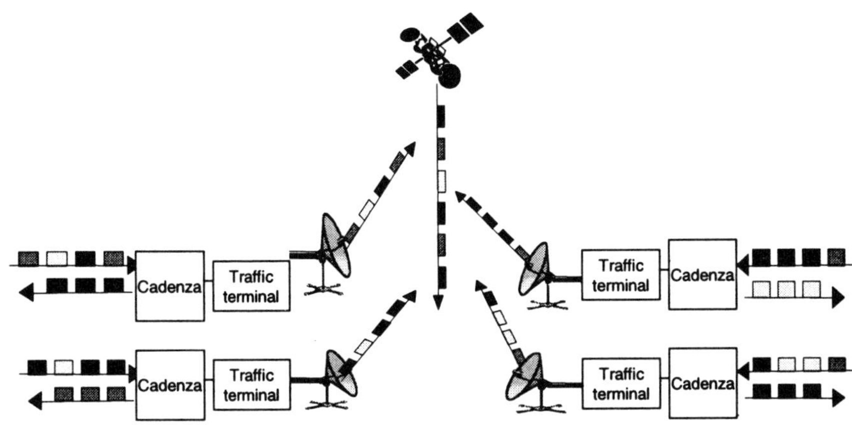

Figure 5.37 VANTAGE distributed switch concept.

A distributed switching, supported by the satellite (space segment) and access-switch units, one in every Earth terminal, performed the ATM switching. Cadenza equipment executed the access-switch units. This equipment implemented all the functions related to cell mapping and OAM management and signaling that was in band through the dedicated VCs.

VANTAGE performed three trials that formed a logical progression from the verification and testing (in trial 1) of the elemental physical functionality which ultimately were required, to the demonstration and use of those functionalities in a fully managed, standards-compliant network with real user applications (in trial 3).

For these evolving configurations, an experimental network was designed. It allowed different access modes to the satellite, FDMA and TDMA, and different bit rates from 256 Kbps to 34 Mbps. The terrestrial links could operate with different transmission technologies (2 and 34 Mbps PDH, 155 Mbps SDH, Ethernet, and so forth).

The trials used a conventional transparent satellite (with free capacity donated by Intelsat) with a variable mix of commercial-standard hardware and software, project-modified hardware and software, and experimental software designed within the project.

The first configuration was a star architecture with a hub station, FDMA access, Ku-band, and 2-Mbps, IDR satellite carriers. In terms of ATM traffic, only permanent VPs and VCs were implemented. The central node at the hub station managed the network. The Cadenza equipment and some routers interfaced the satellite system (see Figure 5.38).

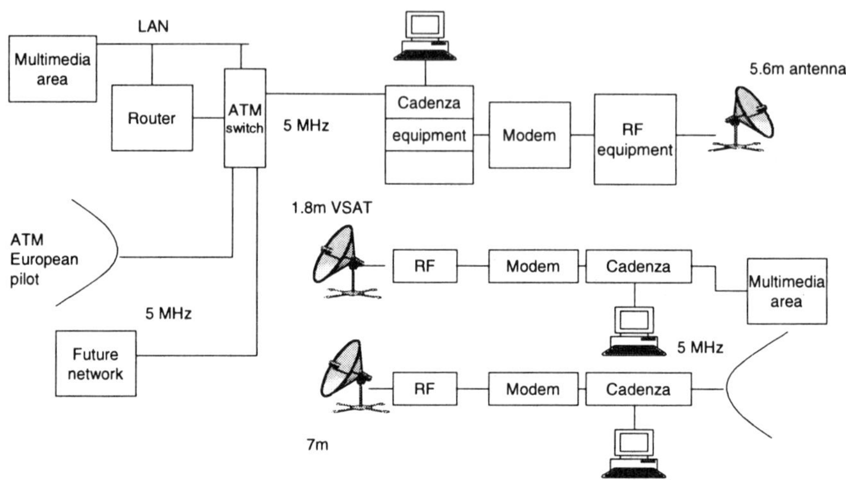

Figure 5.38 Trial-one architecture.

The second configuration was the same network with improved network management to provide the OAM facilities from each remote terminal (see Figure 5.39).

The third trial used a meshed network configuration, TDMA access, Ku-band, maximum 34-Mbps bit rate, and each Earth station was provided with a TDMAX system (already used in other trials) that significantly improved the network management facilities. OAM functions were managed on a TMN concept (see Figure 5.40). Global management was implemented for the setting up and releasing of the operations, for monitoring failure parameters, and for performance monitoring and maintenance through the ATM/OAM functions. Any of the network terminals could monitor network parameters such as BER, CER, CLR, CDV, and so forth. All equipment management systems were integrated in one management system for management of the global network.

VANTAGE's three trials included interworking with terrestrial networks and national hosts [62]. The main purpose for the trials was not the provision of a commercial, operational system, but to demonstrate that the satellite could be introduced into the ATM network without introducing restrictions or requiring significant rework.

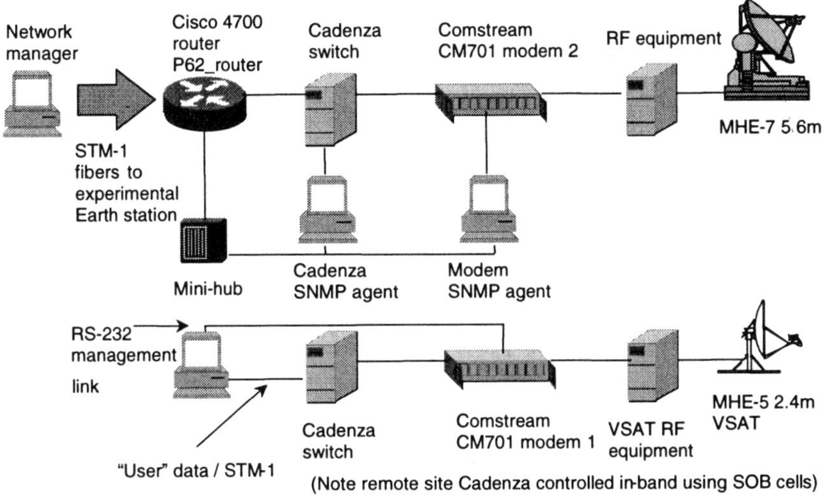

Figure 5.39 Trial-two network-management configuration.

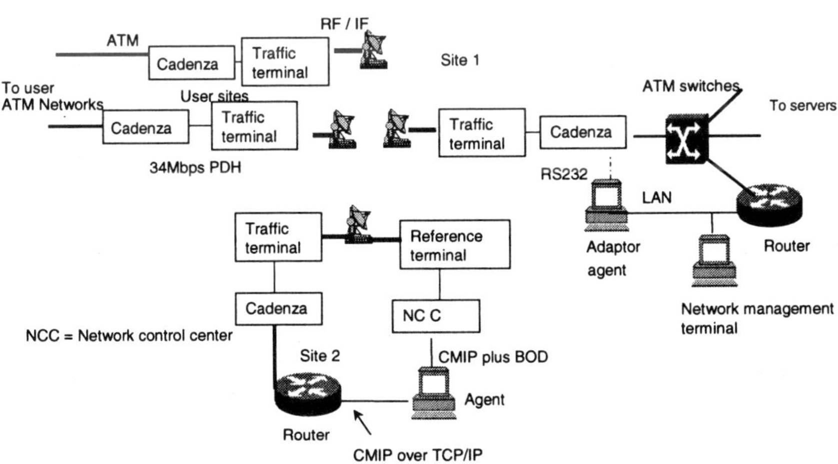

Figure 5.40 Trial-three configuration.

The main objectives of the first trial were the following.

- To set up satellite and ATM network;
- To prove applications and multiplexing;
- To investigate performance.

Most of the trial applications were IP-based (TCP/IP over ATM). VANTAGE proved that IP-based applications could run comfortably over ATM even with a satellite in the loop. The propagation delay effects and the degradation of the transmitted carrier versus cell-traffic parameters were also measured and evaluated. The network management was restricted to the local equipment management systems.

The achievements were the following.

- BTS 5.6 m Earth station to 1.8 m VSAT test;
- Videoconferencing, white-board, file transfer, and MPEG broadcasting applications;
- ATM multiplexing of different service types;
- Cell loss, BER, E_b/N_o measurements.

Trial 1 proved that satellites could offer consistent, reliable, and managed communications in an ATM environment to satisfy all used applications. Trial 2 was more ambitious in developing the management techniques and providing more rigorous analysis of traffic profiles and structured testing of user perceptions.

The OAM functions were identified as essential requirements for proper operability of the VANTAGE network. The provision of effective and versatile architecture for the VANTAGE network management (VNM) was identified as important. The project main objective was to provide management and control of the Earth station and remote site equipment integrated into the core network. The requirements of the VMN for VANTAGE trials were classified into three categories: configuration management, fault management, and performance management. VMN developed an SNMP-based system. The network management configuration developed in the second trial is depicted in Figure 5.39.

The trial included external users. The various applications included telemedicine, videoconferencing, LAN emulation, and multimedia. The main achievements can be summarized as the following:

- *Network management trial*:

    - Incorporation of HP open view network node manager into the VANTAGE network for automatic control of the end-to-end network from a single manager [62];
    - Development of modem and Cadenza MIBs and proxy agents;
    - Remote control and monitoring of satellite Earth station equipment;
    - SNMP traffic intensity measurements.
- *GRAVITAS user trial*:
    - Successfully demonstrated the ability to run geographical information systems (GISs) via ATM over a VSAT satellite link;
    - No adverse effects due to satellite link.
- *School Internet user trial*:
    - The satellite made no difference to user perception or behavior.

Trial 3 developed a more powerful platform with a mesh network, an improved the global network management, providing for higher bandwidth capacity, more flexibility, ATM signaling, and OAM facilities (see Figure 5.40) with objectives as the following:

- To incorporate 25-Mbps TDMA equipment (using larger antennas);
- To use both SNMP and CMIP for network management;
- To integrate SVC functions with a dynamically variable time burst plan to provide per-call capacity allocation;
- To perform extended user trials.

The detailed network architecture for this trial is presented in Figure 5.41. The major trial achievements included advanced network management used CMIP, the setup of SVC in a satellite network, and the allocation of bandwidth in response to SVC setup (BOD).

The project achieved significant results, which were recognized within the existing published standards as the world-firsts. The five world-firsts claimed are listed as follows.

- ATN cell (VP and VC) switching in the sky;
- SNMP-based management of a satellite network;
- CMIP-based management of a satellite network;

- Establishment of SVCs in a satellite network;
- Dynamic satellite bandwidth allocation-on-demand (BOD).

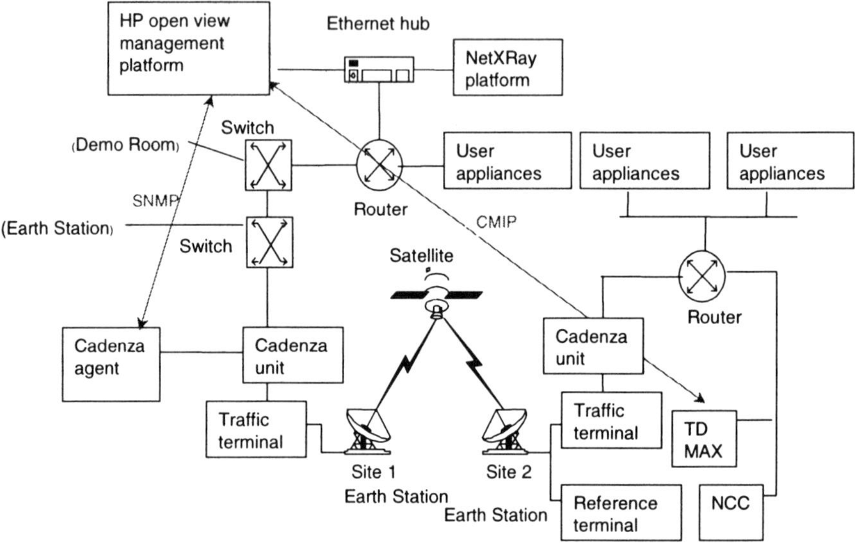

Figure 5.41 Trial-three network architecture.

VANTAGE provided the enabling technology to extend the variety of many known and novel applications and services beyond their urban limits. It contributed to deeper understanding of ATM in the satellite environment and the appropriate operability of management protocols. The conclusions and achievements of VANTAGE contributed to the evolution of both ITU-T network management standards and IETF standards such as SNMP v.3.

VANTAGE demonstrated that seamless integration between satellite and terrestrial networks is possible, and reinforced the new recognized view that satellites and copper/fiber are complementary and not competitive.

5.4 CONCLUSIONS

This chapter presented the importance of terrestrial and satellite network development and standardization. To overcome the variety of network solutions in the present environment, an attempt was made to build up a generic approach for network and interoperability concepts. The project RAINBOW played the most important role in introducing the generic network access approach that identified radio-dependent and radio-independent features and allowed terrestrial and satellite networks to cooperate through unified access points. On top of the RAINBOW concept many other projects structured their architectures, protocol strata, and trials, and tested their concepts with a number of different radio access schemes, including satellite access developed by SINUS.

The functionality of such a complex network demanded powerful planning tools (STORMS), investigation into security aspects (ASPeCT), core network adaptation with IN-based distributed management, and improved signaling protocols (EXODUS). The advantages of optical medium stimulated the introduction of optical support for radio technologies, just before the fiber was deployed to the home (FRANS).

Interoperability between different satellite access schemes, including LEO, MEO, and GEO satellites, in a seamless manner was addressed in the project SUMO. VANTAGE introduced the ATM technology in the terrestrial-satellite partnership, allowing broadband multimedia services to move on scene.

The network concept will continue to develop. The main challenge will be the possibility to seamlessly roam across a range of mobile, wireless, and satellite networks, providing different service support capabilities. GSM, DECT, GPRS, UMTS, and S-UMTS should be complemented by future fourth-generation mobile systems. IP will probably be the new transport technology that will raise new challenges and questions into the mobile telecommunication world. The lower network layer concepts will create network-independent terminals with network-software support. Management issues are critically related to the new arising concepts.

REFERENCES

[1] Da Silva, J., et al., "Mobile and Personal Communications: ACTS and Beyond," *Proceedings PIMRC'97,* Helsinki, Finland, September 1997, pp. 198–203.

[2] Samukic, A., "UMTS/ IMT-2000 Standardization: 3GPP 3G Partnership Project: Development of Standards for the New Millennium," *IEEE Transactions on Vehicular Technology,* November 1998, Vol. 47, No. 4, pp. 1099–1104.

[3] CEC Deliverable: AC015/BELL/CIT/DS/P/032/b1, 1998.

[4] Pereira, J., et al., "From Wireless Data to Mobile Multimedia: R&D Perspectives in Europe," in *Wireless Multimedia Network Technologies,* Norwell, MA: Kluwer Academic Publishers, 2000.

[5] ETSI Web site, www.etsi.org.

[6] ACTS Guideline, BAC-G1, "Access Network Strategies for Broadband Services Provisioning," 1997.

[7] http:/www.infowin.org/ACTS/ANALYSIS/CONCERTATION/bam1-gl.htm.

[8] Berruto, E., et al., "RAINBOW: A Radio-Independent Network Platform for UMTS," *ACTS Mobile Communications Summit,* Granada, Spain, November 27–29, 1996, CD-ROM.

[9] Berruto, E., and G. Colombo, "Definition of Interproject and Interprogram Relations," *CEC Deliverable No. AC015/CSE/PM/DS/P/001/a1,* October 1995.

[10] Jarvis, A., et al., "Migration of DECT/GSM Towards UMTS," *ACTS Mobile Communications Summit,* Granada, Spain, November 27–29, 1996, CD-ROM.

[11] Diaz, P., et al., "Validation of the RAINBOW Network and Service Concept," *AC015/UPC/CIT/DS/P/041/b1 Deliverable,* December 1998.

[12] Mellody, M., et al., "Transport Plane Implementation in the RAINBOW Demonstrator," *4th ACTS Mobile Communications Summit,* Sorrento, Italy, 1999, CD-ROM.

[13] CEC Deliverable, *AC015/NORTEL/CIT/DS/R/022/b1,* "RAINBOW System Concept Studies for the UMTS Network."

[14] Vriendt, J. De, et al., "RAINBOW: A Generic UMTS Network Architecture Demonstrator," *ACTS Mobile Communications Summit,* Aalborg, Denmark, October 7–10, 1997, pp. 285–290.

[15] Casadevall, F., et al., "A Real Time Emulator for the RAINBOW Demonstrator," *ACTS Mobile Communications Summit,* Granada, Spain, November 27–29, 1996, CD-ROM.

[16] Fleming, G., et al., "Architecture and Design of the RAINBOW Mobile Terminals, Base Stations and Mobility Servers," *ACTS Mobile Communications Summit,* Aalborg, Denmark, October 7–10, 1997, pp. 291–296.

[17] Airlie, D., G. Fleming, and M. Mellody, "Macrodiversity in the RAINBOW Demonstrator," *4th ACTS Mobile Communication Summit,* Sorrento, Italy, 1999.

[18] Saidi, A., et al., "RAINBOW Demonstrator Transport Chain," *ACTS Mobile Communications Summit,* Aalborg, Denmark, October 1997, pp. 297–302.

[19] http://www.cpk.auc.dk/asap/rainbow.

[20] CEC Delivererable, *AMD/CNET/SIG2?DS/R/002/b1,* "Detailed Specification of Interfaces for Joint Demonstrations," 1997.

[21] Nikula, E., et al., "FRAMES Multiple Access for UMTS and IMT-2000," *IEEE Personal Communications Magazine,* April 1998.

[22] Richardson, K., et al., "The FRAMES Demonstrator First Implementation Steps," *ACTS Mobile Communication Summit,* Aalborg, Denmark, October 1997, pp. 659–666.

[23] De Vriendt, J., et al., "RAINBOW: A Generic UMTS Network Architecture Demonstrator." *ACTS Mobile Communication Summit,* Aalborg, Denmark, October 1997, pp. 285–290.

[24] ETSI UMTS 30.06, "UMTS Terrestrial Radio Access Concept Evaluation," *ETSI Tech Report* 1998.

[25] Dunlop, J., et al., "Optimization of Packet Access Mechanisms for Advanced Time Division Multiple Access ATDMA," *7th IEE European Conference on Mobile Personal Communications,* 1993, pp. 237–242.

[26] Menolascino, R., and M. Pizarroso, "Software Tool for the Optimization of Resources in Mobile Systems," *STORMS Project Final Report, A016/CSE/MRM/DR/P/091/a1,* April 1999.

[27] Menolascino, R., et al., "A Realistic UMTS Planning Exercise," *3rd Acts Mobile Communications Summit,* Rhodes, Greece, June 8–10, 1998.

[28] Barberis, S., et al., "Planning Methodology for WCDMA and TD-CDMA UMTS Air Interfaces," *4th ACTS Mobile Communications Summit,* June 8–11, Sorrento, Italy, 1999.

[29] http://www.infowin.org.ACTS/ANALYSIS/CONCERTATION/bam1-g1-htm.

[30] Markoulidakis, J. G., et al., "Network Planning Methodology Applied to the UTRA Specifications," *4th ACTS Mobile Communications Summit,* June 8–11, Sorrento, Italy, 1999.

[31] Lyberopoulos, G., et al., "On the Dimensioning and Efficiency of the UTRA Air-Interface Modes (WCDMA and TD-CDMA)," *4th ACTS Mobile Communications Summit,* June 8–11, Sorrento, Italy, 1999.

[32] Tzifa, E., et al., "Software Tool for the Realization of Simulation-Based Studies on the Efficiency of the UTRA Interface," *4th ACTS Mobile Communications Summit,* June 8–11, Sorrento, Italy, 1999.

[33] Galliano, F., et al., "Network Planning Methodology with Adaptive Antennas and Super-Conducting Components," *4th ACTS Mobile Communications Summit,* June 8–11, Sorrento, Italy, 1999.

[34] http://wwwinfowin.org/.

[35] Vanneste, G., et al., "Migration/Evolution Towards UMTS—Security Issues," *Acts Mobile Summit,* Granada, Spain, 1996

[36] Jefferies, N., "ASPeCT—Securing the Future for Mobile Communications," *4th ACTS Mobile Communication Summit,* Sorrento, Italy, June 8–11, 1999.

[37] Vanneste, G., B. Franco, and A. Muller, "Demonstration of UMTS Authentication," *ACTS Mobile Communication Summit,* Aalborg, Denmark, October 7–10, 1997.

[38] Horn, G., et al., "Trialing Secure Billing with Trusted Third Party Support for UMTS Applications," *ACTS Mobile Communication Summit,* Rhodes, Greece, May 1998.

[39] Rantos, K., and C. J. Mitchell, "Key Recovery in ASPeCT Authentication and Initialization of Payment Protocol," *4th ACTS Mobile Communication Summit,* Sorrento, Italy, June 8–11, 1999.

[40] Verrelst, H., et al., "A Rule Based and Neural Network System for Fraud Detection in Mobile Communications," *4th ACTS Mobile Communication Summit,* Sorrento, Italy, June 8–11, 1999.

[41] Burge, P., et al., "BRUTUS—A Hybrid Detection Tool," *ACTS Mobile Communications Summit,* Aalborg, Denmark, October 7–10, 1997, pp. 722–727.

[42] Vinck, B., "A Viable Security Architecture for UMTS," *4th ACTS Mobile Communications Summit,* Sorrento, Italy, June 8–11, 1999.

[43] http:/www.infowin.org.ACTS/RUS/PROJECTS/ac013.htm.

[44] CEC Deliverable Ref. No. AC013/ITAL/WPO/EOPR/98/b1, October 1998.

[45] Tombros, S. L., et al., "Fixed Access to UMTS Service Capabilities," 1998.

[46] CEC Deliverable, AC013/CSELT/TWP1.3/DS/P/023/b1, "Evaluation and Review of Phase I Experiments," 1997.

[47] Morris, C., J. Nelson, and C. Francis, "Mobility Management in the EXODUS," 1999.

[48] Paff, A., "Hybrid Fibre/Coax in the Public Telecommunications Infrastructure," *IEEE Communications Magazine,* 1995, Vol. 33, No. 4, pp. 40–45.

[49] http:/www.infowin.org/ACTS/RUS/projects/ac083.html.

[50] Lane, P., et al., "Fiber Supported Millimeter-Wave Radio—A Route to the Broadband Local Loop," *ACTS Mobile Communications Summit,* Granada, Spain, November 27–29, 1996.

[51] Ellis, R., and M. Capstick, "Performance of Fiber Radio Networks for the Support of Present and Future Mobile and Wireless Communications," *4th ACTS Mobile Communications Summit,* Aalborg, Denmark, October 7–10, 1997.

[52] http://www.infowin.org/ACTS/RUS/PROJECTS/ac345.html

[53] AC345 TNR FOU DS P 102-B1, "Joint Trials System Specifications," *CEC Deliverable*, 1998.

[54] Mort, R., C. Valadon, and H. M. Aziz, "An Integrated UMTS Testbed with Satellite CDMA/FDMA ACCESS—the SUMO Project," *4th ACTS Mobile Communications Summit,* Sorrento, Italy, June 8–11, 1999.

[55] Schubert, A., "URAN Adaptation of the FDMA MAC Layer in the TOMAS/SUMO S-UMTS Testbed," *4th ACTS Mobile Communications Summit,* Sorrento, Italy, June 8–11, 1999.

[56] Karachos, G., et al., "Application Adaptation for UMTS Services," *4th ACTS Mobile Communications Summit,* Sorrento, Italy, June 8–11, 1999.

[57] Efthymou, L. F., R. S. Sheriff, and Y. F. Hu, "Service-Dependent Resource Allocation in Hybrid Mobile-Satellite Networks," *4th ACTS Mobile Communications Summit,* Sorrento, Italy, June 8–11, 1999.

[58] AC345 MMO SUM DS P 208-B1, "Service Test and Trial Plan," *CEC Deliverable*, 1999.

[59] http://www.infowin.org/ACTS/PROJECTS/ac009.html.

[60] AC009 ABT DS P 019 B2, "VANTAGE-Final Report," *CEC Deliverable,* 1998.

[61] http://www.eeng.brad.ac.uk/Research/VANTAGE/trial.html

[62] Pan, H., Y. F. Hu, and N. Efthymiou, "VSAT ATM Network Management," *3rd International Symposium on Interworking,* Nara, Japan, October 1–3, 1996, pp. 602–610.

Appendix I

Alphabetical List of ACTS Projects (by Project Acronym)

Acronym	Project Number and Title	
ABROSE	AC 316	Agent Based Brokerage Service in Electronic Commerce
ABS	AC 206	Architecture for Information Brokage Service
ACCORD	AC 348	ACTS Broadband Communication Joint Trials and Demonstrations
ACTranS	AC 081	A Transaction Processing Toolkit for ACTS
ACTSLINE	AC 344	ACTS guideline Consolidation/Channeling and Streaming Towards Market Applications
ACTUAL	AC 329	Application and Control of Widely Tunable Lasers in HD-WDM Networks
AMASE	AC 346	Agent Based Mobile Access to Information Services
AMOS	AC 356	ACTS MObile Summit
AMPA	AC 213	Advanced Multi-Media Parallel Accelerator
AMUSE	AC 011	Advanced Multimedia Services for Residential Users
APEX	AC 332	Advanced Photonic Experimental X-Connect
AROMA	AC 327	Advanced Resource Management in Service Integrated and Multi-layered HFC Access Networks
ASAP	AC 358	ACTS Success Dissemination and Promotion in the Mobile Domain
ASICCOM	AC060	ATM Switch for Integrated Communication, Computation and Monitoring
ASIS	AC 303	Alliance for a Sustainable Information Society
ASPeCT	AC 095	Advanced Security for Personal Communications Technologies
ASSET	AC 326	ACTS Satellite Switching and End-to-End Trial
ATHOC	AC 037	ATM Applications over Hybrid Optical Fiber Coax
ATLANTIC	AC 078	Advanced Television at Low Bitrates and Networked Transmission Over Integrated Communication Systems
ATMAN	AC 225	Digital Audio Visual Work Trading by ATM
AURORA	AC 072	Automated Restoration of Original Film and Video Archives
AVANTI	AC042	Adaptive and Adaptable Interactions for Multimedia Telecommunications Applications
AWACS	AC 228	ATM Wireless Access Communication System
BIDS	AC025	Broadband Infrastructures for Digital Television and Multimedia Services
BLISS	AC 065	Broadband Lightwave Sources and System
BONAPARTE	AC 022	Broadband Optical Network using ATM Pon Access Facilities in Realistic Telecommunications Environments

BOURBON	AC 001	Broadband Urban Rural Based Open Networks
BROADBANDLOOP	AC 038	Broadbandloop
BTI	AC 362	Broadband Trial Integration
CA$hMAN	AC 039	Charging and Accounting Schemes in Multi-Service ATM Networks
CABSINET	AC 236	Cellular Access to Broadband Services and Interactive Television
CAMELEON	AC 341	Communication Agents for Mobility Enhancements in a Logical Environment of Open Networks
CANCAN	AC 014	Contract Negotiation and Charging in ATM Networks
CAPITAL	AC 047	Customer Access Photonics: An Integrated Technology for Active, Low-Cost Devices
CATVDC	AC 241	Cable Television Domains and Chains
CICC	AC 017	Collaborative Integrated Communications for Construction
CINENET	AC 062	Cinema Films and Live Events via Satellite and Cable Networks
COBNET	AC 069	Corporate Optical Backbone Network
COBRA	AC 203	Common Brokerage Architecture
COBUCO	AC 031	Cordless Business Communication System
CODID	AC 366	Convergence of DAVIC Internet and DVB
CODIS	AC 218	Clip On Demand Interactive System
COIAS	AC 363	Convergence of Internet-ATM-Satellite
COMIQS	AC 322	COmmerce through MPEG-4 on the Internet with Quality of Services
CONVAIR	AC 234	Consensus and Validation in ACTS Results E ploitation
COVEN	AC 040	Collaborative Virtual Environments
CRABS	AC 215	Cellular Radio Access for Broadband Services
CustomTV	AC 360	CustomTV
DAM	AC 087	DAVIC Accompanying Measures
DEMON	AC 354	Demonstrating Evolution of a Metropolitan Optical Network
DIANA	AC 319	Demonstration of IP and ATM Networking for Real-Time Applications
DIANE	AC082	Design, Implementation and Operation of a Distributed Annotation Environment
DICEMAN	AC308	Distributed Internet Content Exchange with MPEG-7 and Agent Negotiations
DIFFERENCEII	AC 207	Dissemination and Facilitation for European Research in Selected Chains
DIGISAT	AC 061	Advanced Digital Satellite Broadcasting and Interactive Services
DIPLOMAT	AC 222	European Charter for Telework
DIVINE	AC035	Deployment of Interpersonal Videoconferencing Systems on IBC Networks
DOLMEN	AC 036	Service Machine Development for an Open Long-Term Mobile and Fixed Network Environment
DOLPHIN	AC 210	Cooperation in Advanced Communications in Europe
DOLPHIN 2	AC 357	Dissemination of IS&N Results
DVBIRD	AC 108	Digital Video Broadcasting Integrated Receiver Decoder
DVP	AC 089	Distributed Video Production
EIES	AC 075	European Information E change Service for the Communication Between Harbours
ELISA	AC310	European Experiment on the Linkage between Internet Integrated Services and ATM
EMERALD	AC 107	European Multimedia Services for Medical Imaging
EMPHASIS	AC 105	Architectures Software and Hardware for MPEG-4 Systems
EPRI-COM	AC 352	European Parliaments Research Initiative Communicating with CEEC
EPRIwatch	AC 079	European Parliaments Research Initiative
ESTHER	AC 063	Exploitation of Solution Transmission Highways in the European Ring
ETD	AC 223	European Telework Development
EURORIM	AC 219	European Research and Consensus on Interactive Multimedia
EXODUS	AC 013	Experiments on the Deployment of UMTS
EXPERT	AC 094	Platform for Engineering Research and Trials
FACTS	AC 317	FIPA Agent Communication Technologies and Services

FAIR	AC 093	Forecast and Assessment of Socio-Economic Impact of Advanced Communications and Recommendations
FAST	AC 059	Fluoro-aluminate Amplifier for Second Telecom window
FIRST	AC 005	Flexible Integrated Radio Systems Technology
FLEXIMACS	AC 311	Bitrate Flexible Code Division Multiple for ATM Access
FlowThru	AC 335	Co-operative Secure Management of Multi Technology and Administrative Domain Network and Service Management Systems
FRAMES	AC 090	Future Radio Wideband Multiple Access Systems
FRANS	AC 083	Fiber Radio ATM Network and Services
GAIA	AC 221	Generic Architecture for Information Availability
GAMMA	AC 239	Global Architecture for Multimedia Action
GESTALT	AC 367	Getting Education System Talking Across Leading-Edge Technologies
GINA	AC 220	Guidelines for Interoperability in Networks and ATM Deployment
HIGHWAY	AC 067	Photonic Technologies for Ultra High Speed Information Highways
HORIZON	AC 058	Horizontal Action on Optical Networks
HYPERMEDIA	AC 361	HYPERMEDIA
IACS	AC 369	Interoperability Assessments in Cable Systems
IBCoBN	AC 101	Integrated Broadband Communications on Broadcast Networks
IMMP	AC 023	Integrated Multimedia Project
IMPACT	AC 324	Implementation of Agents for CAC on an ATM Testbed
InfoBridge	AC 359	The Bridge from ACTS to the Outside World
INFOWIN	AC 113	Multimedia Information Window for National Hosts
INSIGNIA	AC 068	IN and B-ISDN Signalling Integration on ATM Platforms
INSURED	AC 229	Integrated Satellite-UMTS Real Environment Demonstrator
INTERACT	AC 086	Interactive Television Return Channel Standardization and Trials
ISIS	AC 103	Interactive Satellite Multimedia Information System
IthACI	AC 337	Internet and the ATM: Experiments & Enhancements for Convergence and Integration
iTTi	AC 321	Interactive Terrestrial TV Integration
ITUNET	AC 309	International Trials with Users and Networks from European Testbeds
JAMES	AC 111	Joint ATM Experiment on European Services
KEOPS	AC 043	Keys to Optical Packet Switching
KIMSAC	AC 030	Kiosk-Based Integrated Multimedia Service Access for Citizens
LEVERAGE	AC 109	Learn from Video Extensive Real ATM Gigabit Experiment
LOIS	AC 306	Large Optical Integrated Switches
M2VTS	AC 102	Multimodal Verification for Teleservices and Security Applications
MAESTRO	AC 233	Maintenance System Based on Telepresence for Remote Operators
MARINE	AC 340	Mobile Agent Environments in Intelligent Networks
MARINER	AC 333	Multi-Agent Architecture for Distributed-IN Load Control and Overload Protection
MEDIAN	AC006	Wireless Broadband CPN/LAN for Professional and Residential Multimedia Applications
MEMO	AC 054	Multimedia Environment for Mobiles
MEPHISTO	AC 209	Management of Photonic Systems and Networks
METON	AC 073	Metropolitan Optical Network
MIAMI	AC 338	Mobile Intelligent Agents for Managing the Information Infrastructure
MICC	AC 088	Mobile Integrated Communication in Construction
MIDAS	AC053	Multigigabit Interconnection Using Dispersion Compensation and Advanced Soliton Techniques
MIDSTEP	AC 214	Multimedia Interactive Demonstrator Telepresence
MIRADOR	AC 302	MPEG-4 Intellectual Property Rights by Adducing and Ordering
MIRAGE	AC044	Manipulation of Images in Real-Time for the Creation of Artificially Generated Environments
MISA	AC 080	Management of Integrated SDH and TM Networks
MODEST	AC 304	Multimedia Object Descriptors Extraction from Surveillance Tapes
MOMENTS	AC 002	Mobile Media and Entertainment Services
MOMUSYS	AC 098	Mobile Multimedia Systems

MONTAGE	AC325	Mobile Intelligent Agents in Accounting, Charging, and Personal Mobility Support
MOON	AC 231	Management of Optical Networks
MOSQUITO	AC 334	Management of Service Quality in Television Operations
MOSTR IN	AC 104	Mobile Services for High Seed Trains
MOTIVATE	AC 318	Mobile Television and Innovative Receivers
MOVE	AC 343	Mobile Middleware with Voice Enabled Services
MULTICUBE	AC 012	Efficient Multipoint to Multipoint Broadband Switched Network Services for Distributed Multimedia Applications
MULTI-MEDITOR	AC 096	Multimedia Publishing Brokerage Service
MULTIPORT	AC 007	Multimedia Portable Digital Assistant
MUSICIAN	AC 232	Multimedia Services Integration Chain in Advanced Networks
MUSIST	AC 010	Multimedia User Interface for Interactive Services and TV
NEWTEST	AC 202	High Performance Neural Network Signal Processing Schemes for Wireless Terrestrial and Satellite Transmissions
NICE	AC 110	National Host Interconnection Experiments
OCEANS	AC 323	Open Communication Environment for Agent-Based Networked Services
OCTALIS	AC 242	Offer of Content Through Trusted Access Links
OKAPI	AC 051	Open Kernel for Access to Protected Interoperable Interactive Services
ON THE MOVE	AC 034	Application Support Services for Distributed Mobile Multimedia
OPARISoD	AC 076	Open Architecture for Interactive Services on Demand
OPEN	AC 066	Optical Pan-European Network
OPTIMUM	AC 226	Optimised Network Architectures for Multimedia Services
OSC	AC 342	Evolutionary Optical Approach for Intersatellite Space Communication Systems
OSM	AC211	An Open Service Model for Global Information Brokerage and Distribution
OSM PLUS	AC 365	Open Service Model for Global Electronic Commerce
PANEL	AC 205	Protection Across Network Layers
PANOR MA	AC 092	Package for New Operational Autostereoscopic Multiview Systems and Applications
PELICAN	AC 355	Pan-European Lightwave Core and Access Network
PeterPan	AC 307	Provision of an Enhanced Transport by Exploiting Reservation in iP and Atm Networks
PHOTON	AC 084	Paneuropean Photonic Transport Overlay Network
PHOTOS	AC 046	Photosensitive Technology for Optical Systems
PLANET	AC 050	Photonic Local Access Network
PLATO	AC 041	Photonic Links in ATM and Optical Systems
PRAXIS	AC 301	Platform Supporting the Exploitation of ACTS Project Trials
PRIME	AC 370	Promoting Interoperability for MultiMedia Communication in Europe
PRINT-IT	AC240	Publishers Reusable Integrated Network Toolkits for Information Technologies
PRISM	AC 349	Photonic Routing of Interactive Services for Mobile Applications
PROSPECT	AC 052	Prospect of Multi-Domain Management in the Expected Open Services Marketplace
QUANTUM	ep29212	Quality Network Technology for User-Oriented Multi-Media
QUOV DIS	AC 056	Quality of Video and Audio for Digital Television Services
R INBOW	AC 015	Radio Access Independent Broadband on Wireless
REFORM	AC 208	Resource Failure and Restoration Management in ATM-based IBCN
RENAISSANCE	AC100	Integration of High Performance Services for Interactive Vocational Training
REPEAT	AC 305	Regeneration of Pulse Shape, Amplitude and Timing
RESOLV	AC 021	Reconstruction Using Scanned Laser and Video
ReTIN	AC 048	An Industrial-Quality TINA Compliant Real-Time DPE
S3M	AC 313	Digital Interactive Satellite Terminal for Master Antenna Systems in the Third Millennium

SAMBA	AC 204	System for Advanced Mobile Broadband Applications
SCALAR	AC 077	Scaleable Architectures with Hardware Extensions for Low Bitrate Variable Bandwidth Real-Time Videococommunications
SCAN	AC 330	Secure Communication in ATM Networks
SCARAB	AC 339	Smart Card and Agent Enabled Reliable Access
SCREEN	AC 227	Service Creation Engineering Environment
SECOMS/ABATE	AC 004	Satellite EHF Communications for Mobile Multimedia Services
SEMPER	AC 026	Secure Electronic Marketplace for Europe
SETBIS	AC 032	Set Top Box for Interactive Services on Demand
SHOW	AC 350	Showing ACTS Results
SICM	AC 071	Scaleable Interactive Continuous Media Server Design and Application
SINUS	AC 212	Satellite Integration into Networks for UMTS Services
SMARTS	AC 114	SME and Regional Telecom Support
SMASH	AC 018	Storage for Multimedia Application Systems in the Home
SOMMIT	AC 033	Software Open Multimedia Terminal
SONATA	AC 351	Switchless Optical Network for Advanced Transport Architecture
SORT	AC 315	Software Radio Technology
SPEAR	AC 368	Standards Progression Supported by European Ad-hoc Researchers
SPECI L	AC091	Service Provisioning Environment for Consumers Interactive Applications
SPEED	AC049	Superhighway by Photonically and Electronically Enhanced Digital Transmission
SQUALE	AC 097	Security, Safety and Quality Evaluation for Dependable Systems
STORit	AC 312	Storage Interoperability Technologies
STORMS	AC 016	Software Tool for the Optimisation of Resources in Mobile Systems
SUCOMS	AC 115	Superconducting Systems for Communications
SUMO	AC 345	Satellite -UMTS Multimedia Service Trials Over Integrated Testbeds
SUNBEAM	AC 347	Smart Universal Beamforming
SUSIE	AC320	Charging for Premium IP Services in the European Information Infrastructures & Services Pilot
TALISMAN	AC 019	Tracing Authors Rights by Labelling Image Services and Monitoring Access Networks
TAPESTRIES	AC055	The Application of Psychological Evaluation to Systems and Technologies in Remote Imaging and Entertainment Services
TEAM	AC 070	Team-Based European Automotive Manufacture
TECODIS	AC 064	Teleworking in Cooperative Development of Industrial Software
TEESURA	AC 230	Techno-Economic Evaluation and Sectorial User Requirements Analysis
TELEBORG	AC 216	Telepresence Supporting Human-like Presence in Real Remote Areas
TELE-SHOPPE	AC 099	Tele-Shopping Services Using Virtual Reality and Interactive Multimedia
TERA	AC 364	Techno-Economic Results from ACTS
THESEUS	AC 008	Terminal at High Speed for European Stock Exchange Users
TOBASCO	AC 028	Towards Broadband Access Systems for CATV Optical Networks
TOMAS	AC 201	Inter-Trial Testbed of Mobile Applications for Satellite Communications
TOSCA	AC 237	Tina Open Service Creation Architecture
TRADE	AC 328	Trials in the Domain of Electronic Commerce
TRUMPET	AC 112	Inter-Domain Management with Integrity
TSUNAMI II	AC 020	Technology in Smart Antennas for Universal Advanced Mobile Infrastructure—Part 2
UMPTIDUMPTI	AC 027	Using Mobile Personal T l Communications Innovation for the Disabled in UMTS Pervasive Integration
UPGRADE	AC 045	High Bitrate 1300nm Upgrade of the European Standard Single-Mode Fiber Network
USECA	AC 336	UMTS Security Architecture

USINACTS	AC 224	Usability in ACTS
VALIDATE	AC 106	Verification and Launch of Integrated Digital Advanced Television in Europe
VANGUARD	AC 074	Visualisation Across Networks Based on Graphics and the Uncalibrated Acquisition of Real Data
VANTAGE	AC 009	VSAT ATM Network Trials for Applications Groups across Europe
VERTICAL	AC 024	Vertical Cavity Laser Technology for Interconnection and Access Links
VIDAS	AC 057	Video Assisted with Audio Coding and Representation
VISAVIS	AC 314	Fitness-for-Purpose of Videotelehony in Face-to-Face Situations
VISEUM	AC 238	Virtual Museum International
VITAL	AC 003	Validation of Integrated T l Communication Architectures for the Long Term
VPARK	AC 353	Virtual Amusement Park
WAND	AC 085	Wireless ATM Network Demonstrator
WA TT	AC 235	WWW Window for ACTS Trials and Testbeds
WISDOM	AC 331	Wideband Satellite Demonstrator of Multimedia Services
WOTAN	AC 029	Wavelength-Agile Optical Transport and Access Network

About the Editor

Ramjee Prasad was born in Babhnaur (Gaya), Bihar, India, on July 1, 1946. He is now a Dutch citizen. He received his B.Sc. (Eng.) in 1968 from Bihar Institute of Technology, Sindri, India, and his M.Sc. (Eng.) and Ph.D. from Birla Institute of Technology (BIT), Ranchi, India, in 1970 and 1979, respectively.

He joined BIT as a senior research fellow in 1970 and became an associate professor in 1980. While with BIT, he supervised a number of research projects in microwave and plasma engineering. From 1983 to 1988 he was with the University of Dar es Salaam (UDSM), Tanzania, where he became professor of telecommunications in the Department of Electrical Engineering in 1986. At UDSM, he was responsible for the collaborative project "Satellite Communications for Rural Zones" with Eindhoven University of Technology, The Netherlands. From February 1988 until May 1999, he was with the Telecommunications and Traffic Control Systems Group at the Delft University of Technology (DUT), The Netherlands, where he was actively involved in wireless personal and multimedia communications (WPMC). He was the founding head and program director of the Center for Wireless and Personal Communications (CEWPC) of the International Research Center for Telecommunications-Transmission and Radar (IRCTR). Since June 1999, Dr. Prasad has been with Aalborg University, Denmark, as codirector of the Center for PersonKommunikation (CPK) and holds the chair of wireless information and multimedia communications. He was involved in the European ACTS project Future Radio Wideband Multiple Access Systems (FRAMES) as a DUT project leader. He is project leader of several international industrially funded projects. He has published over 300 technical papers, contributed to several books, and authored, coauthored, and edited six books: *CDMA for Wireless Personal Communications*, *Universal Wireless Personal Communications*, *Wideband CDMA for Third Generation Mobile Communications*, *OFDM for Wireless Multimedia Communications*, *Third Generation Mobile Communication Systems*, and *WCDMA: Towards IP Mobility and Mobile Internet*, all published by Artech

House, Norwood, Massachusetts. His current research interests lie in wireless networks, packet communications, multiple-access protocols, advanced radio techniques, and multimedia communications.

Dr. Prasad has served as a member of the advisory and program committees of several IEEE international conferences. He has also presented keynote speeches and delivered papers and tutorials on WPMC at various universities, technical institutions, and IEEE conferences. He was also a member of the European cooperation in scientific and technical research (COST-231) project dealing with "Evolution of Land Mobile Radio Communications" (including personal communications) as an expert for The Netherlands, and of the COST-259 project. He was the founder and chairman of the IEEE Vehicular Technology/ Communications Society Joint Chapter, Benelux Section, and is now the honorary chairman. In addition, Dr. Prasad is the founder of the IEEE Symposium on Communications and Vehicular Technology (SCVT) in The Benelux, and he was the symposium chairman of *SCVT'93*.

In addition, Dr. Prasad is the coordinating editor and editor-in-chief of the Kluwer international journal *Wireless Personal Communications* and a member of the editorial board of other international journals, including the *IEEE Communications Magazine* and *IEE Electronics Communication Engineering Journal*. He was the technical program chairman of the *PIMRC'94 International Symposium* held in The Hague, The Netherlands, from September 19–23, 1994, and also of the *Third Communication Theory Mini-Conference* in conjunction with *GLOBECOM'94*, held in San Francisco, California, from November 27–30, 1994. He was the conference chairman of the *50th IEEE Vehicular Technology Conference* and the steering committee chairman of the *2nd International Symposium on Wireless Personal Multimedia Communications (WPMC)*, both held in Amsterdam, The Netherlands, from September 19–23, 1999. He is the general chairman of *WPMC'01*, to be held in Aalborg, Denmark, from September 9–12, 2001.

Dr. Prasad is also the founding chairman of the European Center of Excellence in Telecommunications, known as HERMES. He is a fellow of the IEE, a fellow of the IETE, a senior member of the IEEE, a member of The Netherlands Electronics and Radio Society (NERG), and a member of the IDA (an engineering society in Denmark).

Index

Recent Titles in the Artech House Mobile Communications Series

John Walker, Series Editor

Advances in Mobile Information Systems, John Walker, editor

CDMA for Wireless Personal Communications, Ramjee Prasad

CDMA Mobile Radio Design, John B. Groe and Lawrence E. Larson

CDMA RF System Engineering, Samuel C. Yang

CDMA Systems Engineering Handbook, Jhong S. Lee and Leonard E. Miller

Cell Planning for Wireless Communications, Manuel F. Cátedra and Jesús Pérez-Arriaga

Cellular Communications: Worldwide Market Development, Garry A. Garrard

Cellular Mobile Systems Engineering, Saleh Faruque

The Complete Wireless Communications Professional: A Guide for Engineers and Managers, William Webb

The Future of Wireless Communications, William Webb

GSM and Personal Communications Handbook, Siegmund M. Redl, Matthias K. Weber, and Malcolm W. Oliphant

GSM Networks: Protocols, Terminology, and Implementation, Gunnar Heine

GSM System Engineering, Asha Mehrotra

Handbook of Land-Mobile Radio System Coverage, Garry C. Hess

Handbook of Mobile Radio Networks, Sami Tabbane

High-Speed Wireless ATM and LANs, Benny Bing

Introduction to 3G Wireless Communications, Juha Korhonen

An Introduction to GSM, Siegmund M. Redl, Matthias K. Weber, and Malcolm W. Oliphant

Introduction to Mobile Communications Engineering, José M. Hernando and F. Pérez-Fontán

Introduction to Radio Propagation for Fixed and Mobile Communications, John Doble

Introduction to Wireless Local Loop, Second Edition: Broadband and Narrowband Systems, William Webb

IS-136 TDMA Technology, Economics, and Services, Lawrence Harte, Adrian Smith, and Charles A. Jacobs

Mobile Data Communications Systems, Peter Wong and David Britland

Mobile Telecommunications: Standards, Regulation, and Applications, Rudi Bekkers and Jan Smits

Personal Wireless Communication with DECT and PWT, John Phillips and Gerard Mac Namee

Practical Wireless Data Modem Design, Jonathon Y. C. Cheah

Radio Propagation in Cellular Networks, Nathan Blaunstein

Radio Resource Management for Wireless Networks, Jens Zander and Seong-Lyun Kim

RDS: The Radio Data System, Dietmar Kopitz and Bev Marks

Resource Allocation in Hierarchical Cellular Systems, Lauro Ortigoza-Guerrero and A. Hamid Aghvami

RF and Microwave Circuit Design for Wireless Communications, Lawrence E. Larson, editor

Signal Processing Applications in CDMA Communications, Hui Liu

Spread Spectrum CDMA Systems for Wireless Communications, Savo G. Glisic and Branka Vucetic

Understanding Cellular Radio, William Webb

Understanding Digital PCS: The TDMA Standard, Cameron Kelly Coursey

Understanding GPS: Principles and Applications, Elliott D. Kaplan, editor

Understanding WAP: Wireless Applications, Devices, and Services, Marcel van der Heijden and Marcus Taylor, editors

Universal Wireless Personal Communications, Ramjee Prasad

WCDMA: Towards IP Mobility and Mobile Internet, Tero Ojanperä and Ramjee Prasad, editors

Wireless Communications in Developing Countries: Cellular and Satellite Systems, Rachael E. Schwartz

Wireless Intelligent Networking, Gerry Christensen, Paul G. Florack, and Robert Duncan

Wireless LAN Standards and Applications, Asunción Santamaría and Francisco J. López-Hernández, editors

Wireless Technician's Handbook, Andrew Miceli

For further information on these and other Artech House titles, including previously considered out-of-print books now available through our In-Print-Forever® (IPF®) program, contact:

Artech House
685 Canton Street
Norwood, MA 02062
Phone: 781-769-9750
Fax: 781-769-6334
e-mail: artech@artechhouse.com

Artech House
46 Gillingham Street
London SW1V 1AH UK
Phone: +44 (0)20 7596-8750
Fax: +44 (0)20 7630-0166
e-mail: artech-uk@artechhouse.com

Find us on the World Wide Web at:
www.artechhouse.com

Fax Back Card

Free information on other Artech House books...

To ensure that you stay up-to-date on the newest Artech House books of interest to you, fill in the address details below so that we may add you to our mailing list. Please fax or mail it to one of the locations listed on the other side of this card.

Name: ____________________

Position: ____________________

Company: ____________________

Address: ____________________

____

E-mail: ____________________

(If you would like to receive new title information by e-mail.)

Please indicate your areas of interest:

- ☐ Telecommunications
  Wireless
  Networking
- ☐ Computing
  Software Engineering
  Security/E-Commerce
  Database Management
- ☐ Microwave
- ☐ Radar
  Remote Sensing
  Electronic Defense
- ☐ Antennas & Propagation
- ☐ Signal Processing
- ☐ Optoelectronics
- ☐ Technology Management/
  Professional Development